"*The Valuation of Technology* is a timely and thoughtful book on a critical issue in the global business arena. Peter Boer's insights constitute important reading for leaders in all fields."

—Jeffrey E. Garten
Dean
Yale School of Management

"*The Valuation of Technology* fills a critical void for those executives who wish to upgrade technology decision making from an art to a more definable science."

—George B. Rathmann
Chairman & CEO
ICOS Corporation

"*The Valuation of Technology* is valuable for every leader in every organization that invests in R&D."

—Jeffrey M. Lipton
President & CEO
NOVA Chemicals Ltd.

"Dr. F. Peter Boer has been a thought leader in his field for some time. His book demystifies the transformation of science into business."

—Dr. Barry Siadat
Vice President & Chief Growth Officer
AlliedSignal Inc.

"Peter Boer has written a book every R&D manager should not only read, but study."

—Walter L. Robb
President
Vantage Management, Inc.

"This book will serve high level executives as well as product and R&D managers who are working to create great wealth and shareholder value."

—Dr. Nicholas Pappas
Former Executive Vice President & Member of the Executive Committee
DuPont

"*The Valuation of Technology* is a valuable reference work."

—Dr. Lester C. Krogh
Senior Vice President
R&D (retired), 3M

"This landmark book bridges a significant gap of understanding between the technical and financial communities."

—L. Louis Hegedus
Vice President, R&D
Elf Atochem N.A., Inc.

"For everyone interested in how to value and evaluate technology in terms of adding wealth to a corporation and growth in shareholder value, Dr. Boer's book is must reading."

—Stanley V. Jaskolski
Chief Technology Officer
Eaton Corporation

"How to evaluate technology and how to choose the appropriate R&D portfolio for further growth is probably the most important issue confronting 20th century companies who plan to be 21st century companies."

—Mary L. Good
Managing Member
Venture Capital Investors, LLC

"*The Valuation of Technology* is an astute review of how judgments should be made of the value of research programs including quantitative measures and analyses."

—Paul Cook
Chairman & CEO
DIVA Systems Corporation

THE VALUATION OF TECHNOLOGY

Wiley Operations Management Series for Professionals

Published titles in this series are:

Logistics and the Extended Enterprise: Best Practices for the Global Company, 0471-31430-7, by Sandor Boyson, Thomas M. Corsi, Martin E. Dresner, and Lisa Harrington

Total Project Control: A Manager's Guide to Integrated Project Planning, Measuring and Tracking, 0471-32859-6, by Stephen A. Devaux

THE VALUATION OF TECHNOLOGY

Business and Financial Issues in R&D

F. PETER BOER

John Wiley & Sons, Inc.

New York • Chichester • Weinheim • Brisbane • Singapore • Toronto

This book is printed on acid-free paper. ♾

Published by John Wiley & Sons, Inc.

Published simultaneously in Canada.

Library of Congress Cataloging-in-Publication Data:

Boer, F. Peter, 1940–
The valuation of technology : business and financial issues in R&D / F. Peter Boer.
p. cm. — (Wiley operation management series for professionals)
Includes bibliographical references and index.
ISBN 0-471-31638-5 (cloth/disk : alk. paper)
1. Research, Industrial—Evaluation. 2. Research, Industrial—Cost effectiveness. I. Title. II. Series.
T175.B56 1999
658.5'7—dc21 98-39117

Printed in the United States of America.

10 9 8 7 6 5 4 3 2

Preface

The linkage between scientific discovery and the delivery of practical results—what we call technology—is vital to the well-being of individuals and to the wealth of businesses and nations. Technology is said to account for half of the economic growth of developed countries.[1] Yet the link between scientific discovery and the development of new technology is poorly understood.

Why this lack of understanding exists is no mystery: the transformation of science into technology is mediated by business forces and brings together two sets of people whose outlooks, specialized knowledge, and professional languages are very different and often out of touch with each other. Many of those charged with making decisions regarding science and technology—business executives and a host of government and university officials—have little or no understanding of scientific process or the culture in which it operates. On the other hand, many who *do* understand technological development are poorly informed about financial and business matters. A chasm of knowledge and interest divides these two important communities and hinders the progress that each seeks. This should not be surprising. The people who enter the communities of business and science are products of different educational processes, operate with different tools and languages, and are generally motivated by different factors. As a result, communication between them is often superficial or limited to a narrow set of issues on which their interests converge. In too many cases, there is a lack of mutual respect. The destinies of these two groups are inextricably linked, however, and they have no choice but to bridge the gap and deal with each other.

The *valuation of technology* is that bridge, and recognition of its central position accounts for the increasing interest in this once obscure subject. The process of valuation provides an opportunity for dialogue and collaboration. Researchers and businesspeople can bring their special knowledge and skill to this process, learn from each other, and share in its mutual benefits.

The term *valuation* has a specific meaning—it refers to the task of determining the monetary worth of an asset, object, or entity. Valuation seeks to answer a fundamental question: "What is it worth?" The valuation process has a long history. Whether the object is a piece of jewelry, a work of art, or residential real estate, people have always wanted an answer to the fundamental question of value. The valuation of businesses—from corner drugstores to giant corporations and their securities has become an important aspect of the world of commerce and is now the subject of increasingly specialized books. Their methods are highly quantitative, and in areas such as the valuation of financial and business options, mathematically sophisticated.

Determining the inherent value of a share of common stock is difficult, but valuing technology is tougher still. Technology may exist as intellectual capital and not even be visible on the corporate balance sheet; it may be embedded in physical assets that are valued on the basis of historic cost and do not reflect technology's wealth-creating potential.

There is growing recognition that the worth of a business cannot be determined without recognizing the value of its technology. In many industries, from pharmaceuticals to software, proprietary technology has become a firm's greatest asset. At the same time, the people who develop technology are increasingly cognizant that they must estimate the future value of their output if they are to acquire the resources needed to turn proposals into programs. The contrary view is widely held: "Nobody can quantify the link between technological research and commercial payoff; innovation involves many factors beyond research."[2]

Nevertheless, many of the relationships among technology, commerce, and research are understood, the stakes are high, and some organizations consistently perform the innovation process better than others. And while those relationships are complex, and events can be fundamentally chaotic, it is equally true that individuals, institutions, and markets act in ways that are generally rational. In this sense, technology markets are akin to financial markets, and any effort to understand their dynamics is worthwhile. A more balanced view can be cited:

> Given the uncertainty of the process, it is a wonder that any manager ever considers financial analysis of corporate technological matters. But all decision-makers do look at financial calculations, probably because there are no other tools so pervasive. Most managers understand basic financial tools in the abstract, and that makes them the coin of the realm no matter how inappropriate their application.[3]

Aim of This Book

The aim of this book is to explore the linkage between research and development (R&D) and shareholder value in a comprehensive way and, in doing so, to create a common language and set of analytical tools that businesspeople, scientists, and engineers can use as they jointly plan new technology and R&D projects. That common language and set of tools will help these different communities communicate and make more enlightened decisions.

The fundamental viewpoint throughout is that *value-based management* leads to effective decision making.[4] Value-based management, however, is not a management system, but a mind-set. It is based on the established observation that capital moves to where it can earn the highest return—and that in providing a return for shareholders, all other stakeholders (employees, communities, etc.) will also benefit. Other metrics for business excellence, such as earnings growth or return on investment, lead into financial blind alleys if not accompanied by value creation. Focus on value also ensures a clear and common vocabulary for corporate communications.

Given these premises and the fact that the translation of science into commercially viable technology is especially risky, this book does not propose to outline a set of managerial techniques for making an inherently difficult process simple, reliable, and efficient. Instead, its goal is to guide readers in analyzing the technological and financial forces in order that they may be more valuable contributors to their organizations, and further their careers in the process.

The book will also convey my own experiences with technology creation and its valuation in a way that may be useful to others. The term "useful," in my view, not only includes offering advice and identifying pitfalls, but describing methods and metrics that help get the job done. Thus, the book will present and critically discuss tools for analyzing decisions about technology while recognizing that all such tools have important limitations. The individual reader can use these tools to supplement his or her own bank of knowledge about technology, markets, or organizational behavior to make informed choices.

Knowledge creates a basis for successful outcomes. Luck and perseverance are also important.

Who Should Read This Book?

Broadly speaking, valuation serves either of two purposes: (1) internal decision support (i.e., judging project proposals) or (2) transaction support (sale of an asset, negotiation of a license, or determining taxes). Thus, this book will be useful to the following readers:

- *Scientists and Engineers.* This book provides a broad introduction to how R&D creates value for business enterprises. It also reviews some of the pitfalls and hidden issues in measuring value. As a result, scientists and engineers will find it useful in identifying or structuring financially attractive projects, and in justifying those projects to potential investors.
- *Business Managers.* Business managers need tools to determine the level of R&D spending required to meet their growth goals, and how to determine whether R&D is sufficiently productive to create, rather than destroy, value. They will find them in the following chapters.
- *Financial Professionals.* Financial managers and analysts are generally familiar with the management and valuation of risk, and the nature of options. However, R&D assets are often hidden or have special characteristics. Risks are both high and variable. Intellectual capital created through research is often intangible. And technology options are more complex than financial derivatives.

- *Investors and Investment Analysts.* The decision to invest in, or to acquire, a firm often hinges on the valuation of its technology. Investors must weigh the intangible assets in the technology portfolio to determine whether the firm is undervalued or overvalued. Technology must also be valued by sellers to ensure fairness to shareholders. Investment bankers who offer valuation services must also come to terms with the special financial characteristics of technology.

Value in the Marketplace

A narrow financial approach to the valuation of technology ("just the numbers, ma'am") is self-defeating. Real-world valuation is a blend of "soft organizational issues, complex strategic questions, and the analytical methodology of discounted cash flow."[5] Nor are analytically determined valuations conclusive. In terms of being a forecast, a valuation expressed as a single number will *always* be wrong. The valuation that ultimately matters usually takes place in the marketplace, through negotiation between two parties or via auction. Each of the contending parties perceives value through a different set of lenses. Analytical techniques such as those presented in this book provide a range of reasonable values to guide the process and to support proposed valuations in a negotiation.

An art auction can serve as an example. An establishment like Sotheby's publishes an estimate of the value of each item, usually in the form of a range. Each bidder approaches the auction with different assumptions and a different perspective on inherent value. Rational dealers will bid up to a price at which their professional estimate of selling price will support their markup. They need to cover costs and earn a return on their capital. Essentially, they are financial buyers. Decorators and private buyers may see higher value because they have immediate use for the auctioned item; to them, a painting may have unique value because it is of a size or color that fits a specific space in the living room. For other bidders, an item that strengthens a collection has greater value than a similar item that does not. The sale of a single first edition volume of Marcel Proust's seven-volume masterpiece, *Remembrance of Things Past,* would be valued lower by Bidder A, who owns no other pieces of the collection, than by Bidder B, who already owns first editions of the other six volumes. Bidder B is a strategic buyer, and is likely to pay considerably more than the estimate (the valuation). In addition, experts with superior knowledge may spot undervalued assets.

In the technology marketplace, too, a range determined by financial criteria is a useful starting point for both sellers and buyers. Again, the value of a technology can be much higher for strategic buyers than for financial investors. And, different parties (major corporations, venture capitalists, and so forth) have different costs of money, which lead them to unlike valuations for the same technology. In one case, a technology with a historic cost of just over $20 million was valued internally by two alternate methods at $60–$90 million (using the viewpoint of a financial buyer). This same technology was sold for $150 million when two expert strategic buyers engaged in a bidding process for it.

For internal decision purposes, technology valuation hinges on the credibility of the analysis and on the individual analyst. *Metrics* can provide that credibility. This book provides a chapter on metrics through which forward projections of the benefits of R&D are linked to past performance and R&D productivity.

Structure of the Book

The flow of this book is motivated by its goal: to show how R&D creates economic value. This goal necessitates considerable discussion of finance, marketing, strategy, and corporate organization. Parts of the book are intended to review and reinforce financial concepts for the benefit of technologists. Other parts are intended for businesspeople to show how basic financial concepts apply to the unique and high-risk context of R&D. To avoid making this book into a "micro-MBA" course for scientists and engineers, general business subjects are discussed only as they bear on technology outcomes. More extensive sources of information on these subjects are identified through the references and notes.

Exhibit P.1 provides a road map to the approach to valuation of technology presented here. Input data are shown as ovals, whereas

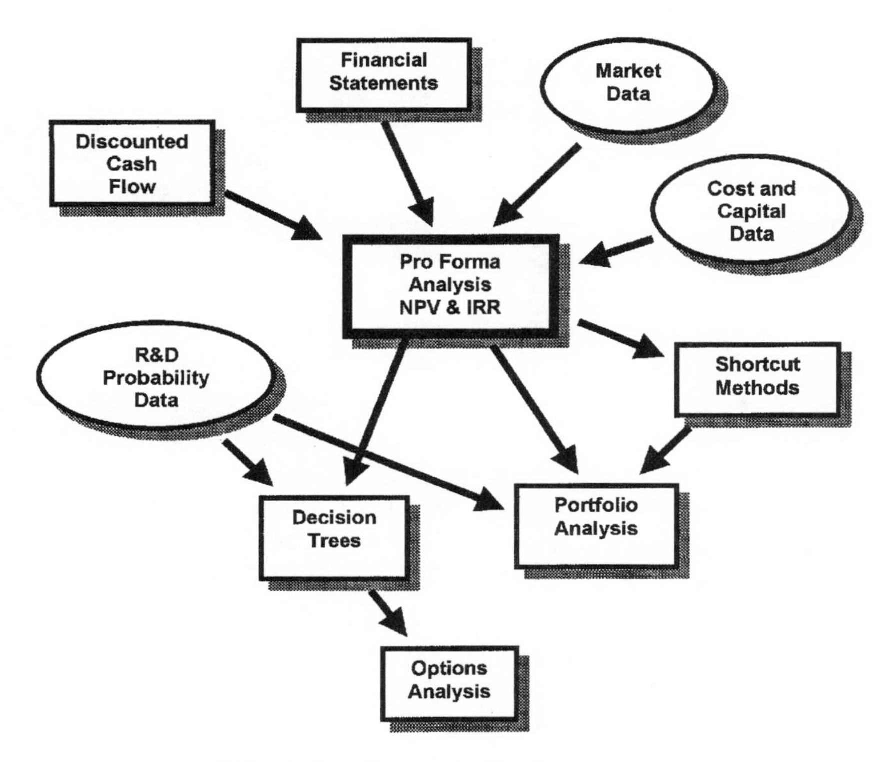

Exhibit P.1 Technology Valuation Roadmap

methods are shown as rectangles. Many of these blocks correspond to chapters or sections in the book, but other chapters include important material needed to keep the methods in context.

Chapter 1 addresses the question, *Why do scientists have difficulty communicating with businesspeople* (and vice versa)? It discusses common cultural and motivational differences that divide scientists and businesspeople and create stumbling blocks to effective decision making. One section of the chapter discusses the different approaches that these two groups take to forecasting.

Chapter 2 is an introduction to *the industrial R&D process.* It describes how value is created at each stage, the typical activities involved, the likely cast of characters, and what business issues come into play as the process unfolds. The chapter is intended for readers who have never directly participated in this process and includes three fictionalized cases drawn from my experience.

Chapters 3 through 5 introduce *key financial concepts* required for valuation to readers lacking a background in corporate finance and examines the applications of these concepts to R&D. Several important valuation pitfalls are presented and discussed. Chapter 3 reviews the principal elements of the income statement, the balance sheet, and the cash flow statement, and explains how they are linked. Chapter 4 is a discussion of operating factors that affect the costs of technology-based projects, such as economies of scale, break-even analysis, and ways to reduce capital-related costs through R&D. In Chapter 5, we introduce discounted cash flow and three closely related subjects: terminal value, the cost of money, and the concept of economic profit.

Growth and shareholder value are the central focus of Chapter 6. Growth is a great motivator for scientists, executives, employees, and investors alike. A quantitative analysis of the value of such growth, what it takes to achieve it, and the trade-offs between long-term growth and short-term profitability can be highly revealing.

Chapter 7 identifies the link between strategy and value, and addresses the mutual relationship between corporate strategy and technology strategy. How do corporate strategists determine the factors that drive value? Assuming that growth is important, how do we get it? And how can we assure that growth will be profitable? The presentation include the mechanics of creating the R&D component of a strategic plan—strategic planning as technologists approach it.

Most valuation is driven by the "top line"—by sales revenues. Chapter 8 indicates how to create a *revenue model,* and how market analysis and marketing tools can be used to estimate revenues and map the road to commercialization. The limitations of market research for making predictions are noted—particularly for new-to-the-world technologies. The use of S-curves and technology forecasting to improve revenue forecasts is also covered.

Chapter 9 reaches the heart of valuation: the *pro forma* project spreadsheet. In its electronic form, the project spreadsheet facilitates the exploration of outcomes through *sensitivity analysis.* "What if" unit sales are less than expected? "What if" the cost of production is higher than anticipated? Subsequent chapters build on the pro forma.

Chapter 10 introduces useful shortcuts and alternative approaches to the full pro forma method. These include *quick and dirty models, economic profit calculations, investment models,* and *licensing models.*

Until this point, the text has considered valuation in terms of the individual R&D project. Chapter 11 moves from the level of the individual project to the dynamic level of the *R&D portfolio*. The portfolio concept is important in estimating risk and minimizing it through diversification. How historical performance data can be used in portfolio situations is demonstrated. Portfolio analysis is essential in determining whether R&D resources are adequate to meet companywide growth objectives. A detailed example of portfolio analysis is presented using the quantitative tools developed in the earlier chapters.

Chapter 12 covers the application of probabilistic methods such as *event trees* and *decision trees* in adding value to the R&D process. It also introduces *Monte Carlo methods.* Because *options* embedded in a technology project are often an unrecognized source of value (and may be the principal value), options are discussed from both a financial and a technology viewpoint.

The important role of *diversification,* in both reducing risk and expanding opportunity, is explored in Chapter 13. This leads to an analysis of why a business must maintain a flow of new conceptual projects to create a high-value portfolio. External technology from a variety of sources can enhance value beyond what an in-house organization can be expected to achieve.

Chapter 14 contains a review and discussion of *R&D metrics* and their use in measuring productivity and in building credibility through metrics that link past, present, and future. The credibility of R&D in forecasting commercial results is frequently an issue. My view is that forward forecasts should be linked to past performance, and any significant discrepancy should be explained.

Some important R&D issues that affect value are the subject of Chapter 15. These include patents, technology transfer, globalization, and the environment.

A brief Afterword comments on the strengths and weaknesses of value management in the context of the mechanisms by which society assigns responsibilities for technological and economic progress, and how the system may be serving us well or less than well.

Origins

This book has three main sources of origin: business experience, teaching, and intellectual curiosity. As an executive, researcher, and board member, I have had the benefit of working with four very different companies. I began my career at Dow Chemical, working as a research scientist at the molecular level. Over a 10-year period, management positions came my way in various research organizations: basic research laboratories, research in manufacturing divisions, and product development (in the field of urethane polymers). This period was rich in business experience, dealing with problems as varied as commercializing radically innovative products (ion chromatography), and positioning a large, technology-driven polymer business in a competitive global environment. Along the way, I absorbed valuable practical knowledge of industrial marketing and of cost accounting. Notably absent from my experience, however, was any real insight into the financial forces driving key investment decisions. It is my impression that many of today's younger scientists, engineers, marketing, and manufacturing personnel experience the same gap in their on-the-job educations.

The next three years at Dow provided an initial opportunity to redress this lack of financial acumen. As head of research for a product line, I became the leader of a cross-functional business team. Among my duties was a requirement to develop a five-year business plan—a daunting proposition in the days before electronic spreadsheets! This job led to the position of business manager for a group of seven businesses, and responsibility for both profit and loss and capital budgeting decisions. The focus, however, was more operational than financial, and the use of discounted cash flow techniques was altogether absent. (A key issue, interestingly, was the proper allocation of cost and capital in a highly integrated business—an issue that seldom goes away.)

For personal reasons, and with serious regrets about leaving an exciting job and a strong company, I took an executive position with the American Can Company, running that company's research and development function and its specialty chemicals business. American Can introduced some very interesting new elements to my business experience. The company was highly diversified among packaging businesses, consumer products, and financial services. Competition among business units for resources was intense, and poor performance by the company's traditional businesses—and major mergers and acquisitions blunders—made resources scarce. I learned that selling healthy businesses to raise cash was, surprisingly, just as real an option as divesting the sick and wounded. Shareholder value was dismal and company management even allowed itself to be "greenmailed" by a dissident shareholder.

As the company struggled to restructure itself during a period of high inflation and high interest rates, I could not help but sense the power of the financial marketplace on my work and career. I knew that I needed to understand the financial dimensions of our work much better. In particular, it became clear that unlocking shareholder value would drive the future of the company, and that businesses and technologies would be bought and sold in the process. Professionally, I learned to use discounted cash flow (DCF) analysis and terminal values to win support for promising research projects such as plastic bottles and cans. The cost of money was an important element in this analysis, particularly in an inflationary period.

My first encounter with valuation occurred at about this time. It was my good fortune to lead a team charged with divesting American Can's specialty chemicals business, which we accomplished at a sale price that was more than twice management's expectation. This useful experience probably would have been unavailable in a healthier company.

It soon became clear that American Can would divest all its technology-intensive businesses in pursuit of a flawed strategy that viewed capital-intensive businesses as inherently bad. So, in 1982, I joined W.R. Grace & Company, which was then enjoying record profits from its oil and gas and chemicals businesses, and seeking to diversify into a number of technology-driven businesses such as health care, biotechnology, separation science, and ceramics. Grace was, like American Can, highly diversified and held a number of non-technology-driven businesses in its portfolio: restaurants, retailing, distribution, and food products.

Grace tolerated, even reveled in, a variety of business styles. It gave its group executives and division presidents great autonomy. However, it united these free-ranging businesses and executives through a culture of detailed financial analysis. Annual business reviews were a major event, involving the creation of thousands of formal numerical charts. These reviews and "the numbers" gave then-CEO, J. Peter Grace, and his key executives opportunities to probe business issues in considerable depth. All major capital projects had to be extensively documented in formal authorization proposals, which, though lengthy and time-consuming, also were superb primers on the business dynamics and financial drivers of the proposals under review.

Historical performance and credible forecasts were the centerpiece of Grace's capital project review. And since the customary penalty for disappointing results was understood to be divestment, managers were not inclined to play fast and loose with their projections. R&D was not exempt from the requirement to account for past performance and to forecast the future returns from its ongoing projects, but to meet this requirement we needed new ways of analyzing the data—that is, new "metrics." We therefore began to assemble data on the success rates of past projects to improve our ability to project future outcomes, to value the projects in the R&D pipeline, and to track our own productivity.

Fortunately, the company's R&D function had its own financial officer and a small staff of MBAs (with technical backgrounds) to perform project analysis. I came to appreciate this capability, both for analyzing project proposals and for communicating to a financially oriented top management group.

Before long, however, it became evident that the level of detail sought by the company was often counterproductive, inhibiting timely reporting, delaying the start of projects, and interfering with the work itself. The planning department was backlogged and holding up research. Faster and simpler analysis was needed if we hoped to keep pace with fast-moving technical programs. This need led to the development of the simplified pro forma DCF methods incorporated in this book as well as systems for tracking milestones and other key project parameters.

In hindsight, it is clear that a serious effort to analyze the research portfolio in business, strategic, and financial terms, and to subject the result to critique by experienced executives, strengthened our technology-related decision making.

My years with Grace (which ended in 1995) and with NOVA, where I have served as a Director since 1991, have reinforced my view that tremendous shareholder value can be unlocked through the value approach, and that researchers are far better served by understanding how their technology contributes to value than by hand-wringing over the myth that focusing on shareholder value is the death of a long-term viewpoint. To the contrary, a careful analysis of modern valuation techniques shows that the single largest element in financial value is the forecast of long-term financial performance, as it is built into projections of long-term free cash flow. But dangerous mistakes can be made if terminal values are handled improperly.

The direct inspiration for this book resulted from a talk I gave in 1992 on the linkage between R&D growth and shareholder value. The setting was a meeting of the Industrial Research Institute (IRI); the audience was a group of top R&D executives representing almost 300 of the country's largest industrial companies. The times were such that many in the audience—particularly those whose firms maintained large, central research laboratories—were feeling the heat from corporate downsizing, outsourcing, and a growing demand that every unit justify itself. Since few of these managers knew how to measure the financial value of their work, especially long-term research, the feeling that "our heads and our departments are on the block" was palpable and pervasive. My talk seems to have struck a chord with this audience, as the IRI asked me to develop a short course on the financial impact of technology for industrial researchers and research managers. The need for the course was predicated on the observation that few of the many "mini-MBA" courses being offered to scientists and engineers related business tools to the actual problems faced by scientists and engineers. Much of the material in this book is an outgrowth of the development of that course.

The most important intellectual source for this work is Simon Ramo's classic book, *The Management of Innovative Technological Corporations.*[6] A distinguished engineer and successful businessperson, Ramo provides a coherent description of the impact of technology-driven growth on a company's fortunes. The book's main drawback is that most of the key discussion is presented in the form of elegant differential equations. These reflect Dr. Ramo's superb intellect, but make the book inaccessible to those lacking the required mathematical background.

The model presented in Ramo's book is one that engineers might call a "steady-state" growth model in which the relationships between revenues, earnings, investment, and cash flow are fixed. In such a model, the corporation maintains a steady growth rate by reinvesting retained earnings while maintaining the same ratios between the various elements of its income statement and its balance sheet. This model was a useful starting point for satisfying my own curiosity about the impact of additional growth on market value.

Books on corporate finance and on financial valuation of companies also influenced my thinking. These discussed event and decision trees, the evaluation of risk, the value of diversification, the valuation of opportunities, and option pricing models.[7] These are powerful concepts. We sense from experience that technology offers ways to create (and destroy) wealth more rapidly than any steady-state model can explain. *Wealth is created and destroyed when the ratios in the steady-state model change.* Because most of these texts are securities-related, where value equates to price, many of the variables of technology decision making and opportunity creation are discussed superficially.

Other, more market-based works on technology valuation have more lately been penned by authors in the fields of licensing, venture capital, and from the emerging profession of technology appraisal.

In directly addressing the issue of technology valuation, from the research laboratory to the securities marketplace, this book bridges a difficult gap in the business literature. It is, in my view, little more than a "rope bridge." I leave it to other authors and to future editions to strengthen and expand on this simple structure. There is much more to be done.

Acknowledgments

This book fulfills my personal belief that each person should make available to others what he or she has learned and experienced, particularly those aspects of learning and experience that are to a degree unique. I am obliged to a great number of people and institutions for these, particularly my former associates at W.R. Grace & Company. Francois van Remoortere and Barry Hotchkies were instrumental in developing metrics to track our research accomplishments and provide credible projections for presentation to top management. Martin Sherwin and Louis Hegedus provided real case studies of enormous interest, and lent their formidable intellects to the task of analyzing the strategic and technology drivers at play. David Seifert and I worked closely in developing simplified pro forma models for individual R&D projects and in exploring their sensitivities to key financial parameters. The analytical atmosphere and strong emphasis on financial controls fostered by the late J. Peter Grace, Jr. provided a rigorous environment in which to test the soundness of valuation concepts. My thanks goes to the many Grace executives who helped develop these concepts in spirited discussions over the years, and especially to Mr. Grace for his steady encouragement.

The support and friendship of IRI Executive Director Chuck Larson and members of his staff deserve special acknowledgment. Bob Burkart worked closely with me in developing the "Financial Impact of Technology" course, and Barry Siadat and Lee Starr helped present the pilot version.

May Adams and Madge Surbaugh, my assistants over most of the past two decades, have been invaluable helpers, not only in the preparation of materials used in this book, but in organizing my activities when the pace was fast. Richard Luecke provided skillful assistance in editing the book and arranging for its publication. Jeanne Glasser and her colleagues at John Wiley & Sons were a pleasure to work with during the final stages of editing and production.

Finally, my wife Ellen has, with patience and good humor, witnessed what all authors' spouses must: long hours at the computer, seemingly pointless research, false starts, and dissatisfaction with early drafts, broken by an occasional glimmer of optimism that something useful is being created. Her vision, support, and enthusiasm are deeply appreciated.

Contents

1 SCIENCE, TECHNOLOGY, AND BUSINESS **1**
Science versus Technology 2
What Is Technology? 4
The Scientist and Businessperson 6
A Value Model for the Corporation 16
In the Same Boat 19

2 THE INDUSTRIAL R&D PROCESS **21**
R&D Stages 22
Stage Zero. Finding and Screening Raw Ideas 24
Stage 1. Conceptual Research 27
Stage 2. Feasibility 32
Stage 3. Development 35
Stage 4. Early Commercialization 38
Summary 41

3 A TECHNOLOGIST'S GUIDE TO FINANCIAL STATEMENTS **45**
The Income Statement 46
The Balance Sheet 52
The Cash Flow Statement 65
Other Key Concepts Relating to Financial Statements 69
Issues in Financial Statements 71
Accounting for Intellectual Property 72
The Value of Technology Is Situational 75
Depreciating Technology 77
Financial Leverage 78
Different Measures of Return 81

4 CAPITAL FROM THE OPERATING VIEWPOINT **84**
Break-Even Analysis 84
Economies of Scale 87

Matching Capacity to Demand 89
New versus Existing Plant 91
R&D Effects on Capital 92
Summary 95

5 Calculating Values Using Discounted Cash Flow 96
Discounted Cash Flow 97
The Cost of Money 103
Discount Factors 112
Risk-Weighted Hurdle Rates for R&D 113
Terminal Value 115
Time Horizons 122
The Pitfalls of Focusing on Cash Flows 124
Summary 126

6 R&D, Growth, and Shareholder Value 127
What Is the Value of 1% in Added Growth? 128
Growth and Profitability—The Trade-Off 130
The Business Life Cycle 132
Portfolio Balance Can Be the Key 134
R&D: Trading Current Profits for Growth 135
Can R&D's Contributions to Value Be Separated from the Other Functions? 137
Other Sources of Growth 138
Two Strategies for Corporate Growth 141
Comments and Caveats 143
Summary 145

7 Strategy: Driving Value in the Competitive Arena 146
The Role of Value Drivers 148
What Is Strategic Planning? 149
The Case for Strategic Planning 151
Resource Allocation 152
Pitfalls in the Strategic Planning Process 153
Developing a Strategic Plan 154
Summary 168

8 Marketing: The Top Line 169
An R&D Perspective on Marketing 170
What Is a Product? 172

What Is a Market? 176
Market Segmentation 178
Consumer Markets 183
Industrial Markets 184
Government Markets 186
Market Research 188
Pricing 190
Delivering Value 196
Distribution Channels and R&D 198
Market Share 200
Technological Performance and Product Life Cycle 201
Product Life Cycle 204
Technology Forecasting 206
Creating a Revenue Model 208
Summary 212

9 BUILDING A PRO FORMA DCF MODEL 213
Purposes of the Model 213
Getting the Assumptions Right 214
Building a Pro Forma Model 216
The Income Statement 218
The Balance Sheet 226
The Cash Flow Statement 230
A Case Study 233
Pro Forma Results 234
Summary 236

10 SHORTCUTS AND MARKET-BASED APPROACHES TO VALUE 237
Sensitivity Analysis 238
Tornado Diagrams 241
Quick-and-Dirty Models 242
Using Quick-and-Dirty Models for Decision Support 247
Project Justification 248
R&D Management as Risk Reduction 250
Treating Sunk Costs 250
Terminating Projects 251
Economic Profit Calculations 252
Market-Based Approaches to Value 254
The Marketplace for Technology 255
Licensing as Valuation 262
Licensing Scenarios 264

Licensing Revenues 266
Technology Appraisal 272

11 MANAGING VALUE AND RISK IN THE R&D PORTFOLIO 274
Corporate Growth Goals and R&D 274
Modeling the R&D Pipeline 276
Success Rates 281
Project Flow in a Model Laboratory Unit 284
Costs, Productivity, and Value in the Portfolio 286
Summary 289

12 DECISION TREES AND OPTIONS 290
Decision Trees 291
Results 293
Probability-Weighted Project Outcomes 295
Monte Carlo Analysis 297
The Option to Accelerate 299
How Realistic? 300
Options Analysis 300
Polyarothene Technology Option 307
Other Characteristics of Technology Options 310
Sources of Value in Technology Options 311
Technology Platforms 312

13 CREATING VALUE THROUGH DIVERSIFICATION 314
Efficient Portfolios 314
Diversification versus Productivity 320
A Value Proposition for Early-Stage R&D 320
Sustaining Innovation 322
Accessing External Technology 324
Later Stage Providers 329

14 R&D METRICS 331
Historical Metrics 332
Current Metrics 341
Future Metrics 347
Direct Valuation of Current Project Portfolio 350
Summary 353

15 SPECIAL ISSUES IN VALUE-BASED R&D **354**
Patents as Value Drivers 354
Technology Transfer 361
Globalization 365
Environment 369

AFTERWORD **376**

CHAPTER NOTES **379**

INDEX **389**

ABOUT THE DISK **401**

CHAPTER 1

Science, Technology, and Business

The business of valuation is about quantification. When we ask the value of this or that, we are seeking a numerical answer, usually in dollars, marks, yen, or other currency. This is true whether the object of our attention is a wealth-producing piece of equipment, such as a machine tool, an intangible generator of wealth, such as a patent on a drug with proven therapeutic value, or an aesthetically pleasing item with no particular commercial utility, such as a fine painting. Each can be, and regularly is, valued in monetary terms. Even the value of human life is routinely quantified in legal cases involving damages and wrongful death.

The value of technology and R&D is also quantifiable, as the following chapters demonstrate. Being able to quantify the value of technology and R&D projects—even those still on the drawing board—is essential for many people: accountants, buyers and sellers of licensing agreements, investors, patent holders, financial managers, and executive decision makers. Before we get into the numbers, it is appropriate to give some attention to the softer, human side of valuation. Value is, after all, very much like beauty—it is framed in the eye of the beholder. And every beholder is different. Because of their experiences, educations, prejudices, and goals, people see the world in many different ways. On passing a new Jaguar automobile in the parking lot, for example, an industrial designer would probably be struck by how elegantly and seamlessly the manufacturer has brought together form and function. A 22-year-old surfer, on the other hand, would probably be disdainful: "Now here's a useless car—how could I get my board on top of this thing?" A U.S. trade negotiator walking past the same car would likely grimace, mentally calculating its contribution to the nation's balance of trade deficit.

Scientists, engineers, business managers, financial specialists, and investors likewise bring different outlooks and expectations to the valuation of technology and R&D. They do not think alike and have diverse backgrounds and motivations; nor do they necessarily understand each other. My own experience indicates that most nonscientific people barely understand the difference between science and technology—and may not even consider that difference to be important. On the other hand, many industrial scientists cannot interpret their employer's financial statements, much less see how the work they do every day finds its way into those statements.

It is not surprising, then, that all these people have trouble discussing the value of the technology and R&D activities that represent the future of their companies. Things would go more smoothly if they understood each other better and had a common language for representing their different outlooks. When businesspeople try to assign a monetary value to a particular technology, it is important that they know what technology is and isn't, and why scientists interpret the facts as they do. When scientists seek funding for their ideas, they need to understand the processes involved in the creation of value and how their own work will benefit or *not* benefit the company—in terms that businesspeople will understand.

Science versus Technology

Though the patterns of our daily activities are heavily shaped by contrivances like telephones, fax machines, computers, electrical appliances, automobiles, temperature controls, and hundreds of other human inventions, few concepts are so broadly misunderstood as the distinction between science and technology, much less the relationship between them. The confusion is driven in part by the fact that the same people (i.e., scientists) work in both areas, and both areas are driven by research. But science and technology are not the same.

And the distinction is important because judgments made about the one do not necessarily apply to the other.

Science is first and foremost a process of discovery. It has a unique characteristic: once a scientific discovery is made and reported, it is made for all time. There is no need for anyone to discover that same fact again (although there may be a need to duplicate it for verification). This characteristic is an enormous motivator for scientists. Published scientific discoveries, whether profound or trivial, are essentially immortal and forever associated with the names of their discoverers. Newton's laws, Maxwell's equations, Avogadro's number, Einstein's theory of relativity, Heisenberg's uncertainty principle, and Planck's constant are among the better known examples. And because scientific discoveries are recorded and referenced in multiple archives, there is little chance that significant work will be forgotten. The creators of technologies, corporations, and physical monuments have far fewer opportunities for leaving their marks. The best artists, writers, and musicians may see their works broadly recorded and preserved, but with the distinction that these works are to be appreciated for themselves rather than as part of a coherent and advancing whole.

Science has a unique culture defined in part by the "scientific method," a process for relentlessly scrutinizing the relationships between observed facts to establish a theory. Once a theory is established, it is used to predict and discover new facts. Peer review assures the integrity of scientific output and the culture that supports it, providing some assurance that inadequately supported facts or false claims will be rigorously censured. A notable example of this censure was seen not long ago in the premature public announcement of the discovery of cold fusion.

The scientific culture is also deeply skeptical of claims made by laypersons when such claims have not yet met standards of scientific verification—UFO sightings and cancer cures are good examples. This innate skepticism of unverified facts is deeply resented by many outside the fraternity of science, who equate it with arrogance and close-mindedness. Individuals with little direct experience with science, including politicians, periodically attempt to capitalize on this resentment and seek to introduce legal and adversarial processes into science in the name of "scientific integrity." Others seek funding and legitimacy for "alternate science" (e.g., creationism), or "alternate medicine" (e.g., acupuncture and homeopathy).

Scientists, however, feel strongly about the value of their culture and methods and, with good reason, fiercely resist attempts to graft the culture of the press and the courtroom to the culture of the laboratory. In so doing, however, they sometimes dismiss legitimate concerns and alienate politically or economically powerful groups.

Scientific culture is learned primarily in graduate school through a process whereby doctoral students serve as apprentices to university professors.[1] There, they learn the standards of scientific scholarship in the course of their thesis research and their participation on research teams. This ivory tower environment is surprisingly competitive and has unique rules and standards, enforced by the system of peer review. Ph.D.s produced through this system dominate the culture of academia, industrial research, and some units of government.

The nonscientist usually gets a dose of the scientific culture in course work, but this brief exposure, and a lack of hands-on research experience, rarely sinks in, and generally leaves the layperson uninformed about the difference between science and technology.

What Is Technology?

Technology is the application of knowledge to useful objectives. It is usually built on previous technology by adding new technology inputs or new scientific knowledge. For example, the innovators of the electric typewriter simply grafted an electromechanical system to the existing keyboard, carriage, and inking systems of the standard manual typewriter.

Technology may even involve little or no science, as scientists define the term. New financial software is surely technology, but creating it requires virtually no new science beyond what is already embedded in computer hardware. And it will almost certainly incorporate software (technology) written by previous generations of programmers.

The general preconception that technology is created *from* science is encouraged by many spectacular examples of which there is broad public awareness: the discovery of penicillin produced the technology of antibiotics, the theory of relativity led to the atomic bomb and atomic power, and the discovery of the nature of DNA is producing new drugs based on modern biotechnology. However, much technology is created by trial and error and optimization by nonscientists and nonengineers. Thomas Edison was neither a scientist nor engineer, and his invention of the incandescent lightbulb involved no new science. It was the outcome of Edison's dogged trial-and-error approach to finding a filament material that would glow brightly and over long periods when electrified in a vacuum.

New technology is also created through the combination of two or more technologies without much true scientific intervention—the portable transistor radio, which combined the technologies of the radio, the battery, and the transistor, broke no new scientific ground. But it was enabled by the low power requirements of the transistor, which incorporated advanced solid-state physics and *was* a scientific novelty.

The criterion for successful technology is usefulness, as defined in commercial, military, social, or medical terms. Usefulness does not, in many cases, require that the user understand the technology or what makes it work. We observe this when a violinist buys a personal computer loaded with musical notation software, and when an Arab sheik purchases a refinery. Chances are that neither the violinist nor the sheik has any clue as to how his technology works. In contrast, science is not very useful to people who lack scientific training. This criterion is helpful in distinguishing science from technology.

Technology also differs from science in the fact that it becomes obsolete. The vacuum tube was a very useful technology in its time, but has been displaced almost entirely by a successor technology, the transistor. Science, in contrast, is never obsolete.

Ironically, advances in science are often dependent on advances in technology, which many view as its stepchild. The early scientific discoveries of astronomy were largely driven by the technology of lens making. Some of that technology was created during the Renaissance, long before the rigorous scientific theory of optics was formalized. Today, our scientific theories of the universe and its origins are being reshaped by data supplied by the Hubble telescope, which was placed in orbit high above the earth's soupy atmosphere by aerospace technologies.

The relationship between technology and the science of astronomy is not unique. The technology of metallurgy preceded the science of metallurgy by at least three millennia. More recently, the improving technology of X-ray diffraction allowed scientists such as Watson and Crick to elucidate the first structures of DNA, and later the detailed structures of many proteins, and create the basis for a new science—molecular biology.

If we accept that science is only one factor driving technology and that new science is itself often mediated by new technology, it remains that the organized translation of science into technology is an exceedingly important and fruitful activity. Furthermore, it is primarily conducted by scientists and engineers in a process called "research and development." In the technology context, research means something quite different from the discovery of basic scientific knowledge; it is the endeavor to use scientific training and skills to produce new and useful products and processes in a timely and reliable manner.

In summary, the linear model that science creates technology in an orderly and predictable fashion is simplistic and generally wrong. An investor cannot be assured a good return or even any return by investing in basic scientific research. Technology has dynamics of its own. Its creation and application will always employ scientists and engineers, even absent new scientific discoveries. And an investor can be assured a good return, at least for some time, by a knowledgeable investment in technology. Science, however, has everything to do with the pace of technological innovation, and hence with how large future technological opportunities will be. And the structure of the scientific community—industrial, academic, and governmental—will matter a great deal in determining which companies and which nations will reap the rewards.

The Scientist and Businessperson

Decades ago, British novelist C.P. Snow referred to a gulf in understanding between scientists and people who lacked technical understanding, and he explored the social implications of this difference as a new breed of individuals rose to prominence in a tradition-bound British society.[2] The society of the modern industrial corporation has changed greatly from the society described by Snow. It is not at all unusual for businesspeople, and even financial professionals, to be technically literate. Many large industrial corporations, in fact, make a technical degree a prerequisite for employment: a chemical salesperson without a degree in chemistry, or a refinery supervisor without a knowledge of engineering, is a fish out of water. Increasingly, traditional service-oriented companies are also hiring technically trained individuals who can master and design complex business systems. The combination of an engineering degree and an MBA is considered powerful training for aspiring executives in manufacturing industries.

Motivations

If so many business executives have adequate scientific education, why does the gap in understanding between scientists and businesspeople still loom so large? The answer is cultural, and can be found in the motivations that these two cultures instill in their members.

The motivation of scientists engaged in industrial research is twofold and may shift with time. First, while they expect to benefit financially from their employer's business success, most find even larger psychic rewards in seeing their work realized in products used by millions and in factories employing hundreds or thousands of workers. Second, they are motivated by the norms and expectations of their scientific peers; here, status and visibility go hand in hand with patents, technical publications, and the quality of one's research. For many, recognition from the scientific community carries more weight than any stock options, bonuses, or corner offices bestowed by their employers. And allegiance to the corporate employer often takes a backseat to the brotherhood of science and to the development of one's individual knowledge and skills. For some members of the scientific fraternity, the corporation is simply a convenient environment in which to do research—a warm, dry, well-lighted place with a budget, a prestigious corporate name, and lots of first-rate equipment, where one need not be bothered with teaching or chasing after funding grants.

The motivations of businesspeople are generally different from those of scientists and vary widely, but there are three key elements. First, businesspeople are more highly motivated by monetary success. This is not just a matter of being rich—it can be a belief that money is how the score is kept. Nor does it mean that scientists are wealth-averse. Still, few of the scientists I have known are as motivated by money as the typical business executive; few would pursue an R&D project with the single-minded expectation of getting rich. The same cannot be said of most businesspeople.

A second major motivator for the businessperson is power and the perquisites of the corner office. Here, there is little difference between the cultures of science and business. University professors scramble as much as anyone else for choice offices and good parking places, and most will avail themselves of as many graduate assistants and postdoctorate fellows as their positions and reputations can muster. Businesspeople, with the exception of some technical specialists, see themselves as managers, and naturally seek to influence if not control the world around them.

The third major motivator for businesspeople is the desire to build a lasting legacy—to build a great institution bearing their personal stamp. Jeffrey Sonnenfeld and others have pointed out that this dream is often unrealistic in view of the inherent fragility of business enterprises, but the emotional need to build monuments is powerful nevertheless.[3] This motivator is particularly strong among CEOs, who have much greater difficulty in separating their personal and corporate identities than do managers at lower levels. And it appears to be strongest among those least satisfied with their lifetime achievements.

Legacy-building *sometimes* leads to irrational behavior. For example, a number of otherwise credible senior executives I have known have taken major gambles on long-shot projects in the years just prior to retirement. These adventures could not have been financially motivated since some of the cost was borne out of present income, whereas the reward (or disaster) would only occur on the successor's watch. These last shots at greatness were made by individuals dissatisfied with their final level of achievement.

At the other extreme, some business executives rapaciously seek to maximize their personal interests as retirement draws near, even at the expense of the firm's long-term interest. Loading up on bonuses, option payouts, and other forms of compensation, they echo the sentiment of the French king who said, "Après moi, le déluge [After me, the deluge]."[4]

Ironically, the scientist, not the businessperson, has the best opportunity to leave a legacy and gain a measure of immortality. Today, we remember Archimedes, Pythagoras, Copernicus, Kepler, Newton, Galileo, and DaVinci, but how many masters of commerce can we recall from those early periods of science?

Vision

The popular business literature of the past two decades has acquainted us with many contemporary executives whose visions of the future were both successful and transforming: Fred Smith with his guarantee of next-day delivery via Federal Express. Lee Iacocca, who very publicly committed to rebuilding the shattered Chrysler Corporation—and succeeded! Steven Jobs in the early days of the PC era, whose challenge to John Sculley of PepsiCo is famously remembered as "Do you want to spend the rest of your career selling sugar water, or do you want to change the way people live?" Jack Welch, who forced giant General Electric through several transformations in succession. John Young, who laid down a corporatewide stretch goal for Hewlett Packard to reduce new product time-to-market to one-tenth its then-current pace.

Businesspeople like these think big and act big. They have the self-confidence to stand at the edge of the unknown, say, "Let's do it," and then take the leap. And if they enjoy sufficient credibility, others will leap with them.

Scientists don't operate this way. Their peer review process encourages skepticism. The reviewers are anonymous, are expert, and are engaged to be devil's advocates. Most are fair-minded and will praise a good piece of work. However, they will scrutinize any work for potential flaws, and if necessary ask the scientist to prove any assertions not adequately demonstrated. A scientific paper with exaggerated claims is likely to be received with biting commentary. Surviving this process builds credibility—a highly prized trait in the scientific community.

Vision for the scientist is a much trickier item. The best scientists have clear and far-sighted vision. However, they have two good reasons for keeping their visions under a barrel. The first is that unsupported theories provoke the skepticism of their peers and undermine their potential for advancement, particularly in academic circles. The second is that good ideas are a scientist's most valuable property, and airing them prematurely may give rivals a chance to beat her to the actual discovery.

Nobel Prize winning chemist Linus Pauling was an exceptional case of a visionary in the field of science. Indeed, as a "second generation student" of Pauling, I was greatly influenced by his ideas and very aware of his style. His provocative text in physical chemistry, *The Nature of the Chemical Bond,* was based on visionary thinking of the highest order for its time, and was replete with original theories and hypotheses. It set the stage for two decades of further research. On careful reading, however, that book had as many near misses as direct hits. But Pauling was a risk-taker. Later in his career, he used his Nobelist stature to advocate the cause of peace (and gained a second Nobel) and a medical hypothesis that massive doses of Vitamin C would cure many human illnesses. The latter idea could never be demonstrated scientifically and tarnished some of Pauling's scientific credibility.

The bottom line is that the scientific culture is skeptical of visionary leadership, and this makes the Linus Paulings a very rare breed.

A clear exception has emerged on a border between science and business, within the domain of science-based start-up companies. To raise money from investors who demand high returns, the scientist/businessperson must declare a vision if not outright promises. In some cases, promises give way to hype—whether it is hype about the potential of his or her technology in the face of other technical approaches to the objective, or the potential to dominate a marketplace with strong, entrenched competitors. Visionary leaders thrive in this environment, and investors are advised to maintain a healthy skepticism, perhaps reversing roles and playing the part of peer reviewers.

Of Style, Precision, and Forecasts

Some cultural differences between scientists and businesspeople emerge in their approaches to data. An example can be seen in their respective preferences for presenting numbers. Scientific presentations have a bias for graphs. Indeed, in companies where technical people are strongly represented in management, graphs dominate the discussion. Financial executives, and businesspeople, in contrast, usually favor numbers. Tables are often the preferred means of communication. Well-constructed tables permit a concise presentation of key ratios and trends in these ratios that graphs cannot convey as efficiently.

A second major difference is related to the precision of data. Accounting data are precise; financial forecasts are not because they are based on fuzzy assumptions, which become even fuzzier in the more distant years of the forecast. Nevertheless, there is a tendency among businesspeople to think of financial forecasts as precise, since they are formatted exactly like historical data, which *are* accurate.

To a scientist, relationships (such as the law of gravity) are precise, but numbers are not. Scientists are rigorously trained to associate each number with an "error flag" relating to random uncertainty imposed by inaccuracy in measurement. They also recognize the potential for systematic errors in measurement; indeed not to recognize it is tantamount to incompetence.

To scientists, the difference between a five-year profit forecast of $100 million and $110 million is insignificant, given the economic uncertainties affecting the forecast. For them, both numbers are "in the same ballpark." Accountants would be inclined to view the difference literally—as $10 million, which is a lot of money—and seek the reason for the difference. The differences between scientists and businesspeople in this area are such that old hands advise scientists to avoid showing quantitative projections, since their business counterparts will remember these and hold the scientist to them. I do not subscribe to this view, but note it to highlight the culture gap.

Despite their differences, businesspeople and technologists must make forecasts for planning and decision-making purposes. And, in terms of valuing technology and R&D, forecasts play a crucial part. The main issues are the methodology employed and the planning horizon. The simplest and most common forecast is a linear or pseudolinear extrapolation of the current trend. A linear extrapolation in its crudest form may consist of simply extending a line drawn through the last two points. A more sophisticated approach is the "least-squares" method, extending a line fitted by an established mathematical procedure to a long series of data points. Although the mathematics of each approach is precise, each begs the questions of whether 2, 3, 5, 10, or 50 data points should be used, and whether some of the data is unrepresentative and best excluded (which will affect the forecast).

At the risk of introducing a "hard" example in a "soft" chapter, I have included a hypothetical case of this forecasting problem. Exhibits 1.1 and 1.2 are based on a series of earnings for a hypothetical business in years 1 through 10. The business has been growing for a decade, and has current earnings of $115 million. It suffered depressed earnings in year 6, due to a major strike in that year. The task is to forecast earnings 3 years out (i.e., in year 13).

Five of many possible forecasts are offered. The least-squares extrapolation based on 10 years of history would give earnings of $122.4 million in year 13. This is the most conservative number in that it fits a trend line to *all* the data points—both good and bad, current and noncurrent. If the strike year is excluded, the forecast becomes $128.6 million. A forecast based on the past 3 years (which were all healthy) is even more optimistic and gives $143.3 million. All of these could be

	Year	Earnings (M$)
Historical Data	1	$ 50.0
	2	55.0
	3	65.0
	4	66.0
	5	80.0
	6	40.0
	7	80.0
	8	95.0
	9	100.0
	10	115.0
Three-year forecast based on:		
10 years of history	13	122.4
10 years of history, excluding year 6	13	128.6
5 years of history	13	171.0
3 years of history	13	143.3
2 years of history	13	160.0

Exhibit 1.1 Hypothetical Earnings Forecast

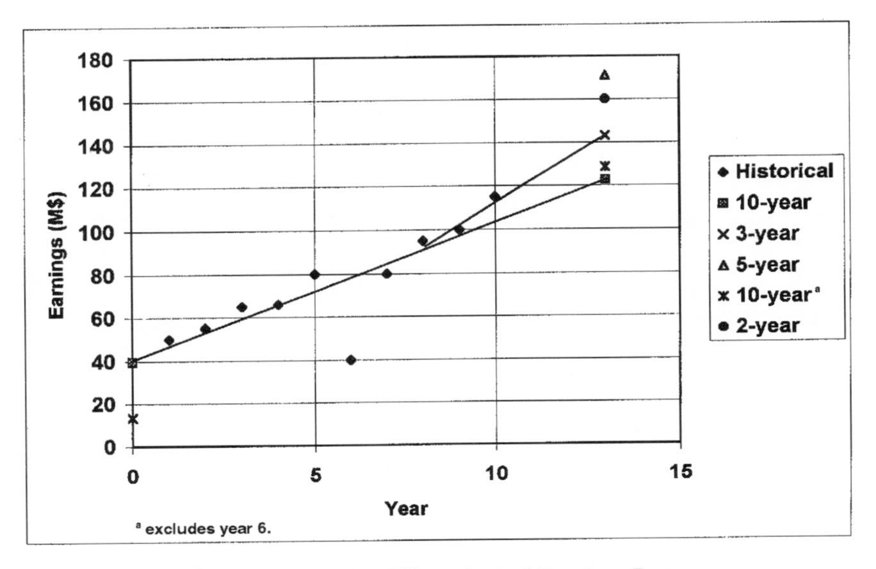

Exhibit 1.2 Graph of Hypothetical Earnings Forecasts

considered reasonable forecasts absent business information to the contrary. An extrapolation from the past two years seems overly optimistic and risky, since on a percentage basis we are projecting average annual growth of 11.6% versus the historical average of 9.7%.

Has the business changed in some fundamental way that would justify this optimism? Or are the past two years a pleasant aberration from the long-term trend of earnings growth?

The most pernicious forecast in this case is the five-point forecast, since it is built off a very weak base year in year 6 and projects an illusory growth rate. In this case, the projected earnings are $171 million. Many annual reports include only 3 to 5 years of data, making management look very good when the initial year is at trough conditions (and, to be fair, bad when it was a peak).

A natural-born optimist could create an even rosier forecast from the same data. The average annual growth from years 6 through 10 was 30.2%! Extrapolating this growth by a constant annual percentage (this is a common type of pseudolinear extrapolation), would give a 3-year earnings projection of $254 million. This type of biased forecasting is, regrettably, not unknown.

In my corporate experience, businesspeople are inclined to base their plans and decisions on the most recent few years of data and a "seat-of-the-pants" extrapolation of the current trend, especially if that current trend casts their performance and the prospects for their pet projects in a favorable light. Their natural optimism leads them to the most favorable case. Statistical tools are seldom used. Why this happens is no mystery. With the exception of financial personnel, selling and convincing are the everyday stuff of life for most businesspeople. And they are always being measured against goals, quotas, and the budget-to-date.

Scientists who look at the same data are inclined to see a different future. They are dismayed by the businessperson's selective use of data and approach the facts with greater objectivity and more attention to the variance in the numbers. For them, forecasts should be approached through the same scientific method that guides their other work. Hypotheses are made, and if verified by statistically convincing experimental evidence, may take on the status of scientific theories. These in turn may be used to forecast phenomena. If the phenomena are current, verification may be straightforward. If a time horizon is involved, and the underlying facts are not scientifically convincing, neither is the forecast. The current scientific controversy over global warming, for example, which is focused on forecasts of global temperatures, amounts to a debate as to what is hypothesis and what is fact.

Analyzers and Synthesizers

The different mind-sets of scientists and businesspeople are also observable in the extent to which they see—or fail to see—linkages between projects and technologies. The classic financial mind-set holds that maximizing net present value is the best criterion for making investment decisions when given a limited amount of cash. This is how most executives are trained to deal with capital budgeting decisions and—by extension—to seek value: place bets on the projects with the highest calculated net present value, starting with the highest, and working downward. Projects are treated independently. Indeed, this approach often goes one step further: it recognizes that project proposals often contain many subprojects, each with its own net present value. Within a promising project, one or more of its component subprojects may have negative net present value. Therefore, the analyst reasons, shareholder value can be increased by doing only those subprojects that maximize value.

Part of the financial analyst's job is to separate the total project into its component parts to ensure that the *all* of company's limited capital is used productively. This mind-set, while narrow, is not always inappropriate.

However, pieces of a project are often strongly linked, and pursuing some but not others may lead to a bad end. Consider a management proposal to build a styrene plant and a polystyrene plant; the latter will use styrene from the former as a feedstock. Assume that the styrene plant does not earn an economic profit when styrene is priced on the merchant market, but the combination of the two plants does earn an economic profit. The business analyst is trained to see this as a "make-or-buy" problem: "It would be better to simply buy styrene in the open market as a feedstock for the polystyrene plant we *will* build." This is what the financial textbooks would suggest. For a number of reasons, however, projects may not be independent. Enough styrene may not be available on the merchant market to supply the proposed polystyrene plant. As a result, customers, knowing that the company's styrene supply is not secure, might be reluctant to enter purchase agreements for polystyrene.

Linkages between technology projects can be extremely strong and difficult to quantify. For example, the future of electrical vehicles will be driven by linkages between fuel cells, batteries, hybrid vehicle technologies, lightweight materials, and energy distribution networks. Likewise, the future configuration of personal computers will be driven by the evolution of both competing communications technologies (telephone lines, cable, and direct broadcasting) and data storage technologies (DVD-ROM, erasable CD-ROM, off-site storage, etc.). Even with advanced decision and risk analysis tools (Chapter 12), it is doubtful that anyone could effectively quantify the uncertain relationships between these evolving technologies.

In fact, technology is all about linkages and scientists are trained to think about them from the onset of their careers. Every good research project begins with a literature search whose aim is to establish where the technology to be developed fits into the context of all previous work. The scientific context may be scientific papers, patents, and research results within one's own company. The technology context is about which technologies are available to build on, which technologies may use the proposed development, and what competitive technologies will be faced in the marketplace.

In this sense, everyone's technologies are linked—those of customers, suppliers, competitors, and one's own. The technology decision maker must assess this information and make astute choices as to which technical developments will create promising positions, and which would be highly compromised. His or her creativity, experience, and training will largely govern the choice of which linkages should be exploited and the much larger set that should be ignored. The independent model simply doesn't fit.

Financial analysts are analyzers, comfortable in dissecting projects into their component parts. This is a narrow but useful discipline. The best technologists, in contrast, are synthesizers. They think broadly, often in domains where there are no quantitative tools. In this game, gut feel, a sense for the future, and *connectedness* to the larger technical community count for as much as technical competence. These two different outlooks explain some of the problems that businesspeople and scientists experience when they work together and attempt to determine the value of R&D.

A Value Model for the Corporation

Many of us grew up with the idea that the major corporations of the world were long-lived, highly integrated, and stable institutions that would prosper indefinitely. We also imagined that our own corporate careers could contribute to the building of these noble edifices. This image was reinforced by decades of steady growth by most of America's largest corporations and relative passivity and tolerance by investors. In hindsight, this image rested on shallow foundations, since few corporations trace their roots even to the nineteenth century, a short period compared with governmental, religious, and academic institutions.

The past two decades have done much to change this image, and the American corporation, as an institution, is rapidly evolving. The perception that the true role and justification of the corporation is to create value—as opposed to jobs or glory—has a great deal to do with the change.

To set the stage for the broad discussion of value that follows, and to put it in the context of an R&D environment, it is useful to think about a corporate value model as shown in Exhibit 1.3.

In this model, corporate value is derived from four elements: its *operations* or business units, its *financial structure* (assets and liabilities), its *management*, and its *opportunities*. To one degree or another, all these things can be bought and sold, removed or acquired, often independently.

The independence of a company's operations from its financial structure is now widely accepted in the science of valuation. The most common valuation approaches determine the values of the operating businesses *based on their earnings or cash flow*, subtract the liabilities, and add back the value of any nonoperating assets. This approach, which is a purely financial model, has been used for all manner of mergers, acquisitions, leveraged buyouts, and other forms of financial restructuring.

This view separates ownership from operations: even if a corporation is hit with a billion-dollar damage award, the value of its underlying

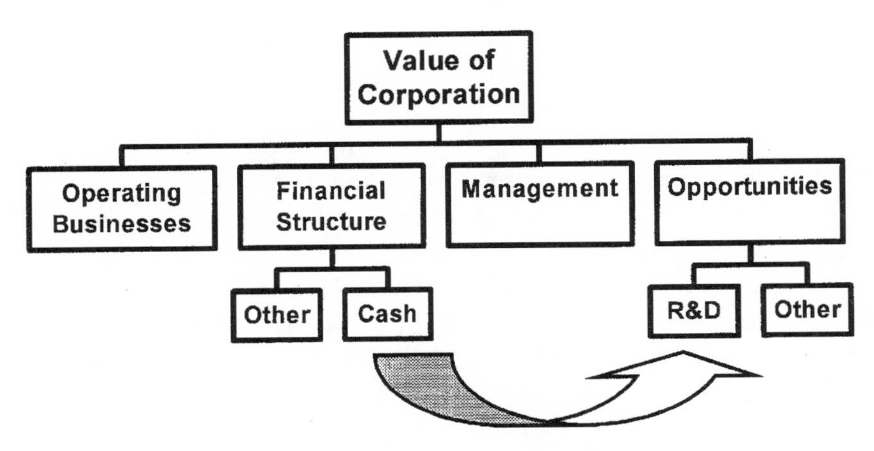

Exhibit 1.3 Corporate Value Model

operating assets is unchanged—only its owners are out a billion dollars. A corporation such as Dow Corning may be legally bankrupt, yet contain healthy operating businesses that will continue to thrive under new owners. To the operating people, rich new owners are far preferable to those who cannot raise the capital needed to sustain growth!

Management (and here we are talking primarily about corporate management) is a somewhat more complex case. Dissatisfied shareholders routinely throw out CEOs whom they perceive to have destroyed value, and more often than not the whole top team goes soon after. When this exercise succeeds, enormous shareholder value is unlocked, as has occurred at IBM and AT&T. Management can, of course, be synonymous with the creation of value: look at business heroes such as Gates, Buffett, Iacocca, and Welch.

Opportunity can be defined either strategically or economically. A common economic model of an enterprise states that value should be based on its cash flows plus the net present value of its opportunities.[5] (The latter is defined narrowly as the ability to earn more than the cost of capital.) In strategic terms, opportunity can be the ability to capture or corner a market, which in the end must be valued by investors in financial terms. We will defer the details for later, but there is a powerful clue here.

R&D is a mode of converting cash into opportunity (see Exhibit 1.3 again). Productive R&D creates options for the corporation to grow in profitability and size. It makes sense to consider this when one is cash rich and opportunity poor, although it must be evaluated against other alternatives for the cash, such as stock buybacks or acquisitions. Conversely, if one is cash poor and opportunity rich, such as a biotechnology start-up, it may make sense to sell some of your opportunity to someone that is in the opposite situation, for example, a cash-rich pharmaceutical company whose R&D pipeline is dry.

Incidentally, it makes no more sense to keep valuable research results under wraps inside a corporation than it does to keep lots of cash on the balance sheet. Neither are earning the cost of capital, and the dollars that created that research were a real investment. Idle R&D assets should be licensed or sold in the opportunity marketplace.

Some may object that R&D should be viewed as part of operations, since this is where new technology will be translated into commercial reality. This is not true in a financial sense, however, for if the value of operations is based on cash flow, R&D's contribution is arithmetically *negative*. It only applies if the operations are valued in a strategic sense, with their full panoply of opportunities.

It further follows that the *sum* of management and opportunities must add value, as measured in market capitalization. Takeovers will be instantly attractive to financiers when market capitalization falls below the asset value determined by operational cash flow net of the liabilities and takeover costs. In this scene, stock price is absolutely critical, for if it drops to that level, the company will be put "in play."

In a value-based marketplace, the research enterprise exists for a reason: to create opportunity and growth. Market forces ensure that research and researchers migrate to wherever the opportunities are best. In commerce, competition is largely a competition for valuable ideas. The past century has shown that the research enterprise has played a role in that competition and has continued to thrive, even as the institutional structures in which it is carried out are transformed.

In the Same Boat

Despite their differences, the destinies of scientists and businesspeople are closely linked. To the degree that they can overcome misunderstanding and appreciate the outlook of the other, they will produce a better future for their corporations and their communities.

Many executives, particularly those who lack technical backgrounds, are inclined to look on R&D and its practitioners with ambivalence. On the one hand, they know that the organization must continually renew itself with new products, new and more effective processes, and new technologies. But lack of scientific knowledge and direct experience with R&D inclines them to view this function as a mysterious black box. "If I put money into it and shake it around," they tell themselves, "something worthwhile *might* come out of it. It has before. It might happen again." At the same time, many who counsel these executives say, "Why bother developing our own stuff when we can buy it?" Indeed, much of the merger-and-acquisitions activity of the past decades has been driven by companies with plenty of cash but either no ideas of their own or no faith in their own ideas. Licensing is another way to "buy" instead of "make."

R&D is, indeed, a cash drain, and its outcomes are never certain. But proper management and good decision making can take much of the mystery out of this black box and make it a productive vehicle for converting cash into opportunity and options for the corporation. The valuation methods described in the following chapter can aid scientists and businesspeople alike in these important tasks.

CHAPTER 2

The Industrial R&D Process

The valuation of technology must begin with some understanding of the process that creates it. This chapter is intended for readers who are interested in the valuation of technology, but who have either never participated in industrial R&D, or who want to broaden their perspectives on this critical endeavor.

For men and women whose training and experiences have been limited to business, the R&D process often seems mysterious, slow, and unaccountably expensive. "What are they doing down there besides spending money?" they ask. "Why can't we get this new technology developed sooner?" "What kinds of risk are we up against?" "Ten million for a prototype production line? Why do we need it?" Businesspeople should not hesitate to ask these important questions. However, they cannot expect productive dialogue with the R&D community unless they first understand how R&D work transforms "good ideas" into commercially viable technologies and products.

To appreciate the magnitude of the potential misunderstandings between businesspeople and researchers, consider this example. An executive with one of W.R. Grace's operating divisions—a man with a reputation for good judgment—called the company's research division to request that the laboratory immediately address an urgent technical issue. Once he explained the problem, the research managers agreed that the issue was urgent and assigned one of their top scientists—a person with a reputation for intellect, responsibility, and timely results—to work on it immediately. Two weeks later, the executive called back, this time to complain that the designated scientist was demonstrating no sense of urgency. "He hasn't begun one experiment!" the executive said. "Instead, he's wasting valuable time in the library!" It took a call from the company's top research executive to persuade this action-oriented manager that experiments are not always the best route to a solution. More often, a thorough understanding of the problem and its potential solutions is the surest and fastest approach, beginning in the library, with a literature search. This search is the first activity in a number of stages that lead to commercial success.

R&D Stages

Research and development typically passes through different *stages.* Managing projects by stage is a widely accepted and useful practice in industry. Companies have different names for these stages, and the activities found in each stage differ greatly by project, by company, and by industry. Stages also differ materially by the level of risk, the level of spending, and to a large degree, by the skills of the personnel conducting the R&D. Typically, the cast of characters changes as a project passes from one stage to another. Most technical people function optimally in only one or two project stages and may spend most of their careers in those stages; only a few move with a project from start to finish.

Stages serve important project management purposes, and each has business and financial implications that both businesspeople and R&D workers need to understand: estimating time to project completion, developing optimal schedules through the use of *milestones,* and tracking R&D productivity (see Chapter 14). Some companies have instituted formal *stage gates* between stages as opportunities for multifunctional teams to review status, ensure consensus on objectives, and approve plans for the next stage. As two authors have described it: "Each gate is a trial by fire in which product teams must justify their work to management review committees and make a case for continued funding. Projects that survive one gate proceed to the next, where they meet another set of inquisitors."[1] Critical review at each gate serves a winnowing process, eliminated unpromising or strategically inappropriate projects. The many product and technology ideas that enter the process dwindle in number as they pass through successive gates, and only a few make it to the end of the process. Some have called this the "R&D funnel," as illustrated in Exhibit 2.1. There is some artistic license in this diagram, however; the reductions in number are most severe at the earliest stages in a well-managed R&D portfolio.

We shall see in Chapter 12 that this process creates value both because it exploits the option to terminate and it reduces risk through a widely diversified portfolio.

To illustrate the stages of R&D, three partially fictional examples are used in this chapter: a medical breakthrough, an innovative laboratory product, and an automotive part. They impart a sense of the diverse practices that exist among businesses and across industries. Assemblers of manufactured goods (autos, aircraft, machine tools) follow different practices than do those whose products are the result of continuous processing (refining, papermaking, chemicals). Time cycles are much longer and the value of patents is much higher in pharmaceuticals than in consumer electronics. Consumer products differ markedly from industrial products in the ways in which customer expectations drive the development process.

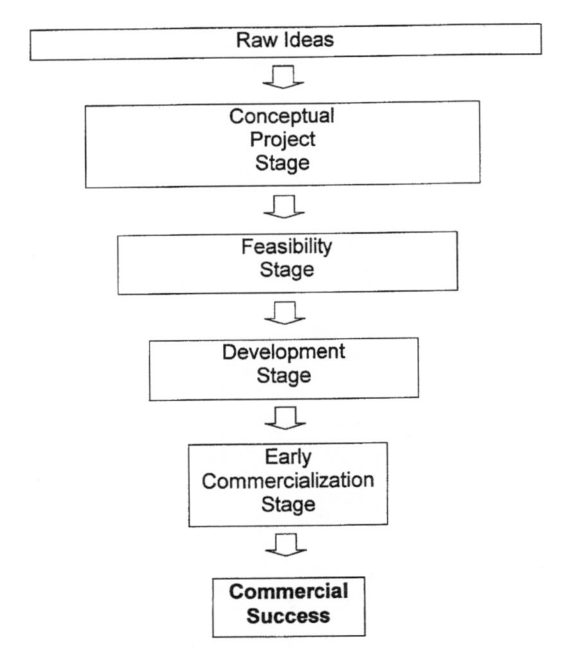

Exhibit 2.1 R&D Stages

Although the stages of R&D described in this chapter are in principle sequential, research may begin at an advanced stage. This is common when a new market opportunity is identified and most of the technology is already in hand. If a major brewer such as Anheuser-Busch approached a packager such as American National Can to produce 10.5-ounce necked-down aluminum beverage cans (which might nominally be a new product line, but technologically would be very similar to 12-ounce cans), the project could skip the earliest stages and move rapidly to commercialization. It is also possible for a project to revert to an earlier stage when a new issue is discovered that is better resolved in the laboratory than in a pilot plant ("Back to the drawing board").

Stage Zero. Finding and Screening Raw Ideas

The task of Stage Zero is to generate a field of commercially promising ideas and identify those that can be transformed into viable research projects. Ideas for research projects can come from just about anywhere—marketing, universities, consultants, top management—even from the weekend experiences of company employees. A Procter & Gamble researcher recognized the commercial potential of disposable diapers while babysitting for his newborn grandchild and handling the messy business associated with that chore.[2] Lore at 3M maintains that the inspiration for its now ubiquitous "Post-it" products came from a researcher in the company's adhesive division who needed a way to attach notes to the musical scores used by him and his fellow choir members.

Many research opportunities are stimulated by external events: a new scientific development, a sudden market need. The discovery of the double helix and the role of DNA laid the foundations for enormous opportunities in modern biotechnology. The oil and AIDS crises stimulated research in new energy sources and new drugs. The development of the microprocessor opened entirely new markets for computers.

Customers, too, are key sources of ideas for new products or product variations. Some customers—so-called lead users—alter or enhance off-the-shelf products in ways that their manufacturers fail to anticipate.[3] A customer may suggest a new product concept to a salesperson, and that suggestion may result in a query from marketing to research. Or a researcher may hear of a new technology at a scientific conference, and recognize how it could be combined with in-house technology to create a new opportunity. In other cases, the researcher may recognize the new technology as a potential threat. Some R&D organizations explicitly assign "gatekeeping" responsibilities to individual scientists to ensure that new developments are not missed and are properly evaluated.

Because the world performs several hundred billion dollars of R&D annually, there are always enough new developments for scientists to address. Information overload is therefore more likely to be a problem than lack of opportunity. Even so, organizations occasionally feel that creativity is at a low ebb, perhaps because they have been overly focused on immediate business priorities. They may then use brainstorming sessions or other forms of creativity stimulation to supplement the idea generation process.

Whatever their sources, raw ideas must be screened to determine their fit with the company's strategies and capabilities, for potential economic significance, and for uniqueness and originality (patentability).

Stage Zero Personae

The intrinsically challenging task of transforming raw ideas into viable research projects is usually performed by scientists with industrial experience. Most have doctoral degrees. Because the work is largely intellectual, laboratory technicians are less involved, though information specialists can play an important supporting role. The 80/20 rule seems to apply here—80% of the best projects are originated by 20% of the scientific staff. A few individuals have been exceptionally skilled in this role. The legendary Thomas Edison, Alejandro Zaffaroni in biomedical science, Victor Mills of Procter & Gamble (the P&G employee mentioned earlier who spearheaded the disposable diaper project), and Raymond Kurzweil in audio engineering are prime examples.

Initial screening is best done informally, through discussions among experts, until the key questions have been answered and a

Stage Zero Examples

Genetic vaccines. Company A holds patent rights to a "gene gun" capable of mechanically inserting DNA into live cells to give them new properties. One of its researchers reads about a new concept called genetic vaccines and proposes that this device could also be used to deliver DNA vaccines; the DNA would cause foreign antigens to be expressed, triggering an immune reaction.

Ion chromatography. A research scientist in Company B discovers a new analytical technique through a novel method of separating the signature of target ions from an ionic background. He calls it ion chromatography. He confirms the principle in some preliminary experiments.

Catalytic converter. A market need is identified to lower tailpipe emissions from vehicles in California, where smog levels are intolerable. An inventor in a company that produces metal catalytic converters for the automotive aftermarket suggests preheating a metal converter electrically to reduce "cold-start" emissions of pollutants. Environmental officials from California and other states are enthusiastic about the concept and willing to support legislation to require lower levels of emission if the technology is shown to be feasible.

reasonable case for proceeding with further investigation can be developed. Managers, engineers, and market experts can be helpful in these roles.

Stage Zero Financial and Business Implications

The success rate for raw ideas in Stage Zero is very low—one source puts them at 1 in 3,000,[4] but this ratio begs the question, "What is a raw idea?" Although the risks (i.e., uncertainties) in an undeveloped new idea are very high, the highest return projects may be hidden in a company's portfolio of "half-baked"—unquantifiable and poorly documented—new ideas. Businesspeople, who are inclined to focus on risk, must be equally open to the return side of the coin. Managers must be wary of dismissing what researchers view as promising ideas. The annals of modern business are replete with cases of researchers who moved to new pastures and created economically dynamic new companies out of the half-baked ideas that their bosses were too quick to reject or too risk-averse to encourage.

Typically, the investigation of a promising idea takes between a few days and a few weeks, and costs only a few thousand dollars. In most organizations, it makes very little sense to track or control such costs since they are limited and the projects themselves are short-lived. Misapplied control can stifle creativity. Generally, it is better to budget a pool of money for such exploratory activities, say up to 10% of the R&D budget. Alternatively, a company can create a policy like that of 3M, which allows researchers to use 10% of their time on their own original ideas. Experience indicates that funds budgeted for exploratory work are usually *underspent* because defined projects and unbudgeted situations compete for the same "surplus" resources. In any case, the costs of these activities are generally borne in laboratory overhead and allocated against defined and approved activities. Accountability is usually satisfied by observing which researchers bring forward successful new projects over time.

The screening process also produces valuable negative information by clarifying what should *not* be worked on and why. For example, a patent search may discover that a competitor has had the same idea and that any project in this area would have to navigate a patent minefield. This could be extremely valuable new information, even though the project as proposed is dead in the water.

Stage 1. Conceptual Research

The conceptual stage involves an effort to understand the full scope and the limitations of new ideas through laboratory research—to explore under what conditions the technology will and will not perform. It also aims to identify any "fatal flaw" in the idea—a potential project killer that is far cheaper to address here than down the road.

The conceptual stage typically begins when a researcher seeks formal authorization to spend time and other resources on a project. Most companies that do substantial R&D have policies and procedures for seeking and receiving such authorization. Generally, the proposer must answer some basic questions:

1. What is the initial target market?
2. Who will commercialize it?
3. How is the proposal distinguished from competing approaches?
4. What is its technical advantage?
5. Is patent coverage likely and will it be broad?
6. Could the technology become a platform for other initiatives?

The more searching the questions, the more time the researcher needs to develop the answers. I have used a policy limit of three months of researcher time, though there are valid arguments for both tighter and looser controls; it is basically a matter of risk tolerance and good judgment.

Far from being "full steam ahead," a conceptual project seeks a balance between opportunity creation, defined as attractive project opportunities for the firm, and risk reduction, defined as avoidance of the needless waste of money. However, the balance should be tilted strongly to opportunity creation; all too often firms fail to see the real payoff from their inventions because they moved ahead to the obvious and familiar. Initial research ideas are explored and results are compared with alternative solutions to the same problem. Researchers read the literature and consult with technical and industry experts to avoid reinventing the wheel and to capture all the important issues. These activities require time. Time must also be spent in obtaining materials, facilities, and equipment. These tasks may involve negotiating with external parties and creating strategic alliances to expedite progress. Good experimental design (a scientific specialty in itself) accelerates results and reduces laboratory costs. On the patent front, "reduction to practice" (i.e., performing successful experiments based on the inventive concept) greatly strengthens the patent position, and becomes a high-priority activity because of the importance of dates of invention and of filing in the patent process.

A project's value changes after each experiment. If it goes *exactly* as planned, confidence increases. If an experiment goes *better* than planned, more resources and a faster track may be in order. More commonly, researchers discover new problems and opportunities that require time and effort to address. Their solution, however, adds to the value of the technology and the company's competitive position. In the case of the genetic vaccine project, new conceptual projects aimed at additional diseases may spin off from it even as the lead project moves toward the feasibility stage.

Finally, a work plan and milestones are needed for the next stage of research, which may require arranging access to larger-scale equipment and addressing new safety and regulatory problems. Questions need to be answered in each stage to support the plan for the next.

Stage 1 Personae

By this stage, each of our example projects has become interdisciplinary, requiring multiskilled research teams. Skilled technicians have joined scientific personnel to run experiments, build prototypes, handle animals, and so on. Patent attorneys are now involved and may suggest additional experiments to strengthen coverage. Marketing input is becoming vital in sorting priorities.

Stage 1 Financial and Business Implications

The scope of the work required to define the concept can be surprisingly broad. As a result, the costs of an individual project can be substantial, involving several professional man-years at a per man-year cost of $100,000–$300,000. These costs are much higher when aggregated over a balanced research portfolio, because high project mortality is normal in the concept phase, and because the time value of

Stage I Examples

Genetic vaccines. The genetic vaccine concept has vast potential, including possible immunization against HIV (an enormous unmet need) and superior performance against a host of other diseases for which conventional vaccines already exist. However, the development of any vaccine is a long shot, and potential competitors have access to methods other than the gene gun. The first milestones will be to demonstrate and patent the concept, which will require negotiating access to a number of genes and to an animal model. Given the number of possible applications, considerable thought must be given to prioritization and strategy, and to whether the project warrants a major commitment of resources. The risks are fairly high that "blocking patents" from competitors will surface down the road as the project progresses.

Ion chromatography. The primary goals for the ion chromatograph at this point are to obtain a strong patent and to define the scope of its potential application. These will require a program of testing the instrument with a wide variety of artificial samples to define the scientific limits of its application, and to test it against difficult real-world samples, such as wastewater and blood, to judge its practicality. Various practical problems are found and mostly solved in Stage 1, adding to the company's technology base.

Catalytic converter. The conceptual stage of the electrically heated catalytic converter program begins with the construction of a laboratory prototype and testing its performance in the lab. Mechanical and electrical skills and expertise in catalytic science are required. A patent application must also be written and filed. This last task is complicated by (1) the large

Stage 1 Examples (Continued)

number of design possibilities, and (2) the crowded field of patents on improvements for conventional catalytic converters. Manufacturing will be challenging, because continuous production of a coated metal sheet must, after cutting, be integrated with piecemeal assembly into individual modules.

Comment: The genetic vaccine project runs an obvious risk of getting bogged down in Stage 1. The ion chromatography project is on a fast track. The catalytic converter is moving quickly now, but faces large obstacles down the road.

money places greater weight on R&D costs incurred today than on commercial revenues obtained several years in the future.

In aggregate, conceptual stage research is an important part of the total R&D budget, as will be discussed in Chapter 11. Expenses are normally allotted to a project budget, and tracked monthly by recording hours spent on the project and multiplying by an appropriate hourly rate to cover direct labor plus an allocation for laboratory overhead, supplies, and support costs.

Project slippage and missed milestones are common in the conceptual stage. Researchers run up against impasses that could not have been anticipated in earlier planning. Business sponsors are frustrated by slippage and often infer a lack of focus. Fortunately, researchers generally learn from the unanticipated technical problems in this stage, and their solutions often strengthen a company's competitive position and its future profit margins.

Because the research teams are already making critical decisions with long-term strategic implications in this stage, management involvement is important.

Stage 2. Feasibility

The main tasks of this stage are to resolve the known issues and to generate the cost and performance data that engineers and marketers need to undertake development. To do this efficiently, a tentative target market must be identified; in most cases, this requires an investment in market research by internal staff or outside consultants. When a technology has many individual applications (such as the genetic vaccine example), a single "lead" application may be brought forward to the feasibility stage, while other applications are slowed or put on hold.

A milestone system gives definition to timelines and responsibilities. It also ensures that the most critical problems are addressed early, and that all feasibility issues are resolved in the time designated for the onset of the development stage. This approach minimizes risk. Other functions, notably marketing, may share these milestones.

Entry into the feasibility stage implies that both R&D and business management are convinced that the project should move forward to development. However, they want any technical "show stoppers" (toxicity, competitor patents, etc.) identified, the scope of the project and its resource requirements defined, and the full path to commercialization outlined. Confidence should be high that any remaining regulatory and patent issues will be solved before moving forward to development.

The feasibility stage typically addresses a number of generic technical issues. A preliminary set of specifications must be drawn up for the lead product, based on what is so far known about the market need, about performance and about cost. Test methods must be established for critical specifications if they are not already in hand. A conceptual process for manufacturing must be envisioned. For example, if a continuous process is to be used for producing the product (where prototypes and samples had been made so far by hand), the feasibility of continuous production must be shown. The proposed manufacturing process must be sufficiently refined to not require excessively expensive materials or generate unacceptable waste. Experimental activity may move up a level from laboratory tests to "bench scale" equipment. The first estimates (± 30%) of manufacturing capital should be made,

Stage 2 Examples

Genetic vaccine. The feasibility stage for the genetic vaccine project begins with the decision to follow up on promising preliminary results on a target disease in an animal model. There is a need to optimize the gene and the delivery system (consisting of disposable gene cassettes, a prototype hand-held gene gun, and a protocol for their use). Plans must be formulated for producing gene products under the Food and Drug Administration's (FDA) Good Manufacturing Practices (GMP) guidelines, and for recruiting respected clinical investigators for human testing. Consultants and vendors must be engaged to support these activities.

Ion chromatography. A decision is made to build a few prototype ion chromatographs and to develop validated analytical methods for several promising applications. These instruments are *beta tested* in several internal laboratories to get user feedback regarding performance and reliability. A local engineering contractor is hired to assemble prototype units, an exercise that provides excellent cost data. It is clear that there is a business opportunity to sell both instruments and associated disposables, and that costs will be in line with other chromatographic instruments. Now that a patent application has been filed, papers will be presented at national scientific meetings to stimulate outside interest in the concept.

Catalytic converter. Although the concept of reducing vehicle emissions through an electrically heated converter has been proven, the requirements for gaining product acceptance by the automotive original equipment manufacturers (OEMs) are formidable. A major technical issue has been the time required to preheat the converter prior to ignition. This problem is solved with a series of improved and ingenious designs that reduce

(continued)

Stage 2 Examples (Continued)

time to zero. Deploying the converter in an actual automobile raises systems issues regarding the location, size, and shape of the device and the source of electrical power. The device must have a life of 100,000 miles and must pass strenuous heat and vibration tests, which are different for each OEM. A prototype production line must be built and operated and a source of noble metal catalyst formulations engaged. A few cars are custom equipped with converters to get actual use experience. Competitors are emerging and promoting other technical solutions to the problem.

based on a preliminary manufacturing process. This latter exercise is likely to suggest avenues for process and cost improvements, no doubt requiring further R&D.

Stage 2 Personae

All our illustrative projects now include engineering personnel and consultants to design and assemble prototype equipment. The staff and spectrum of skills required for technical work have expanded considerably. Scientists are now generating the data required for design purposes, and switching over to team-oriented problem solving. Project staff are also heavily engaged with marketers in communicating and promoting their concepts to potential customers and collaborators, and in evaluating feedback. A multidisciplinary team and team leader may be named to manage the project.

Stage 2 Financial and Business Implications

By this point, the project's financial burn rate has increased substantially. Charges for man-hours, equipment, consulting fees, focus groups, and supplies are beginning to hit the CFO's desk with no revenues yet in sight. Management is anxious for more assurance of success before committing additional funds. The emphasis has changed markedly from *opportunity creation* to *risk reduction*. Tracking of costs and technical progress is more frequent and more intense.

As timelines, costs, and resource requirements are determined, valuation becomes less fuzzy. An investment in market research, if it has not been made during the concept stage, is now imperative and will clarify potential revenues in various volume and pricing scenarios. If a project looks too risky, this is the time to pull the plug, regardless of sunk costs and bruised egos.

Stage 3. Development

The development stage seeks to define the specifications of the product and the manufacturing process for producing it. As discussed in Chapter 8, product specifications are much broader than scientific, technical, and performance specifications: they involve channels of distribution, marketing, packaging, legal concerns, environmental issues, and many other considerations.

There are two reliable indicators that the development stage has been reached. One is the exposure, or imminent exposure, of the technology to outsiders, especially customers. This step is usually not taken unless there is confidence that the product and its technology will reflect well on the company and that there is an internal commitment to produce the product if customer feedback is positive. The second indicator is the appearance of extraordinary expenditures: the authorization and construction of a pilot plant or pilot line, the contracting of expensive testing programs to outside organizations, and so forth. Serious money and the company's reputation are now on the line.

Stage 3 Personae

It is highly likely that project leadership will change in Stage 3 as decision making and project management pass from laboratory scientists to marketing, sales, engineering, and manufacturing personnel. A team leader with business skills may well be designated at this stage.

Stage 3 Examples

Genetic vaccine. The development stage of a medical product can be thought to begin with the preparation of a request to the FDA to begin human trials, referred to as *clinicals*. (Animal tests are called *preclinicals*.) Submission of this request requires voluminous data regarding safety, efficacy, manufacturing practices, and experimental controls. This stage can be lengthy, since a small set of human safety trials (Clinical Phase I) are required before permission is received to proceed to Clinical Phases II and III, where efficacy must be proven. Laboratory personnel will be heavily engaged in dialogue with the FDA and supporting the trials. Their work may include experiments that add to the scientific database created in the feasibility stage. Engineering and manufacturing personnel will also be engaged with FDA issues, support of the trials, and arrangements for commercial scale manufacturing. This stage will not be complete until FDA approval is certain.

Ion chromatography. The development stage of the ion chromatograph also requires parallel efforts supporting market development and the creation of one or more commercial products. Potential customers need more than an instrument, they need validated methods for applying the instrument to their samples. They will also want technical support. The R&D organization must address these issues through the development of an operating manual.

Engineering must work with industrial designers to create a model chromatograph that is attractive, versatile, and reliable. Customers will want product options; these bells and whistles must be mated and tested in the context of ion chromatography. Also, a supply of disposable materials must be arranged—an add-on business opportunity.

Stage 3 Examples (Continued)

Catalytic converter. Since the electrically heated catalytic converter will be a new product for OEM customers, company marketers engage them in extensive dialogue with regard to specifications and potential price. The latter involves not only the price to be received by the supplier, but the total cost implications *on a systems basis* for the OEMs, who are bringing purchasing managers to the now frequent meetings. The OEMs can be counted on to demand the lowest possible price and to take no risk of a recall.

The technical staff is heavily involved in the assembly and testing of numerous prototypes. Since both long-term durability and dynamometer testing are required, the tests are expensive and time-consuming. At the same time, engineers are working with outside contractors to detail a prototype automated production line, which will replace much of the hand assembly involved in prototyping. This is needed to validate manufacturing cost projections for bidding purposes, to demonstrate credibility of supply to customers, and to prepare a capital request document for management approval.

The development stage can be frustratingly slow—customer evaluation and feedback must await the customer's priorities, and pilot plant operations may be delayed by the long lead times needed to acquire and install specialized equipment.

Stage 3 Financial and Business Implications

At this stage, projects have recognized future economic value. Revenue and cost projections are now more certain, their anticipated ranges narrower, and estimates are based on fewer assumptions. This economic value is not simply in the eyes of project champions, but can be perceived by customers, competitors (who see a viable threat), investors, and potential licensors or joint venture partners. These perceptions are confirmed when important customers make conditional commitments to adopt the product or technology under development.

Now management may evaluate alternatives to internal commercialization (licensing, a joint venture, spinout, sale, etc.) to determine which course maximizes value. Their negotiating position with outside parties is much stronger than it was in the conceptual or feasibility stages. At the same time, the advantages of strategic partners in reducing project risk and gaining speed to market become ever clearer.

Stage 4. Early Commercialization

There are two key tasks of the early commercialization stage:

1. To establish a beachhead in the market.
2. To resolve lingering design, quality, and production issues.

If there are few such issues and time to market is critical, this phase may be bypassed.

For industrial products, early commercialization is the stage at which customers begin to make purchases, even though uncertainties in final specifications, quality, and delivery dates continue to exist. For consumer products, local or regional test marketing indicates entry into the early commercialization phase. In many or most cases, the shots are called by the business unit responsible for marketing the product. Construction and start-up of a full-scale manufacturing facility can take 24 months or more if preliminary engineering has not already been initiated in the development stage. On the other hand, when a new product can be produced in existing production facilities, early commercialization may be reduced to a few months.

Early commercialization generally represents a dangerous transition for the project and its backers. Because economies of scale have not been realized, unit manufacturing costs likely exceed unit revenues. Yet output volume is high enough that start-up losses are significant. The "bleeding" associated with this stage is significant and highly visible. It is best not to enter it until the risks and issues are

Stage 4 Examples

Genetic vaccine. While the development stage for a new biomedical product can be lengthy, the early commercialization stage, which begins with FDA approval to market the product, should be short. If Phase III clinicals have gone well, much of the work to accelerate commercialization will be done while awaiting the FDA's decision. For products with technically or medically sophisticated applications, however, it may be desirable or necessary to begin commercialization more cautiously with a few prestigious medical teams in a university hospital setting. A second tier of such hospitals can be added later, and in time, the product will be marketed broadly. Technical support for the medical teams will be initially an R&D responsibility. Serious thought is given to a joint venture of this technology in view of the parent company's lack of expertise in some of these areas.

Ion chromatography. The early commercialization stage for the ion chromatograph may involve a manufacturing agreement with an established producer of analytical instrumentation. This mitigates the risk associated with substantial fixed investments in full-scale manufacturing at a time when marketplace success is uncertain; the main capital costs will be limited to the inventory. Inventories can be further minimized if orders can be filled as produced. Since there are no equivalent instruments, this strategy is both attractive and feasible. Of course, the contract manufacturer's requirement for a return on capital diminishes operating margins. The option of self-manufacturing when sales reach an appropriate level will be evaluated. Financially, alternatives are considered to license this technology to an established instrument manufacturer, or to spin it off since it is not strategic to the parent. The latter alternative is actually adopted.

(continued)

Stage 4 Examples (Continued)

Catalytic converter. Early commercialization for the electrically heated converter begins with negotiation of a contract with the lead OEM customer for converters produced from the prototype production line. The primary purpose of this line is to manufacture commercial quantities for market development and gain manufacturing experience (a *semi-works* in the parlance of the chemical industry).

The OEM contract triggers management authorization to construct the line (long-lead-time items may have been ordered in advance to accelerate the schedule). Delivery and acceptance of the product would usually initiate transfer of the business to operations and the end of the early commercialization phase. Success with the first OEM venture would trigger R&D involvement in new, related, development projects for other OEMs and vehicle models. With the demonstrated success of the technology, new conceptual and feasibility projects on emissions from stationary sources of air pollution are being proposed.

defined (although market pressure may dictate otherwise) because errors of judgment or omission in the previous stages will be magnified. Managing entrance into and exit from the early commercialization phase is the ultimate test of any development team.

One may fairly ask: "Why is early commercialization a stage of research and development?" The answer is that early commercialization *consumes significant R&D resources,* regardless of the cost center charged. Even when these resources take the guise of "start-up costs" absorbed by the larger business, hidden subsidies from the R&D budget to the profit center are common. The reputation of R&D management is squarely on the line, and its troops will be there to put out any fires.

Stage 4 Personae

The Stage 4 project has become a minibusiness, with a leader, a sales group, a manufacturing facility, and a set of books. It may have its own R&D staff to provide incremental performance advances and to support customer needs. The original R&D organization will likely be called on to develop additional products and process improvements aimed at expanding profit margins.

Stage 4 Financial and Business Implications

From a financial viewpoint, this stage can mitigate the financial risks associated with a full commercial rollout of the product. It also generates important commercial information. If confidence in success is high and time is critical, however, the early commercialization stage may not contribute to value creation. This would likely be the case with a blockbuster new drug that has just won FDA approval, or a new line of vehicles or aircraft.

Summary

Sound industrial research management balances opportunity creation against risk reduction. Although R&D stages may seem arbitrary (because projects are so different), they are a critical tool to managing this balance intelligently. Many aspects of technology risk and its management are discussed throughout this book, but a look at the big picture here is in order. The key relationships are shown in Exhibit 2.2, where "certainty" is a proxy for the absence of risk. Certainty and expended costs increase as R&D proceeds through the stages, but the opportunity set and diversification, so abundant in the raw idea stage, contract at the same time.

Robert Cooper has effectively summarized five key principles of risk management as it applies to new product development.[5] The first rule is that if uncertainties are high, keep the stakes low. As uncertainties decrease, the relative amounts at stake may be increased (Rule 2). Rule 3

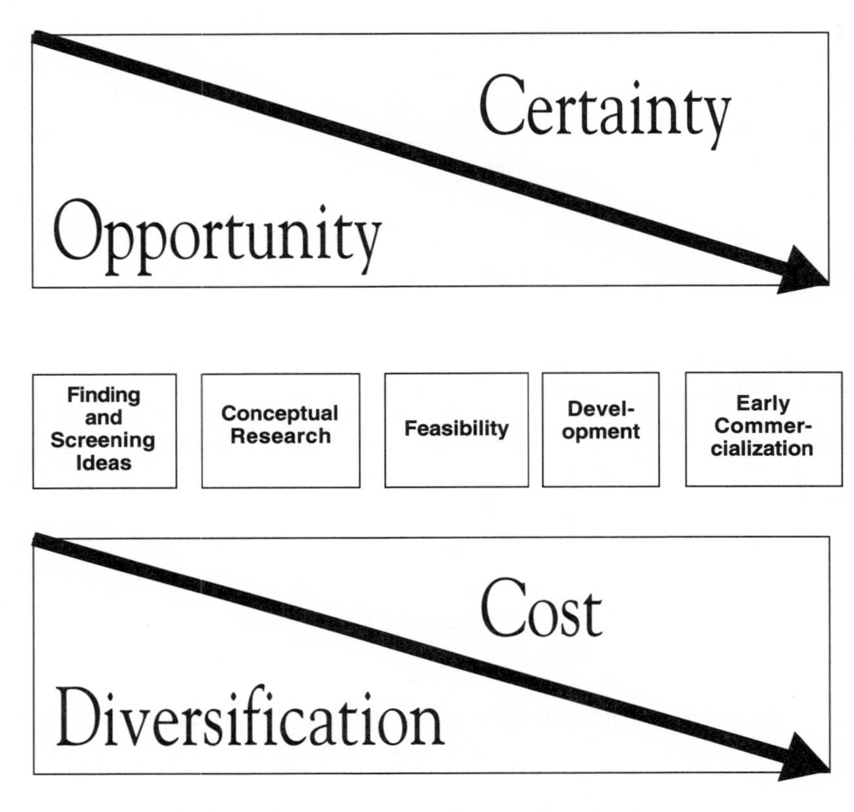

Exhibit 2.2 Key Relationships in the R&D Process

is to make incremental decisions and avoid all-or-nothing gambles. Rule 4 is to be willing to pay for the information required to reduce risk. Rule 5 is to create a process with an abundance of exit points.

In the early stages of research, the focus is on opportunity creation. Risk is managed through the well-known principles of diversifying the portfolio over many projects, and keeping the bets small. So there are lots of low-cost projects. The objective is to find potentials to earn exceptional returns.

In the middle phases, the company needs to eliminate the principal causes of risk so it can safely increase the bets. These include market risk (inadequate market research, changes in the marketplace, competitor reaction, and regulatory risks), business factor risks (sensitivities to value drivers such as volume, cost, capital intensity, and cycle time), resource risks (having the required core competencies, organizational issues, capital availability), and technical risks (meeting performance specifications, moving targets, patents, etc.). Cooper's admonition that one must be willing to pay for information is important here—expenses should not be spared on essential items such as market research or patent studies.

In the late phases, more and more of the project specifications are frozen, and changes become very expensive. The company's financial exposure has grown tremendously. Hence, it is crucial that certainty be high. There are many fewer projects, but they are costlier, and the company is counting on them.

These rules are not just common sense. Diversification is a financially proven method of reducing risk, and the *option value* of creating termination points in a research program can be determined quantitatively. Another technique for mitigating excessive risk is to share it, by seeking strategic or financial partners—an increasingly common practice in R&D management.

At the same time, one must avoid the mind-set that risk is bad. Because riskless projects tend to earn no more than the cost of capital, they create no value. Risk must be managed, not avoided. The price of increasing certainty can be lost opportunity. Exhibit 2.2 summarizes the changing relationships of certainty, opportunity, cost, and diversification as a project moves down the R&D track.

The role of senior management in this process is both important and counterintuitive. Since new products and new technology represent the future of any company, their development has strategic implications. And strategy is the prerogative of senior management. How will a new technology enhance the company's current strategy? Will it support or savage the company's existing strategies? Will its combination with other technologies produce something with remarkable market power? These are questions that all employees have a duty to ponder, but only senior executives have the authority to allocate the R&D and capital resources that will convert strategic concepts to marketable products and technologies.

Generally, senior managers—particularly those with little scientific or engineering expertise—do not participate in the early stages of R&D, only entering the process when the big-ticket decisions are on the table, which is usually just before the early commercialization stage. On the surface, this seems logical; top people only get involved in the large issues. However, this approach has two drawbacks. First, the strategic insights of senior managers can have an important influence on R&D choices as early as Stage 1, where they can filter out undeveloped concepts that lack strategic fit and give a nudge to others that do. Second, entering the process in the late stages where the plan is largely frozen leaves the executive with few if any choices other than "Go ahead" or "Kill the project." By that point, market and cost estimates are sufficiently developed that any number of knowledgeable company officials could make the right decision.

The tools that researchers and managers can use in making decisions about technology at any stage of the R&D process occupy the remaining chapters of this book.

CHAPTER 3

A Technologist's Guide to Financial Statements

There is no escaping the basic vocabulary of finance if one is to engage in the valuation of technology. It is the "coin of the realm" whenever businesspeople and researchers parley over plans and the commercial prospects of their mutual work. This chapter will familiarize you with the essential elements of financial statements and some important measures of return used by business managers to evaluate performance. It will also define four key numbers on which valuation is commonly based: net income, EBIT (earnings before interest and tax), EBITDA (earnings before interest, tax, depreciation, and amortization), and free cash flow. Experience indicates that these are as mysterious to many scientists and engineers as thermodynamics and kinetics are to their business-trained colleagues.

Of course, the reader has an alternative to this chapter and the subsequent chapter on discounted cash flow. He or she can curl up with a good book on corporate finance, most of which are at least 700 pages long. Indeed, doing so will provide additional dimensions to the reader's understanding. Some readers will have already done this reading, or have taken a course on the subject. Many scientists and engineers find that finance has more value than any other course in the business curriculum. Among other things, it helps them see their own work with new eyes—its constraints and its effects on the value of their companies; and more than any other discipline it provides insights into the business world and the lingua franca of its practitioners. Finance is to practitioners of modern commerce what Latin was to the intellectual life of the Middle Ages. Practically any business problem, anywhere in the world, can be viewed through the lens of finance.

The business reader may be saying, "Oh, finance. Been there. Done that." Most business readers will be familiar with the concepts of finance through either training or experience. Generally lacking in that training and experience, however, is the application of financial concepts to the unique environment of R&D and the commercialization of its output. An example is accounting for intellectual property; navigating the issues associated with this accounting is critical to intelligent technology valuation. A significant portion of this chapter is dedicated to these issues.

But first a caveat. This chapter is limited to those elements that appear in corporate financial statements, which have serious shortcomings and are not in themselves adequate for valuation purposes.

All corporate annual reports include three financial statements: an *income statement*, a *balance sheet*, and a *cash flow statement*. These can be viewed as a sum of the corresponding statements for each of the corporation's profit centers (i.e., business units), cost centers (i.e., headquarters and the central laboratory), its financial assets, and its liabilities. As such, each profit center can be viewed as an enterprise with its own set of three financial statements. And conceptually at least, each R&D project can be viewed as a mini- or microenterprise with *its* own set of financial statements. These individual statements can therefore be summed up with financial assets and liabilities into a book valuation for the corporation.

The format of these financial statements can be used as we forecast future results. If projects are in a proposal stage, they are purely forecasts, but can be summarized as *pro forma financial statements*. These are the takeoff points for the discussion of valuation of R&D projects in Chapter 9.

The Income Statement

In its simplest form, the income statement is where revenues are stated and where costs and taxes are subtracted to arrive at profit—the *bottom line*. The income statement represents these additions and subtractions for a stated accounting period, usually a quarter or a year. Exhibit 3.1 describes the basic elements of the income statement—which many organizations call the "profit & loss" or P&L statement.

Sales and Operating Expenses

For most manufacturing operations, the top line represents sales but may include many other forms of revenue, such as rents and royalties. Expenses directly related to the creation of these revenues are called *operating expenses.* These include costs incurred in manufacturing (often called, somewhat confusingly, "cost of sales"), selling costs, general overhead, and R&D (these are discussed in more detail in Chapter 9).

Operating Income (EBIT)

The net of sales and operating expenses is called *operating income* or *operating profit.* Financial analysts and investment bankers often refer to it as *EBIT*, earnings before interest and tax.[1]

EBIT is a very important number in valuing and running a business because it is independent of how the organization is financed (i.e., using borrowed money or its shareholders' own capital) and taxed, either of which can result in very different bottom line profits for two otherwise identical companies. Exhibit 3.2 shows a simple

Sales/Revenues
- Cost of goods sold
- Selling expense
- General and administrative expense
- R&D expense

Operating Profit
- Interest expense
+ Interest income
± Nonoperating items

Pretax Income
- Taxes

Net Income

Exhibit 3.1 Basic Income Statement

	Company X	Company Y
Operating profit (EBIT)	$ 100	$ 100
Less: Interest expense	0	(15)
Pretax income	$ 100	$ 85
Less: Taxes (40%)	40	34
Net profit	$ 60	$ 51
Value of business @ 10 × EBIT	$1,000	$1,000
Less debt	None	(200)
Value of company	$1,000	$ 800

Exhibit 3.2 Calculation of Value from EBIT (in $millions)

example of how one can separate the value of operations from the financing structure.

Example

Corporation X has operating income of $100 million from its U.S. superconductor operations. Because it operates with no debt, X pays no interest. It does pay 40% in taxes, however, so its bottom line profit is $60 million.

Company Y is in the very same business and also produces $100 million in operating profits. Because it uses lots of borrowed money to finance its operations, it pays $15 million in interest on debt of $200 million. That interest is deductible from its taxable income, but its bottom line profits are knocked down to $51 million after taxes at 40%. One of the simpler approaches to the valuation of an individual business is as a multiple of EBIT. Thus, if Company X has operating profits of $100 million, and other similar businesses have been sold for 10 times EBIT, Company X could be considered fairly valued at 10 times EBIT, or $1.0 billion. Company Y's identical business has the same EBIT. *But the companies have different values.* Y must be valued less its debt, or at $800 million.

A company's EBIT is a good indicator of operating effectiveness—how well it can generate revenues relative to costs. It ignores the differences caused by financing costs and taxes, which are operations-neutral. To a potential buyer, who will refinance the business in the act of purchasing it, EBIT is then a useful starting point for evaluating the wealth generating power of a business and, by extension, determining its value.

Nonoperating Expenses

The next tier in the income statement typically addresses nonoperating expenses and income. If a company finances its operations with debt—typically bonds—interest payments can represent a major expense. That expense may be offset by some interest income. Also, the corporate accountants may choose to define other transactions as nonoperating expenses (or income) and include them in this nonoperating category.

Writing off nonperforming assets is popular among corporations as a way of enhancing future earnings and hence shareholder value. These write-offs, although often related to operating businesses, are often made to show the continuing fundamental earnings power of the businesses in the most favorable light. As will be discussed, write-offs also translate to a reduction of shareholder equity on the balance sheet.

Start-up costs are sometimes included among nonoperating expenses. Nonoperating gains, such as the sale for profit of the corporate art collection, would also be taken here. The net of these operating and nonoperating items is referred to as *pretax profit* or *pretax income*.

The final adjustment on the income statement is for taxes, which is usually dominated by the federal corporate income tax rate. Federal income taxes can change appreciably from year to year, especially in cyclic businesses, as a result of carrying forward previous operating losses, write-offs, and so on. In valuing profit-making projects, it is often useful to look at an *average* corporate tax rate absent such extraordinary tax items.[2]

After deducting taxes, the bottom line is referred to as *net income* or *earnings*. For valuation purposes, taxes attributable to operating profits (i.e., EBIT), should be separated from taxes attributable to nonoperating income.

Exhibit 3.3 shows a consolidated income statement from a corporate annual report.[3] The term "consolidated" implies that some partially owned subsidiaries, such as joint ventures, are included in this simple and clear statement for "Company E." In actual reports, terminology differs considerably, as does the level of detail presented. Sometimes, large "other" items are included and the reader is referred to detailed footnotes for their understanding.

	2000	1999	Percent Change
Net sales	$6,822	$6,052	12.7
Costs and expenses			
Cost of products sold	5,028	4,397	14.4
Selling and administrative	927	890	4.2
Research and development	227	213	6.6
	$6,182	$5,500	12.4
Income from operations	$ 640	$ 552	15.9
Other income (expense)			
Interest expense	$ (86)	$ (83)	3.6
Interest income	6	7	−14.3
Other income—net	32	12	166.7
	$ (48)	$ (64)	−25.0
Income before income taxes	$ 592	$ 488	21.3
Income taxes	193	155	24.5
Net income	$ 399	$ 333	19.8
Per Common Share			
Net income	$5.13	$4.40	16.6
Cash dividends paid	1.50	1.20	25.0
Average number of common shares outstanding	77.8	75.6	2.9

Exhibit 3.3 Company E Consolidated Income Statement for the Year Ending December 31, 2000 (in $millions)

The terminology used by Company E is slightly different from that used in the previous exhibit, and Company E has elected to combine selling expenses with administrative expenses. Research, however, is broken out separately. Two years of data are shown to facilitate year-to-year comparison.

Below the net income line, it is common to calculate *earnings per share* by dividing net income by the number of shares outstanding. The number of shares most commonly increases slightly from year to year as a result of grants of options to employees, but may decrease as a result of stock buyback programs. If the company has elected to raise capital by selling newly issued shares, the number may increase substantially.

Company E's net income per share is $5.13. When combined with the current price of the stock (found in the newspaper, not in the annual report) this figure can be used to calculate a very closely watched number, the *price to earnings* or *PE ratio*. If the current share price is $100, the PE ratio will be $100/$5.13, or 19.5. The reciprocal of this number, 1/19.5, is the earnings a shareholder investing today would obtain, expressed as a percentage—in this case 5.1%. Because this return is less than the investor could obtain elsewhere (e.g., from corporate bonds), it is clear that the market is anticipating earnings growth.

Another per share number reported on this page is the *dividend*, $1.50 per share. If divided by the current share price, it gives the *dividend yield* ($1.50/$100) or 1.5%.

Earnings can be either paid as dividends to shareholders or retained for reinvestment. Young, fast-growing companies often pay no dividends because they need all the cash they can get their hands on to finance expansion. These companies generally retain all earnings. As will be shown, *retained earnings* are transferred to the balance sheet as shareholder equity, a key link between the income statement and balance sheet.

Neither the price earnings ratio nor the dividend yield appear in the annual report, since they vary from day to day with share price changes.

It is useful at this point to mention the concept of *market capitalization*. "Market cap" represents the total nominal value of all shares held by investors and is calculated by multiplying the number of shares outstanding by the current share price. For Company E, if the share price is $100, the market cap is $7.78 billion.

By simple arithmetic, the ratio of market cap to corporate net income is identical to the PE ratio, and the total dividends paid by the corporation divided by the market cap is the dividend yield.

Income Statement Analysis

Let's analyze Company E's income statement for some first impressions, based on the numbers alone. These impressions might be confirmed, or otherwise explained, by reading further in the annual report.

Company E seems to have had an excellent year. Revenue gains were 12.7%, indicating healthy growth. Somewhat worrisome is that the cost of products sold grew even faster, 14.4%, indicating erosion of profit margin. Is this due to competition, product mix shifts, or increased plant level costs? Nevertheless, management was able to improve the operating bottom line to 15.9% by holding the line on selling and administrative costs, which only increased 4.2%. R&D expense has not kept pace with sales growth, increasing 6.6% in absolute terms, but decreasing on a relative basis to 3.3% of sales.

Good as these operating results are, a manager or investor might have grounds for concern. A continuing trend in margin erosion cannot be offset long-term by holding the line on selling costs and R&D—that would require improvements in productivity out of line with historic experience. But one year does not make a trend. Savvy managers would look back over a few years to determine the long-term direction of these costs, and they would monitor *current* costs closely.

Nonoperating income for Company E was up smartly owing to one or more "other" items. (We would have to turn to the footnotes for explanations of these.) As a result, net income was up even more sharply than operating income, by 19.8%—an impressive number.

Net income per share, however, increased only 16.6%, owing to *dilution*. The company issued 2.2 million more shares and increased the number of shares outstanding by 2.9%—quite a large number. Shareholders were rewarded with a 25% increase in the dividend.

The Balance Sheet

If the income statement is a measure of profitability during a stated period, the balance sheet (some call it the "statement of financial position") indicates all the assets owned by the company and all the claims against those assets at a stated moment in time (Exhibit 3.4). This picture of the company is summarized in a simple equation:

Assets = Liabilities + Shareholders' equity (or Net worth)

Assets	Liabilities and Shareholders' Equity
Current	**Current Liabilities**
Cash	Short-term debt
Marketable securities	Current portion of long-term debt
Accounts receivable	Accounts payable
Inventories	Accrued liabilities
Total Current Assets	**Total Current Liabilities**
Gross property, plant, and equipment	Long-term debt
– Accumulated depreciation	Other long-term liabilities
Net property, plant, and equipment	Deferred income taxes
Goodwill	**Total Liabilities**
Investments	Common stock and paid-in equity
Other assets	+ retained earnings
	Total Shareholders' Equity
Total Assets equal	**Total Liabilities + Shareholders' Equity**

Exhibit 3.4 Corporate Balance Sheet, as of (Date/Year)

The fact that the two sides of the equation must be equal explains the term "balance sheet."

The assets in this equation are the things the company "owns," including monies owed to it. Liabilities are claims against those assets: what it "owes" to others. Shareholders' equity is a residual figure, whatever value is left over after all claims have been satisfied, or put another way:

Assets – Liabilities = Shareholders' equity (or Net worth)

Thus, the balance sheet is an accounting measure of shareholder wealth at a stated moment.

It is perfectly possible for a company to be very profitable, but poor or even bankrupt; it is also possible to be very wealthy but unprofitable. To the mathematically inclined, profitability is the first derivative of wealth. Short-term, the two are weakly correlated; long-term, the relationship is strong. Dow Corning was bankrupted by huge legal liabilities incurred by a small product line, yet it is a very profitable specialty chemicals company. On the other hand, IBM in a dark period during the early 1990s was enormously rich in terms of its assets, but was making returns far below its cost of capital. To be both profitable *and* wealthy is what every manager and shareholder hopes for.

Accounting versus Economic Value

There are fundamental differences between how accountants and individuals (and economists) value the assets on the balance sheet. As an individual, I value my assets—my house; my investments in stocks, bonds, and mutual funds; and my bank accounts—at their *current market values.* The difference between the current value of my assets and my liabilities is my personal net worth. If my assets exceed my liabilities, my net worth is positive, and I am in good shape. As a technologist, I am unrepentant about this commonsense calculation, but I am not thinking like a corporate accountant!

Corporate accounting differs from the commonsense approach in one major respect. "Generally accepted accounting principles" value most assets at *historic cost:* the price at which the asset was purchased. Even though my house may have doubled in value, the rules of accounting would dictate that it be represented on the balance sheet at its historic cost, less depreciation. Some of my stocks have tripled in market value since I acquired them, but they too would be valued by the accountants at what I originally paid for them. Naturally, this "noneconomic" approach to valuing balance sheet assets would cut deeply into my net worth and that of just about every reader.

So why do the accountants' rules insist that most balance sheet items be listed at historic cost, or "book value?" The primary reason is that the true market values of many, if not most, corporate assets are difficult to determine, are even more difficult for outsiders to verify, and could be easily manipulated to make a company appear to be in better shape than it is. (Imagine the cost and effort involved in having professional appraisers come into a corporation every year to determine the market value of its buildings, furnishings, inventories, delivery vehicles, and on and on.) So, if historic cost is a poor measure of economic value, it may be the lesser of two evils. As one source put it, "Accountants would rather be consistently wrong following a uniform method than allow suspicious values to find their way into the balance sheet."[4]

And why is equity (which seems to be an asset) on the right side of the balance sheet along with the liabilities? To understand this, we must mentally separate the corporation from its owners. To the owners, their equity *is* an asset, but to the corporation, it is an obligation, just as the corporation's bonds are an obligation to the bondholders. In effect, if the corporation for some reason were to be liquidated at book value, the proceeds, after all other liabilities were paid, would go to the shareholders of common stock.

ASSETS

Both assets and liabilities are normally classified as either current or long-term. "Current" assets are cash or anything else that can be turned into cash in a short time—usually one year: inventory that will be sold; bills that will be collected; securities that can be sold. Everything else is "long-term."

Cash and Marketable Securities

Among a company's typical current assets are cash and *marketable* securities. The term marketable implies that they are liquid. These are normally minimized since they are economically unproductive—earning much less than the company's cost of capital.

Accounts Receivable

Accounts receivable is usually the larger portion of current assets. It represents monies due from products already sold and shipped. Typically, customers receive terms of 30 to 60 days, but these may be longer or shorter depending on the business. Big money can be tied up in accounts receivable, which must be financed.

Inventory

This balance sheet category represents finished goods ready for sale, work-in-process, raw materials, and supplies. Companies go to great lengths to minimize inventories since, like most other current assets, they are nonproductive of revenues and must be financed with either borrowed money or shareholders' equity. Because equity money is usually more expensive than secured debt, some companies borrow against, or "factor" their receivables and inventory.

The Concept of Working Capital

Financial people generally call the money represented in current assets *working capital.* It represents the capital used for the everyday business activities of paying wages, converting raw materials into products, products into sales, and sales into customer payments in hand.

On the right-hand side of the balance sheet is a category called "current liabilities." These are short-term financial obligations that usually (like current assets) must be met within one year. The difference between current assets and current liabilities is *net working capital.* A company that has lots of net working capital generally has no trouble paying its bills in a timely way. If the ratio of current assets-to-current liabilities is less than one, however, it cannot, and is "insolvent."

Property, Plant, and Equipment

These represent a company's most visible physical assets—factories, land, process equipment, computers, vehicles, and so on. They are normally shown in the balance sheet first as *gross fixed assets,* which represents their original cost. An adjustment is then made for *accumulated depreciation,* giving the value of *net fixed assets.* Typical depreciation rates are 3 to 6 years for computers, office equipment, laboratory equipment, and vehicles; 5 to 15 years for plant equipment; and 15 to 30 years for structures and buildings.

Depreciation is a noncash charge used to match the cost of a productive fixed asset with the revenues that it generates over its useful life. Because depreciation schedules vary—from a few years for vehicles and computers to a few decades for buildings—depreciation will vary with the mix of assets. Depreciation schedules also affect valuation, especially when the valuation method is based on reported earnings.

Although it is not itself a cash expense, depreciation nevertheless finds it way into the income statement (and then into the cash flow statement) because it can be used to reduce taxes.

> *Example*
> Corporation XYZ paid $10 million for factory equipment that it could depreciate over a period of 10 years at an even (straight-line) rate. Thus, XYZ included $1 million each year for 10 years among the expenses on its income statement.

Slower depreciation improves reported accounting profits in the short-term, since the initial depreciation expense charged against revenues is less. (With any depreciation schedule, the asset is fully depreciated in time, so we are basically concerned about the time value of money.) However, slower depreciation worsens shorter-term cash flow because the business initially pays out more in taxes. Because IRS rules allow some flexibility on how fast assets can be depreciated, corporate managers and accountants must judge whether it is better to state higher profits (and pay more taxes), or reduce stated profits through greater depreciation (and pay less taxes). To make things more complicated, depreciation for tax purposes is different from that used for management accounting, which has given rise to two sets of books: the so-called tax and management books, one used for tax purposes and the other for reporting to shareholders and making decisions. (Since this book concerns valuation, the perceptive reader will understand that the difference between valuation methods based on cash flow versus methods based on reported earnings will be sensitive to choice of depreciation schedules.)

Goodwill

Goodwill is another type of long-term corporate asset. Though the name suggests something intangible, goodwill is well defined in a financial sense. Goodwill is created when one acquires an asset and pays a price in excess of the amount on the seller's books. Ironically, intangibles as we normally think about them, such as a corporation's reputation for quality or its brand franchise, do not show up on the balance sheet unless the company is sold to another. Under American tax law, goodwill can be amortized only very slowly (typically 40 years) and is a very unproductive asset. As explained later, goodwill is described appropriately as "excess of cost over net assets of businesses acquired."

Intellectual property may also show up after a transaction as goodwill, although it is wise to formally value acquired patents and, if possible, depreciate them over their remaining life. If so, patents will be found within the "other" category of balance sheet assets. In accordance with generally accepted accounting principles, internally generated patents have no asset value, no matter what their market value. However, if a powerful patent is *acquired,* it can be assigned a value and amortized over its lifetime. This happened in the case of the aspartame composition of matter patent that Monsanto obtained when it acquired Searle. The patent was valued on Monsanto's balance sheet at over a billion dollars and amortized over 5 years. This was far better for tax and cash flow purposes than a 40-year amortization for goodwill.

Investments and Other Assets

Finally, major corporations over time often create a network of partially owned entities, such as joint ventures and affiliated companies. Their results (revenues and earnings) may or may not be consolidated with the parent, usually depending on the degree of ownership. If they are not consolidated (that is, the ownership of the asset is relatively small), the value of the parent's share of these assets, at cost, will be shown on the balance sheet as either "investments" or included among "other assets." If the results are consolidated, an entry on the balance sheet for "minority interests" representing the equity of the minority shareholders will be shown as a liability.

LIABILITIES

Like assets, liabilities generally have current (payable within one year) and long-term components. And they are listed on the balance sheet in that order. Current liabilities generally include *accounts payable, accrued liabilities* such as wages and taxes accrued but not yet paid, short-term borrowings, and the current amounts owed on long-term debts. Long-term liabilities generally include long-term debts and reserves or contingent liabilities, such as possible losses due to pending lawsuits.

Accounts Payable

This represents the amount owed to vendors for supplies and services. You can think of it as the flip side of accounts receivable by the corporation. Vendors will often sell on standard terms, such as 30 days; these represent an interest-free loan to buyers and a working capital offset against accounts payable.

Accrued Liabilities

Money owed to employees (who are not paid until some time after the actual work performed) is one kind of accrued liability; so too are accruals for taxes due but not yet paid. Both reduce the amount of capital the firm must have to pursue current operations.

Debt

Debt is what it appears to be, monies owed to third parties. It will usually be a mix of short- and long-term. Short-term debt (payable within one year) includes monies owed to banks and long-term debt which must be repaid within one year. Most corporate debt is long-term and takes the form of long-term bonds issued at rates generally lower than short-term funds.

Debt is one of the key determinants of a corporation's cost of capital. Although many individuals and some corporations take pride in stating, "We don't owe a dime to anyone," some debt in the corporate capital structure is generally desirable: it can reduce the cost of money, leverage the shareholders' investment, and provide a tax shield (since the interest expense is deductible from taxable income). Such tax shields can be readily valued in a quantitative sense; the interested reader is referred to corporate finance texts for details.[5] However, there are diminishing returns on how much can be borrowed at attractive rates. Senior debt, especially if secured by a company's current assets, is relatively cheap. Unsecured junior debt is a bet by lenders against the future creditworthiness of the company and, thus, will always be more expensive. Lenders may also impose *covenants* that restrict a company's ability to borrow beyond negotiated limits.

Corporations also typically secure a large line of credit not shown on the balance sheet, called a *revolver*. This credit, which is similar to the line of credit you have on your credit card, is negotiated with a group of banks for a fee and can be drawn on when needed, albeit at high interest rates. It has the benefit of helping a corporation to minimize its cash on hand, since cash is a nonproductive asset. Seasonal businesses whose inventories expand and contract on a regular basis find that the line of credit is a useful and dependable financing tool. A candymaker, for example, would draw on its line of credit to finance the production of big inventories in the months just prior to Easter and Halloween, then pay off what it has borrowed as sale receipts come in. The line of credit also allows a corporation to act quickly when faced with a threat or opportunity that requires swift response. However, this short-term money is usually expensive and leaves the company exposed. It will wisely restructure its balance sheet by issuing long-term debt, issuing new equity, or selling assets to pay off the line of credit after such an opportunistic transaction.

Debt, in any form, makes the corporation and its stock riskier. The corporation has a legal obligation to repay the interest and principal of its debts as they come due. If a business reversal makes timely repayment impossible, the debtholders could drive it into bankruptcy.

Deferred taxes are a form of liability created as a benefit of tax laws when the tax code allows a higher rate of depreciation for tax purposes than the company uses in calculating profits. As a result, the earnings for tax purposes are lower than those reported to the shareholders in the first few years after an investment, and the tax bill is reduced. In later years, the situation reverses and additional taxes are due. In practice, the deferred tax liability is a permanent feature of the balance sheet as long as a growing company continues a capital investment program. As one deferred tax liability is paid down, a new one more than replaces it. This is not a significant issue for nonaccountants. Deferred taxes are a by-product of a tax incentive to encourage capital investment. They are regarded by financial experts as a form of pseudo-equity, since investors expect to make a return on the capital saved by tax deferral.

Some types of deferred taxes can be classified as assets, for example those associated with the creation of reserves or the de facto capitalization of R&D, subjects to be discussed.

Reserves

Reserves are monies held aside to pay for future liabilities; for example, unfunded health care benefits for retirees, remediation of environmental damage, or settlement of product liability suits. Reserves are created when management believes it should disclose a future liability to shareholders. Typically, the company's auditors are consulted in this decision. In conformance with accounting practices, there is a corresponding reduction in shareholder equity when reserves are expanded. For example, if a company were to create a $100 million reserve for personal injury litigation, its shareholder equity (book value) would be reduced by an equal amount.

Shareholder Equity

Shareholder equity, or book value (or net worth), is a residual: to an accountant, it is what is left of corporate assets after all other liabilities have been satisfied, or subtracted. That is a fair way of looking at it, since shareholders stand at the end of the line of claimants to the corporation's assets. However, as stated earlier, the accountant's measurement of shareholders' equity, based on historic costs, may be wide of the mark in terms of real economic value, as many mergers and acquisitions will verify. When you see companies being valued at 2 to 10 times book value, you learn to treat this accounting of shareholder wealth with a grain of salt.

Equity should grow in a profitable company, since all annual earnings not paid out in dividends are reflected in shareholders' equity added to the equity account each reporting period.

Shareholder equity typically is divided into two portions: *paid-in capital* and *retained earnings.* Matters are confused somewhat because accountants separate out a mostly fictitious piece of paid-in capital called the *par value* of the issued shares. Par value is typically a small fraction of the paid-in value of the shares, often a nominal $1. The remainder, naturally enough, can be called paid-in capital in excess of par value. For example, if Company X has a public offering of 10 million shares at $20, with a par value of $1, it will have paid-in capital in excess of par value of $19 per share or $190 million. A typical company may issue equity (stock) a number of times in its corporate life. The first will be called the *initial public offering* (IPO), and subsequent offerings will be described as secondary offerings.

Shareholder equity increases each year with retained earnings, assuming that retained earnings are positive.

To get a better handle on this tricky section of the balance sheet, look at the balance sheet of our example corporation, Company E (see Exhibit 3.5).

The income statement of Company E shows that retained earnings in 1999 were $3.20 per share (earnings of $4.40 less dividends of $1.20), or $242 million based on 75.6 million shares outstanding. This is within rounding error of the reported increase in retained earnings from $988 million to $1.232 billion (the slight difference is

Consolidated Balance Sheet	2000	1999
Assets		
Current Assets		
Cash	$ 56	$ 18
Short-term investments	28	23
Accounts receivable	932	889
Inventories	735	698
Deferred income taxes	150	151
Other current assets	66	67
	$1,967	$1,846
Property, plant, and equipment		
Land	$ 52	$ 50
Buildings	578	539
Machinery and equipment	2,584	2,321
	$3,214	$2,910
Accumulated depreciation	(1,561)	(1,441)
	$1,653	$1,469
Excess of cost over net assets of businesses acquired	$ 895	$ 850
Other assets	$ 538	$ 517
Total Assets	$5,053	$4,682
Liabilities and Shareholders' Equity		
Current liabilities		
Short-term debt	$ 30	$ 14
Current portion of long-term debt	20	22
Accounts payable	486	449
Accrued compensation	168	163
Accrued income and other taxes	62	60
Other current liabilities	379	394
	$1,145	$1,102
Long-term debt	$1,084	$1,053
Postretirement benefits other than pensions	$ 579	$ 573
Other liabilities	$ 270	$ 274
Shareholders' equity		
Common Shares (77.6 in 1995 and 78.0 in 1994)	$ 39	$ 39
Capital in excess of par value	812	806
Retained earnings	1,232	988
Foreign currency translation adjustments	(55)	(71)
Unallocated Employee Stock Ownership Plan shares	(53)	(82)
	$1,975	$1,680
Total Liabilities	$5,053	$4,682

Exhibit 3.5 Balance Sheet, Company E December 31, 2000 ($millions)

probably caused by quarterly vs. annual accounting intervals). For companies that have been essentially self-financing over their history, the greatest portion of the shareholders' equity will be in the form of retained earnings. This is not the case for Company E.

Retained earnings is the key link between the income statement and the balance sheet.

Warning! Retained Earnings Is Not a Piggy Bank

Many newcomers to financial statements make the mistake of thinking that retained earnings is like a savings account, where ready cash is available. Accounting professors love to trap their students with this question, "What would you do if your company needed lots of cash in a hurry?" Anyone who answers, "Take it out of retained earnings," is immediately stripped of his green eyeshade and made to sit in the corner. Every dollar of profit retained in the business is noted under "retained earnings," but the actual dollars are reflected on the left-hand side of the balance sheet, in the form of cash, inventories, equipment, or whatever purpose toward which those retained earnings were directed.

Double Entry Bookkeeping

This concept is important to understanding balance sheet transactions (some have called it one of the great inventions of Western civilization). The principle is simplicity itself—for every credit or debit there must be an offsetting entry. If a product is sold, inventory is debited and accounts receivable is credited. When the bill is paid, accounts receivable is debited and cash is credited. If a plant burns down, net fixed assets are debited, and shareholder equity is debited. In considering a transaction, it is useful to think about where the entries are likely to be made and their effect on the corporation's financial ratios.

Capital Structure and Investment Performance

The right-hand side of the balance sheet tells you at a glance how management has chosen to finance its assets and activities. Most corporations use a mix of the shareholders' money (equity) and other people's money (debt). The sum of these is referred to as *total capital.* Note that *total capital* is not synonymous with *total* assets.

For analysts at financial institutions and bond-rating agencies, such as Moody's and Standard & Poor's, a corporation's *debt as a percent of total capital* and its *debt/equity ratio* are important metrics in determining creditworthiness. And creditworthiness has a powerful influence on a company's cost of borrowing. As noted, the use of borrowed money increases the risk of business failure—or the inability to pay interest and principal in a timely way. That risk is reflected in the rate that the individual corporation will pay on its borrowed funds. When debt rises above 50% of total capital, borrowing rates usually escalate sharply.

Though debt may increase business risk, it creates *financial leverage,* which increases the return on the capital contributed by shareholders (assuming there are profits). We'll get to this important subject later.

Company E has a solid balance sheet. Its long-term debt of $1,084 million is 35% of its total capital of $3,059 million (long-term debt plus shareholder equity). Its interest payments of $86 million (see the income statement) represent an average pretax cost of borrowing of 7.9%. As discussed in Chapter 5, the *book value* of debt and equity counts less than the *market value.*

Two other comments about this balance sheet are worth noting. First, its working capital assets, inventories and accounts receivable, are substantial, in fact slightly larger than its fixed assets, each being about $1.6 billion. In the case of Company E, accounts receivable of $932 million represent 13.7% of sales, or 50 days.[6] Inventories represent assets of $735 million, or 10.8% of sales, or 39 days. Company E's financial department has done very well in managing these hard-to-manage items, since they each increased by only about 5% year to year, whereas sales increased over 12.7%.

Second, deferred income taxes are shown as an asset rather than a liability. Commonly, deferred taxes are a liability arising from accelerated tax depreciation, but in some circumstances they are an (unproductive) asset, among them when a company elects to capitalize some development costs.

The Cash Flow Statement

In the world of consumer spending, "plastic" is king. In the world of financial analysis, cash flow is king. Yet, accounting hocus-pocus makes the meaning of cash flow unclear to many. Cash flow is the after-tax cash from operations that is available to pay dividends to shareholders and interest to creditors.[7] The cash flow statement is a usual companion to our other financial statements in that it tell us where cash has come from and where it has gone, describing the sources and applications of cash over a stated period. The cash flow statement is also the key to most valuation methods.

The typical cash flow statement has three sections: *operations, investments, and financing activities.* This structure is very useful for valuation purposes. The sections are summarized as Exhibit 3.6.

OPERATIONS

Under operations, the first element is operating income. This is identical to the bottom line net income from the income statement. As such, it includes expenses for taxes and interest. However, in calculating net income, we had taken a noncash charge for depreciation. In calculating cash flow, we must *add it back*. Cash flow is, after all, generated as the difference between the price received for goods sold and

Operations
- Operating Income
- Depreciation
- Changes in working capital
  - Inventories, accounts receivable and accounts payable
- Changes in deferred taxes

Investments
- Capital expenditures
- Acquisitions

Financing Activities
- Changes in debt level
- Purchase or issue of stock
- Dividends

Exhibit 3.6 Elements of a Cash Flow Statement

the expenses paid out of pocket to produce them. Without mentioning it at the time, we had included a charge for depreciation in cost of goods sold in discussing the income statement. The actual cash outlay for the plant and equipment being depreciated was made in the past and accounted for as a capital investment. Its cost is now being recaptured for accounting purposes through depreciation schedules. Amortization of goodwill or patents is perfectly analogous to depreciation and is also a cash generator.

The third major item in operating cash flow is the *change* in working capital. For example, if a company is on autopilot and growing at 10% per year, its inventories, accounts receivable, and accounts payable may also grow at 10% per year. In looking at the balance sheet for Company E, which had working capital assets of over $1.6 billion and a year-to-year growth rate in sales of 12.7%, this item had the potential to eat up more than $200 million per year in cash. Fortunately, the company was not on autopilot and the year-to-year increase was kept under $100 million. In general, for a rapidly growing business, working capital requirements are a major cash drain and must be made up from other sources such as operating income or borrowing.

INVESTMENTS

The second section of the cash flow statement is categorized as investments. *Capital expenditures* are a key item here. The business will need a certain amount of *maintenance capital* just to replace equipment and facilities that have outlived their useful lives and which are usually fully depreciated. Holding back on growth or on maintenance is usually a desperate or unwise business decision. Additional capital will be needed to finance *new capacity* to accommodate normal business growth. Such essential capital is typically subtracted from operating cash flow to give what is called *free cash flow*, a key number for valuation purposes.

In addition, operating managers may budget *discretionary capital projects* such as production facilities for new products or high-return cost-savings projects to enhance manufacturing competitiveness. Generally, there are enough such ideas that discretionary capital must be rationed or allocated based on strategic intent or return on investment (ROI). Indeed, these are the high-return investment opportunities that top management typically looks to its R&D laboratories to produce.

Acquisitions are very similar in concept to discretionary capital projects and compete for capital with them. Acquisitions may be used to enter entirely new businesses, or can be used to grow an existing business through acquisition of small or marginal businesses that benefit from the efficiencies of the acquiring company.

Financing Activities

Finally, financing activities are an important part of the cash flow statement. Their primary purpose is to allow ongoing operations to grow and for the company to invest in new opportunities that earn more than the cost of capital. These after all are the underlying aims of the company and why investors buy its stock. If money needs to be raised, it can be done through the issuance of new debt or new equity.

Corporate treasury operations also seek to reduce the cost of capital by restructuring its balance sheet. It may judiciously replace high-cost debt or equity with lower cost financing as opportunities to do so arise in the financial marketplace. Stock buybacks may be part of this picture. These activities, which have no direct connection with operations, can enhance shareholder value.

In addition, the company may elect to pay dividends to shareholders and so, in principle, allow them to reinvest proceeds that the company does not require to finance its own growth.

Company E's cash flow items are described in Exhibit 3.7. Operating income and depreciation are throwing off a cash flow of over $600 million combined. Some amortization is occurring, as we would expect, given the substantial goodwill on this company's balance sheet. Working capital changes have been carefully controlled in 2000, but were high in 1999. Looking beyond operations to investments, a very large amount of acquisition activity, over $1 billion, occurred in 1999 and a smaller but material amount in 2000. Capital

	2000	1999
Operating Activities		
Income before extraordinary item	$ 399	$ 333
Adjustments to reconcile to net cash provided by operating activities		
Depreciation	238	216
Amortization	43	35
Deferred income taxes	1	35
Long-term liabilities	19	40
Other noncash items in income	22	37
Changes in operating assets and liabilities, excluding acquisitions and divestitures of businesses		
Accounts receivable	(20)	(190)
Inventories	(30)	(115)
Other current assets	(11)	(12)
Accounts payable and other accruals	(10)	129
Accrued income and other taxes	(1)	10
Other—net	(3)	4
Net cash provided by operating activities	647	522
Investing activities		
Acquisitions of businesses, less cash acquired	(143)	(1,058)
Divestitures of businesses	11	61
Expenditures for property, plant, and equipment	(399)	(267)
Purchases of short-term investments	(10)	(7)
Maturities and sales of short-term investments	6	252
Other—net	28	9
Net cash used in investing activities	(507)	(1,010)
Financing activities		
Borrowings with original maturities of more than three months		
Proceeds	368	731
Payments	(251)	(609)
Borrowings with original maturities of less than three months—net	(73)	173
Proceeds from sale of Common Shares		252
Proceeds from exercise of stock options	11	18
Cash dividends paid	(117)	(91)
Purchase of Common Shares	(40)	
Net cash provided by (used in) financing activities	(102)	474
Total increase (decrease) in cash	38	(14)
Cash at beginning of year	18	32
Cash at end of year	56	18

Exhibit 3.7 Cash Flow Statement for Company E for the Year Ending December 31, 2000

expenditures for property, plant, and equipment are somewhat higher than depreciation, as is to be expected for a growing company. A relatively small amount of cash was raised through divestitures of businesses. However, Company E during this two-year period used over $1.5 billion in cash, which was offset by less than $1.2 billion from operations.

Financing activities made up the difference. In addition to making up the cash shortfall from operating and investing activities, another $0.2 billion in cash was paid out over the two-year period in dividends. The corporate treasurer needed, therefore, to raise about $0.5 billion.

Let's see how he did it.

On the debt side, Company E extensively restructured its long-term debts with repayments ("payments") approaching $900 million and "proceeds" of about $1.1 billion over 2 years. On a net basis, its total borrowings increased by over $339 billion. On the equity side, Company E raised $252 million by issuing common shares in 1999, but bought back $40 million in shares in 2000.

By raising equity in addition to debt, this aggressive expansion program kept the debt/equity ratio in a very comfortable zone.

As a final commentary, the cash flow statement reveals that corporate growth was heavily fueled by new investment and acquisition activity and did not occur solely as a result of the dynamics of the ongoing operations. The income statement would not reveal this.

Other Key Concepts Relating to Financial Statements

FREE CASH FLOW

As noted, free cash flow equals operating cash flow less related capital expenditures. It can be looked at as a business's true after-tax operating cash flow.[8] It is independent of financing, and in the absence of debt, it is what would be available to the company and its shareholders for investment or distribution. Free cash flow is the most common basis for valuing businesses and will be used throughout this book.

EBITDA

In the world of valuation for transaction purposes, the term of EBITDA—*earnings before interest, tax, depreciation, and amortization*—is used commonly enough that it should be mentioned. It is a measure of pretax cash flow. Operations are often valued as a multiple of EBITDA, say 7 × EBITDA, with the multiple reflecting the "going rate" in M&A transactions. This metric is an alternative to EBIT (pretax operating income). The going multiple of EBITDA for valuation purposes will of course be lower than the going multiple of EBIT, since EBITDA, which includes depreciation, is a bigger number than EBIT.

The importance of EBITDA is illustrated by a typical leveraged buyout transaction,[9] where the strategy of the seller is to get as much cash for a business as the buyer is able to borrow. The buyer, who will focus on reducing costs and paying down the debt, is likely to forgo growth, with its attendant demands for capital, to get the debt down as rapidly as possible. Because of heavy interest payments, taxes will initially be minimal, making the use of *pretax* cash flow a nonissue. Hence, a multiple of EBITDA can provide an excellent metric to financial value within the confines of this strategy.

Cash Flow and Operating Decisions

It is common to continue to operate businesses for their cash contribution even if they are unprofitable on a reported earnings basis since the depreciation can still be recaptured. It is only when the cash margin goes negative—that is, when the selling price has fallen below the variable cash costs of making the product—that you lose money on every item you produce. At this point, shutting down the operation is financially rational. For this reason, manufacturers in cyclic commodity businesses carefully track the cash contribution of their mills in a cyclic trough, since it is assumed that producers with the highest cash costs will be the first to shut down.

Issues in Financial Statements

Broadly speaking, there are two important classes of problems associated with the corporate accounting systems that produce financial statements: allocation and valuation. Readers should be aware of both.

ALLOCATION

Allocation issues in accounting exist at all levels. They have become particularly troublesome with the widespread adoption of the concept of profit centers, which "are treated as mini-companies with allocated corporate expenses so that they can be managed as independent financial entities."[10] However, when costs and resources are badly allocated, value-destroying decisions can be easily made. Each company has its own allocation conventions, some intelligent and other downright stupid. Let's look at some that pertain to R&D.

At the lowest level, individual scientists and technicians must allocate their time to projects, but the formal project definitions may fit the actual uses of their time quite poorly. At the next level, the laboratory may have to allocate some or all of its resources among several profit centers. Headquarters expenses (which may already include an allocation from the corporate laboratory!) may then be reallocated to business units.

In some companies, imputed interest may be allocated to profit-center managers to remind them that capital is not free (although most think it is much better to separate financing from operations).

Other classes of headquarters expenses can be difficult to allocate fairly. Should the cost of a legal settlement be charged to the unit that incurred the problem, be divided among all business units, or kept as a corporate expense? Such questions are endless.

A related issue is that few profit centers are truly independent: if the Latin American subsidiary sells film made in Texas, using equipment owned by the U.S. operation, formulas must be established for allocating cost and capital. Experienced business managers know that some of the thorniest issues in multidivisional companies have to do with "transfer pricing." Most of these issues do not affect the corporation's total financial statement, but do affect the valuation of individual enterprises, the behavior of their managers, and internal investment decisions.

VALUATION

The biggest problems of all in accounting are issues of *valuation* because the values of assets and liabilities can change with time. Accountants like to value things based on historical cost, but this can lead to major problems. The value of a five-year-old plant can be estimated by cost less depreciation, but will that value be realistic? Inventory is traditionally valued at cost, but how long can it sit in the warehouse before it should be declared worthless? Should inventory accounting be based on a last-in first-out or first-in first-out basis? How large a reserve should be created for product-liability litigation or future site remediation work? In this light, the problems in valuation of intellectual property, which we shall deal with next, are hardly unique.

The tools and terminology of the financial statement, and especially the links between the income statement and the balance sheet, are essential both to the valuation of technology and to an understanding of its impact on the corporation as a whole. A thorough understanding of these terms is essential if a technologist is to reach an understanding with, (or effectively debate) the corporation's financial team. At the same time, financial statements may contain serious inaccuracies and, in any case, relate only to book value. Questioning them is okay. And the marketplace may impute quite different values.

Accounting for Intellectual Property

It has been suggested[11] that in the information age a larger and larger portion of a company's economic value is associated with its intangible assets, intellectual property often being the most important.

Intellectual property takes many forms. These include identifiable properties such as trademarks, brand names, patents, and copyrighted works, and far less identifiable property: knowledge about customers, about technology, about how an industry works, about high-quality production processes. There is also a vast corpus of written information—reports, notebooks, drawings, manuals, memos, computer codes, and so on. Each of these is potentially valuable, as evidenced by the existence of employee confidentiality agreements, noncompete agreements, and bitter legal battles over patent and trademark infringements and the rights of former employees.

Movie studios, software houses, biotech start-ups, law firms and the Coca-Cola Company, among others, must surely count intellectual property among their most highly valued assets. Their market capitalizations reflect these values, though their books may not.

Intellectual property in a technology sense is a somewhat narrower subject and it is worth reviewing its characteristics briefly. An employer often hires a researcher or an engineer in part to gain a specific expertise that the employee has acquired through education or experience. For example, the recently hired engineer may be an expert on extruding plastic film. In effect, the employer is *renting her expertise* in the form of a salary that reflects the value of her skills in the marketplace. This type of intellectual property is often referred to as *background* technology. As individuals learn and study on the job, their background expertise, and this portion of the company's intellectual capital, will grow. In addition, the company will gain *foreground* technology. This will be proprietary to the firm. Assume in this example that the engineer develops a technique for running an extruder 50% faster than the competition. This improves the capital productivity of the firm and gives it a big edge in film extrusion. Normally, the intellectual property would be recorded in the engineer's notebooks, translated into drawings, and embodied in physical modifications of the extruders. It will be shared with some other individuals in the firm.

The firm now has a choice in how it will treat this intellectual property. It can patent its new extrusion technology or treat it as a trade secret. Patenting might help prevent others from using the invention, perhaps giving the patent holder a competitive edge. Alternatively, the patent holder could license the technology exclusively to a leading manufacturer of extruders, or license it nonexclusively to all comers.

The other choice, to treat the invention as a trade secret, may be feasible if few knowledgeable outsiders visit the company's manufacturing facilities. It might be possible to restrict knowledge of the invention on a "need to know" basis, reducing the likelihood that the secret would get out through ex-employees. The device itself may be concealed in a "black box" to keep it away from the eyes of visiting repairpersons, tech service representatives, customers, and others.

Valuing this intellectual property *in the marketplace* would be straightforward in principle. Assume that the company had 10 extruders prior to its productivity invention. The 50% improvement is worth at least 5 extruders, less the cost of modifications. It may be worth more if the company has an attractive option to expand its extrusion business based on its new competitive advantage.

If licensed, the invention would be worth the discounted cash flow (see Chapter 5) of the anticipated royalties over the life of the patent, an entirely different calculation. However, one should not double count. The productivity advantages of the invention would not be sustainable, aside from a short period of *lead time*, as competitors will begin to employ the licensed technology. The actual decision to license or not license will hinge in part on the valuation exercise, but also on judgment regarding whether the trade secret can be kept, whether competitors will invent (or worse, patent) something similar, and whether a patent will be granted and be enforceable.

Accounting for this intellectual property and valuing it, however, are two different matters. In most companies, the time spent by the engineer and her associates in developing the technology and procuring the patent will be expensed. Hence, the book value will be zero; patents are not normally shown as assets on the company's books. The real value of the asset might emerge if the extrusion business were to be sold—the enhanced productivity would create a purchase premium, and that premium would be booked as either goodwill or as a patent by the buyer, and as income by the seller.

Conceptually, the alternative to zero valuation is to capitalize R&D, just as the labor involved in constructing a physical facility is part of the capital investment. In principle, management could define a development project to improve extruder productivity, and charge the engineer's time and materials to the project. This approach would treat the project as an investment, just as a production plant is an investment. This sum could then be depreciated over the useful life of the technology, which might be the life of the patent. The book value of this asset would be based purely on cost, which would still *not* reflect the real value of the technology.

The Value of Technology Is Situational

The example above illustrates why the valuation of technology can only be undertaken in the context of a specific business situation. *Technology does not have intrinsic value.* (This actuality accounts for the oft-heard, but sometimes premature, criticism about "technologies seeking a market.")

In the case above, the situational factors that will drive the company's decision as to whether the highest value is obtained via license or as a trade secret may include:

1. How many extruders does the company have?
2. Are there opportunities to expand its business?
3. Is the company large or small with respect to its competitors?
4. Can the technology work on the competitor's equipment?
5. Is the patent likely to be enforceable?

Given the circumstances, this decision may be a "no-brainer" or it may require, and lend itself to, a proforma analysis of alternative scenarios using known facts, reasonable assumptions, and perhaps even estimates of probabilities. Furthermore, the company may elect not to accept the highest value alternative for reasons of overall strategy or to protect and build valuable relationships.

The Pros and Cons of Capitalizing R&D

Capitalizing R&D, as described, would have important pros and cons. As a practical matter, it is not done and is not consistent with generally accepted accounting principles, which ordinarily require immediate expensing of R&D costs.[12] The reason is based on the notion that most

R&D efforts are too uncertain to warrant capitalization and writing them off immediately is the conservative course. However, there are exceptions, including the capitalization of development expenses within major projects[13] or from the capitalization of international R&D.[14] Starting with the "cons," by capitalizing the work:

- Based on the FASB argument, there are no accounting problems if an expensed project fails. But if it has been capitalized, it must be written off. If some of the project's goals have not been met, it should be written down. This accounting requirement, when extended to a broad portfolio of R&D projects, would engage R&D managers in countless discussions with auditors as to which projects to write off and when. It could give management another tool by which to "manage earnings" by electing to defer or accelerate R&D write-downs, a generally unwelcome situation given the large discretionary factor in R&D decisions. Write-offs made by public companies might also need to be reported to investors, giving competitors access to some very interesting information.
- If R&D were capitalized, the company could not expense the R&D costs on the income statement. Normally, legitimate business costs can be used to reduce taxable income from all sources, thereby improving cash flow. Those tax savings take some of the sting out of R&D expenditures. Most financial managers prefer the benefits of better cash flow to better reported earnings. However, since some valuation methods are based on cash flow and others on earnings, this financial decision does affect valuation.
- For capitalized R&D, deferred income taxes are listed as an *asset* (and an unproductive one), since the full tax deduction is yet to be realized (accelerated depreciation is classified as a *liability*).

Here are the arguments for capitalizing R&D:

- R&D costs can be depreciated over time. This may not be as tax-advantageous as expensing costs immediately, but it helps to match the costs of R&D with the income it produces.

- No single year's earnings are torpedoed by huge developmental expenses, which often bunch up in one or two years. This may be no small matter if the added R&D costs of the proposed project reduce operating income below what management considers acceptable or slow the apparent growth rate of operating income.
- The books will reflect that an investment is being made. Confusing investment with expense is probably the biggest mistake that corporate executives make in dealing with R&D. Capitalizing its costs is a visible signal that the organization sees R&D as a bridge to the future, not as a cost center that needs to be limited or reined in.

Accounting methods are inadequate at valuing technology. To the degree that investments are confused with expense, accounting numbers can mislead management and investors about both current and future profit levels. Accounting earnings are particularly pernicious in judging small, rapidly growing businesses, which typically must make above-normal investments in R&D problem-solving, in market development, in the training of personnel, and in plant start-up costs. As the business gets larger, these expenses are likely to be spread over a larger base, thus improving margins. And as growth slows, the need for the infrastructure required to support growth will also diminish.

Depreciating Technology

A consequence of not recognizing technology as a balance sheet asset is that one cannot recognize the phenomenon of technology depreciation. And technology *does* depreciate. Under accounting rules, however, you can't depreciate something that is not on your books. (Some technology depreciation is recognized in the useful life of equipment, such as computers, which has technology embedded in it, but we are speaking more broadly here.)

Technology depreciation is driven by the fact that the performance parameters for most technologies improve continuously, at rates that are characteristic of the technology's position on the S-curve (see Chapter 8). "Moore's Law," that the density of circuits on a chip doubles every two years, indicates just how quickly technology can improve. Speeds of paper machines and of oil-drilling rigs also increase, as does the fuel efficiency of vehicles, but in these examples much more slowly. When radical innovations succeed, the value of existing technologies may not just depreciate—they may collapse.

Technology depreciation has real financial consequences. For businesspeople, eroding profit margins provide the signal that their once-golden technology has depreciated. The market value of their shares is the next to slide. To prevent this slide toward failure, R&D budgets must recognize that the bar of competitive performance is always being set higher. Thus, to stay even with an advancing pack of competitors, a company must continue to invest in R&D, just as it offsets the depreciation of its equipment with new investment. This is intellectual maintenance capital, and its bottom line impact can only be measured against what would have happened if the money had not been spent. The cost of just staying even in a competitive environment is often high. For a group of specialty chemicals businesses with R&D budgets in the range of 4 to 5% of annual sales, one expert told me that he believed that about half of those budgets were required to just stay even, with the other half employed in projects that might actually create shareholder value.

Financial Leverage

Financial leverage refers to the relationship between debt and equity in the company's capital structure. The more debt relative to equity, the greater the financial leverage. Shareholders benefit from leverage to the extent that the returns on debt capital exceed the cost of borrowing (interest payments). The concept of leverage is critical to maximizing profits. To the corporate treasurer, leverage is a way to reduce the cost of money. To the investor, it is a way to concentrate profit in the equity portion of the balance sheet.

In the example shown in Exhibit 3.8, a company is assumed to require total assets of $10 billion to support its operations, and to have an after-tax return of 15% on its assets or $1.5 billion. In Case A, we

	Case A	Case B
Total assets	$10B	$10B
Equity = Book value	$10B	$6B
Debt		$4B
Return on assets	15%	15%
After-tax operating income	$1.5B	$1.5B
After-tax interest on debt @ 6%		–$0.24B
Net income	$1.5B	$1.26B
Return on total capital	15%	12.6%
Return on equity	15%	21%
Shares outstanding	200 million	120 million
Book value/Share	$50	$50
Earnings per share	$7.50	$10.50
PE ratio	15	15
Price per share	$112.50	$157.50
Market capitalization	$22.5B	$18.9B
Price/Book	2.25	3.15

Exhibit 3.8 Example of Leverage

assume it has no debt. Therefore shareholder equity is also $10 billion, and return on equity equals return on total capital, or 15%. In Case B, we assume the company has leveraged its equity and taken on debt equal to 40% of total capital. It must pay interest on that debt, which we are assuming to be 6% on an after-tax basis (equivalent to about 9% on a pretax basis). While net income drops to $1.26 billion, return on shareholder equity of $6 billion is a nifty 21%.

The second tier of Exhibit 3.8 indicates the effect of leverage on shareholder value. In Case B, only 60% as many shares are outstanding as in Case A; this reflects the policy that debt will be maintained at 40% of total capital. In other words, to have the same level of total capital, the corporation requires a lower level of shareholder money. Earnings per share are proportionately higher for the leveraged case, however. If the stock market were indifferent to the degree of leverage on the balance sheet (not a good assumption) and valued each case at a multiple of 15 times earnings, leverage could raise the price per share from $112.50 to $157.50.

Although it is clear that leverage increases return on equity, it is far less obvious how much value it creates. Merton Miller and Franco

Modigliani, two Nobel Prize-winning economists, have argued persuasively that in perfect financial markets the *stock price should be indifferent to leverage* and depend only on the value of the underlying assets. Theoretically, potential shareholders in the leveraged company will demand a higher return, because leveraged companies represent a greater business risk, and hence invest at a lower PE ratio. However, markets are imperfect. To the degree corporate treasurers forgo the tax shield implicit in debt, they will not be maximizing value. In practice, markets do reward a reasonable debt level, and corporate treasurers lever up accordingly. The details of this complex and interesting subject can be found in a number of financial textbooks.[15]

How can one get from Case A to Case B? As a company grows, its equity will increase with retained earnings. If the corporate treasurer elects to maintain borrowings at 40% of total capital as a matter of policy, he will borrow $4 for each additional $6 of retained earnings. Hence the company's total capital will grow at a rate considerably higher than the portion generated by reinvesting profits. Alternatively, if debt levels are low and the company lacks attractive internal investment opportunities, the treasurer may elect to *purchase shares* with borrowed money, simultaneously increasing debt and reducing equity until the target of 40% is reached.

Looking at the transaction in reverse (going from Case B to Case A), raising equity to pay off debt would needlessly *dilute* shareholder interests. However, if debt levels are uncomfortable, the treasurer may issue equity (i.e., sell new shares) and use the proceeds to retire part of the corporation's debt; if the equity markets are uninviting, the company could sell an asset and use the proceeds to pay off debt.

In the real world, the unleveraged corporation is a tempting takeover target, especially if it is consistently profitable. The acquirer can even finance the takeover using the target's own money! Here's how it works: the acquirer borrows money from an investment banker and uses it to buy enough shares of the target company to gain control. He then borrows money again, using the target company's assets as collateral, and uses the money to pay off his original loan from the investment banker. The end result is that the acquirer, using other people's money, has gained control of a company whose capital structure is now leveraged to one degree or another. The acquired company is riskier—

owing to its new debt—but in a position to produce a better return on shareholders' capital due to financial leverage.

Different Measures of Return

The number of different measures of return used in measuring business performance can be highly confusing. Naturally enough, management may add to the confusion by selecting those measures that happen to make their performance look best. Here are some measures that can be derived from financial statements alone.

From the right side of the balance sheet, we can calculate two very important measures. One is *return on equity*—perhaps the most popular measure of performance. Author Robert Higgins notes, "It is not an exaggeration to say that the careers of many senior executives rise and fall on their firms' ROEs."[16] ROE is obtained by dividing net income by shareholder equity. It indicates the return the corporation produced with its shareholder capital *only*. For most investors, this is what really counts.

On the other hand, we have seen that *return on shareholder equity* (ROE) is highly dependent on leverage. The example showing ROE of 21% is readily attainable with businesses achieving a more modest 15% return on assets, and without undue leverage.

But do the shareholders receive a 21% return on their investment. No way! (Unless he or she is an original shareholder and is valuing the original investment at book value.) At a PE multiple of 15, the shareholders' return as measured in earnings per share is only \$12/\$180 or 6.67%. They get the reciprocal of the PE ratio, or 1/15. This example dramatically illustrates the difference between market value and book value—a factor of 3.15 in the leveraged example. ROE is not an attainable return since one cannot purchase the stock at book value—it is only a measure of historical performance.

Indeed, it may not even be a good example of history, because in the event of a write-off of assets (usually associated with bad news and a disturbingly frequent management practice), shareholder equity is reduced and earnings are increased, resulting perversely in an improvement in ROE. Hence, ROE should be taken with a grain of salt because it reflects neither the real return available to shareholders, the real value of assets currently employed, nor the historic costs of these assets (which may have been distorted by write-offs).

The broader measure of return is the return on total capital employed, also known as *return on investment* (ROI). ROI tells shareholders and prospective investors how well a company's managers have performed, given *all* the capital at their disposal. ROI is a useful measure in that it is essentially independent of the leverage in the balance sheet.

The operating *return on assets,* sometimes called return on net assets (RONA), is also a useful measure of operating performance. As discussed, operations can in principle be separated from financial entities in a corporation. Owing to corporate history or philosophy, debt can in principle be 0%, 40% (Cases A and B), or even more than 100% of assets. Conversely, the balance sheet can include a huge cash asset earning interest. Operating performance, and the value of the operating businesses, is independent of these financial vehicles.

Running a business to maximize RONA, however, is by no means the right formula for maximizing shareholder value—indeed it can result in a low-growth strategy where RONA increases each year as the fixed assets are depreciated, and the company forgoes opportunities to invest at returns above its cost of capital.[17]

The preceding considerations lead to another measure: *total shareholder return,* which sounds like return on shareholder equity but is not. It is always defined over a time period (1 year, 5 years, 10 years) and is the actual increase in market value per share plus the dividend yield over the time period, expressed as an annualized percentage. This measure depends primarily on stock market performance.

If earnings grow at 12.4% per year as in the hypothetical firm in Chapter 11, and the price earnings ratio of 15 is maintained, shareholders can expect an annual capital gain on investment of 12.4%. If they obtain a dividend yield of another 2.6%, their total return will be 15%. This number is a lot better than 6.67%, the reciprocal of the PE ratio. But it is also a lot less than the 21% return on equity figure.

Total shareholder return is the underlying reason for buying the stock. One expects to make money not only from the earnings stream itself (which will be reinvested or dividended to shareholders), but from the capital gains associated with growth. The company is charged with reinvesting earnings to achieve this growth. And without growth there will be no capital gain.

An expectation of a 15% return assumes that a constant ratio of price to earnings is maintained over the time period, or at least that the investor buys and sells the stock on days when the PE ratio is the same. This is not unreasonable if the company continues to perform, since historically PE ratios revert to the mean. However, the investor will do even better if he or she is astute enough to buy at a low PE ratio and or sell at a high one.

The effect of a sustainable acceleration in corporate earnings growth on PE ratios and shareholder value is discussed in Chapter 6. In this happy event, average annual returns can become even more attractive.

Finally, there are important performance corporate metrics based on discounted cash flow analysis—including *internal rate of return*, *economic profit*, and *net present value*. These require a determination of the cost of money, covered in Chapter 5.

Net, net, in buying equities growth and return are everything—without growth the returns are mediocre and the premium one pays above book value cannot be justified. But *with* consistent growth in the 10% range, equities far outperform securities offering a fixed rate of return.

CHAPTER 4
Capital from the Operating Viewpoint

Thus far, we have concentrated on capital in financial terms. Return on capital is a dominant consideration in investing. Regardless of which of several performance measures we adopt, one way or another that return is a fraction in which the numerator is profit and the denominator is invested capital. Reaping greater profits from the same amount of capital will improve return. This increases the numerator. Another way to improve return is to reduce the denominator, that is to reduce or control capital, an option that is too frequently overlooked.

This chapter covers several decision-making tools that relate to operating capital: break-even analysis; economies of scale; matching capacity to demand; and new versus existing plant.

Break-Even Analysis

Break-even point analysis is a simple and powerful decision-making tool. It takes into account the fundamental differences between *fixed* and *variable* costs.

Fixed costs are those that do not change over an observable period, even as the level of production changes. And most are capital related. For example, many of the costs of owning and operating a production plant do not change, whether the plant operates at 50% or 100% of capacity. Interest and principal payments on the money borrowed to build it remain the same; so too do depreciation, property taxes, insurance, maintenance, control, security, and supervision. Fixed costs do not also stay fixed, of course. The costs of maintenance and supervisory staff can fluctuate with different levels of plant utilization. At the plant level, fixed costs apart from depreciation may be lumped together in an item called *factory overhead.*

Variable costs are those costs that increase more or less linearly with output. They include raw materials, direct utilities, packaging and shipping, and direct labor. The term "direct" means direct contact with processing equipment. Variable costs are sometimes referred to as *direct manufacturing cost,* or DMC.

The distinction between fixed and variable cost is admittedly simplistic—the maintenance cost of mowing the grass is more or less fixed, while the maintenance cost of extruder wear is largely variable. Direct labor is viewed as variable, but cannot be varied continuously in a plant with many sequential operations; more likely the decision is made to operate with more or fewer shifts. In a plant that operates continuously, such as a refinery, direct labor costs may be relatively fixed. Judgment is needed, therefore, in applying this useful concept.

The excess of revenue over variable costs—the difference between selling price and DMC—is referred to as *gross margin* (GM), or *contribution margin* (or contribution to fixed costs and profits). It happens also to represent cash being received by the business. It increases linearly with sales. As Exhibit 4.1 shows, the business does not break even until the gross margin earned exceeds the fixed costs. That point is called the break-even point. Below that the business is losing money.

Once breakeven is reached, profits begin to accumulate. Sometime thereafter, they will reach another key point, that at which profits have returned the cost of capital. This is economic breakeven. It essentially adds the cost of capital to the other fixed costs. The curve does not extend indefinitely, since eventually the demands exceed the capacity of the plant, and more fixed costs must be added to produce more product. This raises the break-even point.

No matter how small the scale of plant one builds, fixed costs have a practical minimum level. In the chemical business, a product with a market of $2 million that requires a stand-alone facility will have virtually no chance of ever being profitable. Small entrepreneurs

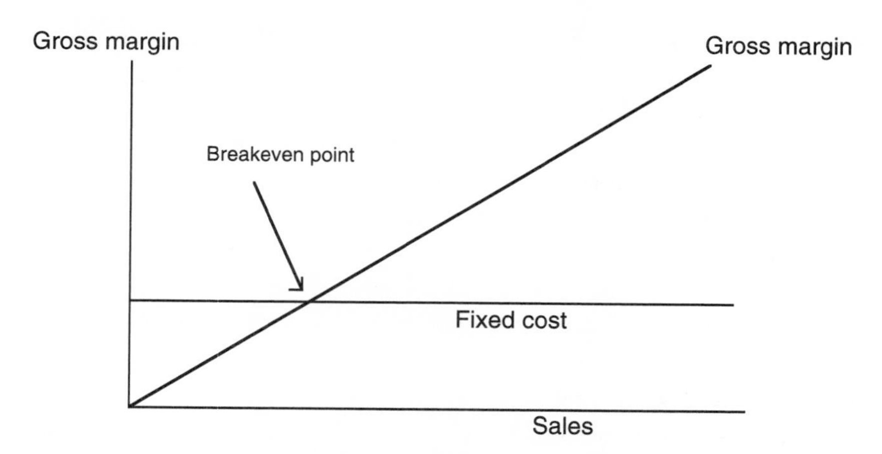

Exhibit 4.1 Breakeven Analysis

operating in garages may be able to do so by avoiding greenfield capital costs, outsourcing services on a variable basis, and taking shortcuts on control, safety, and security, but even this is a difficult proposition. Therefore, R&D projects that fail to meet a minimum size-of-market criterion can be eliminated from the portfolio by break-even analysis alone.

Break-even analysis can also show that projects that require a large amount of initial R&D also become uneconomic below a given market volume. This statement implicitly reflects the idea of capitalizing R&D and then amortizing it over the life of the project. R&D depreciation becomes part of the fixed cost, just like physical asset depreciation. An example might be a new drug, where the initial R&D investment can reach hundreds of millions of dollars. Only a big market can support this investment. In fact, Congress has recognized this reality by creating additional economic incentives for pharmaceutical makers to apply their expertise to the development of "orphan drugs" for treating diseases that afflict a relatively small patient population who would otherwise represent "noneconomic" markets.

When a project or a business appears uneconomic despite decent gross profit margins, break-even analysis points out the road to the commonsense solution: reduce the break-even point by finding a way to reduce fixed costs.

The concept of contribution is important in business tactics because it is rational to keep operating a business as long as its contribution is positive, even if it is not covering fixed costs such as depreciation and is therefore "in the red." This is what business managers mean when they say they are reluctant to shut down a losing business because it is helping to "cover overhead." As long as operations cover variable costs, and then some, it pays to continue. Every dollar earned that exceeds variable costs pays a dollar of fixed costs that would have to be paid in any case. When that contribution drops to zero, liquidation becomes the compelling choice.

Fixed costs are relevant to the investment decision. But once you have made the investment into a business, pricing and strategy will hinge only on an analysis of your contribution and that of your competitors. For example, an excess of industry capacity can cause prices to drop to the point that the contribution of several suppliers may be negative. One of these should close a plant to restore the health of the industry. This can be a game of "chicken." But it is logical for the plant with the highest cash costs to be the first to fold. Winning at this game—knowing when to fold and when to hold—is obviously facilitated by an understanding of everyone's cash costs.

Economies of Scale

The term *economies of scale* refers to the reduction of average unit cost that results from greater output. It is another major factor in capital investment decisions, and has important implications for the break-even point. Economies of scale depend on the nature of the business and on multiple factors within it. Larger plants have lower fixed costs per unit production, but larger fixed costs on an absolute basis.

Simple geometry is one underlying source of economies of scale. Construct a building, a tank, or a chemical reactor, and the volume (or capacity) goes up with the cube of the dimensions, while the surface area (corresponding more or less to the amount of material) goes up as the square. For those elements of a plant that are volume dependent, capital cost would be expected to rise at only the two-thirds power of capacity. Doubling the circumference of a pipe increases its cross-sectional area by a factor of 4: hence the cost of materials tends to go up as the 0.5 power of capacity. Meanwhile, the size of some items associated with a production facility—the control system, the front gate, the plant manager's office—may not need to expand as plant capacity increases. The economics of scale for such items are virtually infinite: on a per unit basis, a plant that is 10 times as large has one-tenth the cost. On the other hand, some items do not benefit from appreciable economies of scale and must be added in direct proportion to capacity; examples might include additional looms in a textile plant or cash registers in a supermarket. Other factors that affect the scale relationships are purchasing leverage (bigger discounts for larger volumes) and transportation costs (which usually favor more, smaller plants closer to customers).

A rule of thumb in the chemical industry indicates that, on average, capital costs rise as the 0.6 power of capacity.

I once had experience with a technology that converted forest slash (branches abandoned after tree trunks were trucked to the sawmill) to charcoal briquettes. An economic analysis of this project showed that one value-driver, economies of scale, was offset by another, transportation cost. The former varied as the 0.6 power of the plant's capacity, while the latter varied as the square root of the capacity, since trucking costs depended directly on the distance from the plant, and the gathering area was proportional to the square of that distance. In this case, one could calculate an optimum plant size using simple high school calculus. Exhibit 4.2 shows the implications of a 0.6 relationship.

When such dominating economies of scale apply, managers should consider strategies to build a big plant and price aggressively to sell out unused capacity quickly, since they enjoy an overwhelming advantage in capital costs per pound over small competitors. These economies lead to tremendous incentives to add large-scale new capacity to service the market. If you don't lead, your competitor will, and grind you into the dust.

Volume (MM lbs.)	Fixed Capital ($MM)	Fixed Capital (per lb.)
10	10	$1.00
20	15	0.75
50	26	0.52
100	40	0.40

Exhibit 4.2 Economies of Scale with 0.6 Relationship

The risks are also high because the break-even point (assuming it is directly proportional to fixed capital) of a 100-million-pound plant is more than 50% greater than a 50-million-pound plant. So if you can't sell out most of your capacity, you are in worse shape! Several competitors playing the economies of scale game is another business equivalent of playing chicken—miscalculations plus a determination to never surrender market share can create massive overcapacity leading to disastrous profit margins for the entire industry. Playing the game adroitly is *the* major issue in some capital-intensive businesses.

Matching Capacity to Demand

"What capacity should we build?" Matching capacity to demand is a critical business decision and must often be made in the face of uncertainty. Unused capacity has an initial economic cost, but one that should in time be more than offset by sales growth. With a growing market, a company must regularly consider building a new plant, or expanding an existing one, if it is to maintain market share. Adding to the complexity of the decision, the risk factor for Plant 2 is likely to be lower than for Plant 1, which bears the costs of the learning processes inevitable in start-up. Also, competitors' decisions with regard to the size, timing, and location of their plants will be, in part, a reaction to one's own decisions, and must be anticipated to the degree possible.

But is additional capacity the sole solution to growing demand? Assuming that the demand is there, can one sell more than one has capacity to produce? Yes. One strategy is to build inventory when capacity exceeds demand, and sell it when the reverse is true. Because this strategy increases working capital (while delaying capital investment in fixed facilities), its utility is limited. For commodities (products that are virtually indistinguishable from those of competitors) swapping or tolling deals with competitors may be the right answer—permitting one to maintain share, defer premature investment, *and* build only plants with world-scale capacity. Increasingly, competitors are actually sharing economies of scale through joint plant ownership.

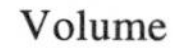

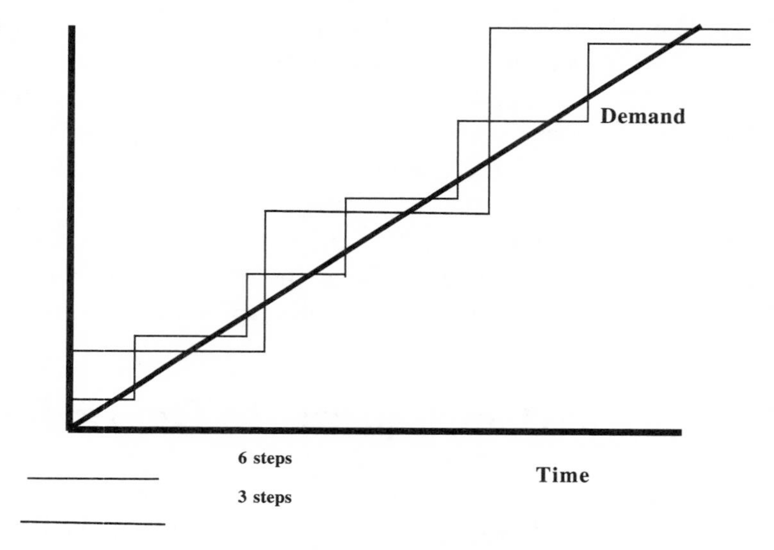

Exhibit 4.3 Matching Capacity to Demand

Exhibit 4.3 shows two approaches to matching capacity to demand. In the first, capacity is added in six small increments; in the second, three larger increments are built. Unused capacity is the area below the step-like supply curve and above the demand curve. It is also assumed that some inventory is built preceding each step, as shown in the areas below the demand curve. Six small steps result in less unused capacity than three large ones, but are handicapped by lower economies of scale. This curve does not consider the multicompetitor case, where the increments in capacity may be added by different parties.

Some unused capacity is inevitable in an expanding market, and the decision on how much to have is a trade-off involving economies of scale and competitive dynamics. In evaluating an R&D project proposal, it is probably unrealistic to anticipate the actual decision-making process and simpler to add extra capital to the project economics to account for average unused capacity. This premium should be larger if greater economies of scale are built into the economics in terms of lower capital charges per unit of production. In addition, further capital resources will be required for *plant construction in progress,* since it must be anticipated that each incremental step will be preceded by one to two years of design expense, deposits on long lead-time equipment, equipment purchases, and installation, construction, and start-up costs before any revenues are earned.

New versus Existing Plant

Building capacity from scratch generally involves high capital outlays. By comparison, any project that can be run out of an existing production facility has considerable advantages even though some capital spending for modifications may be required. The first advantage is the use of depreciated assets versus entirely new plant and equipment. This strategy will reduce the book assets employed by the project by an amount equal to the previously accumulated depreciation. The second advantage is time: retrofitting an existing facility generally takes much less time than building a facility from scratch. And time is money.

Siting a new project within an existing facility, however, raises interesting financial and strategic questions. Should the capital charged to the project be determined on (1) an incremental basis, (2) a book basis, or (3) a reinvestment cost basis? To answer this question, let us consider the case of an existing facility that is partially idle. Should the project be charged only for the incremental capital required to modify it, or for its fair share of the book value of the facility in which it is sited? The answer is a "no-brainer" when the project meets its required hurdle rate with or without a charge for preexisting book capital. However, if the project return cannot stand the charge for preexisting capital, the project should be viewed as opportunistic—a cash generator for the short term, but probably not a good long-term business. Some companies would insist that the project also meet its hurdle rate on a full reinvestment basis; that is, as if it were charged the capital required to equip a new stand-alone site. If it can, the project is a worthy long-term investment and will support investments in future expansions. The use of an existing facility merely adds short-term frosting to the long-term cake.

The issue just described demonstrates the common business dilemma between taking on any projects that earn shareholders more than the cost of capital, versus insisting that scarce resources be spent only on projects that will generate future high-value investment opportunities. If there are alternative uses for an existing plant (e.g., if an existing product is to be displaced by a new product), the decision must consider the return for each alternative use on an apples-to-apples basis. Strategic and marketing considerations may, of course, outweigh the purely financial calculations.

R&D Effects on Capital

Beyond the techniques for maximizing returns through astute timing, scale, and location for investments in operating capital, R&D can also be effectively deployed to decrease capital requirements, often with a very high rate of return on the R&D investment. In some cases, the R&D inputs that make more effective use of capital will be the make-or-break factor in determining whether a project meets the corporate hurdle rate.

Good communications is an important part of the art of minimizing capital. If left to themselves, members of the design engineering team will generally gather inputs regarding process parameters from R&D, and supplement these with additional inputs from manufacturing executives; these inputs will involve special circumstances at the site and their views of operating and design philosophy. In many cases, the design team will include contractors who have little incentive to reduce cost and much incentive to avoid risks. The design team will then set out to design a plant that is safe and reliable. When in doubt, they will err on the side of conservatism (translate as greater expense). There will be a strong tendency to include all features requested by the project sponsors (more expense). Typically, the first estimate of the capital cost of the plant will be made from price lists of the key process units (reactors, heat exchangers, dryers, etc.) marked up by a substantial factor (2 × or more) based on industry experience to cover piping, electricity, structural steelwork, instrumentation and construction. Hence, each "extra" feature comes with a big multiplier.

The antidote to this approach is a cross-functional team with representation from R&D, manufacturing, marketing, and engineering. Its members should review the preliminary design. Cost should be a major focus of its review, but the team should have flexibility to adjust product specifications, operating rates, maintainability, ease of start-up and shutdown, safety, and other attributes that can affect financial results. Team members should contribute both ideas and intelligent compromises on cost/benefit issues. Typically, a marketing representative may agree, once the full cost of providing for all possible customer needs is understood, to streamline the proposed product line and thus reduce manufacturing costs, inventory, and storage requirements.

Here are some areas in which capital cost reduction can be achieved:

- *Materials of Construction.* The use of lower cost materials is one of the most obvious ways to reduce costs. For example, substituting mild steel or plastic for stainless steel or high performance alloys in certain pipes and vessels can have dramatic effects on total costs. However, laboratory or pilot plant data may be required to demonstrate that these changes will not jeopardize safe and reliable operation. These potential cost savings should be considered and the tests performed well before design engineering begins.
- *Throughput.* Throughput takes many forms. If a line can run twice as fast, half as many lines may be needed. If residence time in a chemical reactor is halved, the vessel size can also be halved. Such improvements can be studied by varying processing conditions with an eye on capital. Sometimes a relatively small amount of additional capital on a process unit (e.g., improved mixing or increased cooling) can more than pay for itself by downsizing the process unit as a whole. Or a more costly, but more active, catalyst may pay for itself in capital savings.

  For the plant as a whole, identify the process that is most likely to limit capacity. Increasing the throughput of this process may leverage the capabilities of the rest of the plant sufficiently to give an excellent return on investment. This exercise is referred to as *debottlenecking*. Usually, debottlenecking one unit moves the bottleneck elsewhere, where there may be further opportunities for high-return capital investments.
- *Yield.* Yield refers to the amount of product produced relative to the raw materials used. Yield improvements obviously improve per unit manufacturing costs, but their effect on capital should also be noted. An increase in yield usually translates to

an increase in throughput, and, as such, to capital savings. It will also reduce waste. Less waste means reduced costs and capital in downstream operations that must separate good product from waste and treat or recycle waste streams. Yield pays off in at least four ways, two of which are capital-related.

- *In-line Processing.* To the degree that one process leads directly to another, it is possible to create large gains in yield and capital productivity. Looked at in reverse, if in-line processing is *not* performed, an intermediate storage capability is required. This has three negatives for capital. Most obviously, a storage facility (warehouse or tank) is required for the in-process inventory, adding to fixed capital. Second, a buffer supply of inventory must be maintained, adding to working capital. And third, quality problems are less apparent and more difficult to trace, resulting in lower yield. In-line processing is seldom physically feasible in the research laboratory, and therefore is usually addressed in the development process. Ignoring opportunities for in-line processing can be very expensive.
- *Elimination of Process Equipment and Storage.* Each process step should be carefully analyzed to see if it is essential or has been added to increase process latitude (the ability to continue to run the process outside normal operating parameters). Are three stages of separation required when two will do? Are there other ways to get the needed process latitude? In a similar vein, tankage and warehousing at the plant are always found among the usual suspects. Plant managers, who are typically held accountable more for operating costs than capital employed, like tanks and warehouses because they are good places to store off-spec products (most of which they hope in time to repair or blend away). Available storage tends to be filled—with working capital. It is probably better to insist that manufacturing rent warehouses or railcars to store their mistakes; doing so gives its managers an incentive to resolve problems quickly and avoid rents or demurrage expense.
- *Maintenance.* A plant shut down for maintenance creates no revenue. In addition, shutdown and start-up costs can be expensive in terms of direct maintenance, off-specification products,

and raw material waste. Good plant design anticipates maintenance requirements and maximizes time on-line. R&D can contribute through research on key issues such as lubrication, scaling, and corrosion.

- *Energy Efficiency.* Efficient use of heat and electricity based on principles of heat exchange and insulation reduces capital requirements for the production of steam and power. Cogeneration investments can be attractive when heat requirements are high and energy is expensive.
- *Working Capital.* The ability of technology to reduce working capital tied up in inventories is often overlooked. We have already noted one example, the reduction of in-process inventory through continuous processing. Shipping liquid products in the form of solid or liquid concentrates can also substantially decrease inventories by increasing the economic shipping distance (which tends to be volume-related) and reducing both the number of storage locations required and the volume of storage (additional fixed capital savings) required at those locations. R&D will be required to demonstrate the performance of concentrates and to develop procedures for reconstituting and packaging them.

Summary

Chapter 3 focused on financial terminology and the importance of capital and return on capital in financial decision making. This chapter has looked at the proactive use of R&D and engineering skills to reduce capital requirements and increase returns. Thus, we have begun to create the link between the corporate viewpoint and the scene as it looks to an engineer or research scientist. In the next chapter, we use the concept of discounted cash flow to understand why capital can be so expensive, and the implications of that fact on technology projects.

CHAPTER 5

Calculating Values Using Discounted Cash Flow

The concept of *discounted cash flow* (DCF) is central to the valuation of any asset when any part of its return is captured in the future—whether it is a security, a business, or an R&D project. Financial instruments with fixed returns, such as bonds and mortgage contracts, have relatively straightforward valuations. The valuation of businesses and technologies, whose future returns are uncertain, present a greater challenge, requiring a combination of sophisticated analytical tools and the exercise of critical judgment. If the judgment factor is not present, the tools alone are likely to give the wrong answer: "Garbage in, garbage out."

This chapter introduces the DCF concept and explains two important terms that most scientists and engineers have heard, but that many do not understand: *net present value* (NPV) and *internal rate of return* (IRR). Neither of these terms makes sense, however, without a grasp of another key concept: the *discount rate*. The discount rate is tied to the structure of the individual corporation's balance sheets, covered in Chapter 3, and the corporation's *cost of capital*.

In rapid growth situations, a central focus of this book, most of the economic value is likely to be derived from the *terminal value* of the project, not from short-term cash flows. Mishandling or misunderstanding terminal value is a major pitfall in the analysis of R&D projects. Why this is so is explored in considerable detail, as are several approaches to calculating terminal values.

We will then discuss the conceptual difference between a discount rate and a *hurdle rate*, and why a single risk-weighted hurdle rate is inappropriate for the management of projects *where risk changes with time*—which is generally the case in R&D projects.

Discounted Cash Flow

When I first encountered the DCF concept as a young research manager, the company's financial wizards were applying discount rates of 20% to 30% to my proposals. The effect of these high rates was to kill almost any long-term R&D project that either I or my colleagues could think up. Facetiously speaking, my alternatives seemed to be to either forget these R&D projects or understand DCF analysis more critically. If this predicament sounds familiar, read on.

Financial analysis and technological progress often seem to be in direct opposition. Financial analysis favors projects that will bear fruit in a short time; but many, if not most, of the technical breakthroughs that have shaped the ways we live and work today—and created huge wealth—have had long incubation periods. The time between thinking, developing, commercialization and significant profits has often been decades. Think of your own company's most successful, wealth-producing products: the big cash cows that took shape in R&D many years ago. How many of these would have ended up in the dustbin of unpursued ideas had they been subjected to the same financial tests that your current projects are required to meet?

A backward glance at history of technology seems to show that almost all successful new technologies traced their origins to ideas born many, many years earlier. Historically, project champions overcame significant and expensive setbacks on their way to success, but subsequently created great wealth. Consider Nutrasweet, a speculative new-to-the-world venture. Its patent term of 17 years was about to expire when the FDA finally approved its use as a sweetener. It is unlikely that Nutrasweet's champions would have stood the test of a

30% discount rate for 17 years—yet this product did not destroy wealth, it created mountains of it.

Nutrasweet is no isolated case. Thomas Edison's first-of-a-kind electric lighting business took over 12 years to turn a profit. Corning's investment in fiber-optic technology was in red ink for over a decade. Pilkington Glass spent over 5 years and a small fortune developing the now commonplace technology for producing plate glass through a continuous process. Another 14 months were required to get this breakthrough process to produce salable glass. In the end, the company would not break even on cash flows for 12 years.[1] But from that point forward, it reaped tremendous wealth.

Would these projects have survived in a world of bean counters and discounted cash flow? Surprisingly, the answer is "Yes." Discounted cash flow methods, properly applied, can strongly *support* long-term projects.

NET PRESENT VALUE

The DCF method is simplicity itself. Its premise is that a dollar received tomorrow is worth less than one in hand today. The reason is that one can invest today's dollar, with the result that it will be worth more tomorrow. Conversely, one can invest less than a dollar today and obtain a full dollar tomorrow. The real question is *how much* must one invest today to earn tomorrow's dollar? The answer is the *net present value* (NPV) of tomorrow's dollar. NPV is the present value of a stream of future cash flows less any initial investment. As an analytical tool, it is one of the most useful in the field of finance, making it possible to put cash flows occurring in different time periods—some positive, some negative—on an equal footing that recognizes the time value of money. And positive and negative cash flows, some occurring now and others occurring later, exactly describes the typical R&D project.

Assume that tomorrow's dollar is $100 and that "tomorrow" is next year. Further assume that we live in a happy world free of taxes and risks where there are plenty of opportunities to earn 12% per year on investments. The amount we must invest today to get $100 a year in the future is $89.29. That is, $89.29 plus 12% interest of $10.71 returns $100 next year. Viewed in reverse, the present value PV of $100 (at 12%) is $89.29.

Note that this is not the same as taking a 12% discount on $100, which would yield only $88.00. The correct result is obtained by *dividing* $100 by $(1 + M)$, where M is 12%, the cost of money (in this case as determined by a hypothetical alternate investment at 12%).

Now let us consider a case in which we must wait two years to get paid. The net present value is now only $79.72. If we invested $79.72 at 12% for one year, we would receive interest of $9.57, giving us $89.29 after one year. This amount will earn $10.71 during the second year, bringing us up to $100 at the end of the second year. We obtain the present value in this two-year situation by dividing $100 by $(1 + M)^2$.

Should we wish to consider a third case, where we need to calculate the NPV of both $100 payments combined (that is, we receive $100 in year 1 and another $100 in year 2) we simply add the two values and get $169.01. The general extension of this principle is that the NPV of an investment is the sum of the cash flow in each year "n" discounted by a factor $(1 + M)^n$.

The general formula for present value (PV) is $PV = \Sigma P_n/(1 + M)^n$ where P_n represents cash payments in year n.

NPV derives from the preceding formula. It is used to evaluate situations in which there is an initial investment (I_0) made today (time 0), and a stream of payments or cash flows (P_n) received in future years. The *net* present value would then be

$$NPV = -I_0 + \Sigma P_n/(1 + M)^n.$$

NPV is the present value of a stream of future cash flows less the initial investment.

Discounting a future payment or cash flow to its present value is the *reverse* of the familiar process of compound interest that makes your savings account balance grow year over year. Using the following figures as an example, if you were to put $250 into an investment account paying 12% compounded annually, you would have $351 (a future value, or FV) in the account at the end of three full years.

$$PV \times (1 + M)^n = FV$$

$$\$250 \times (1 + .12)^3 = \$351.23$$

A Discounted Cash Flow Example

Consider this simple question: Would you pay $250 today for a three-year stream of income of $100 per year, starting next year? The typical classroom answer is "It would depend on what interest rate I could earn with that $250 on an alternate investment, or at what rate I could borrow." In other words, it would depend on your *cost of money*. If your cost of money was 12%, each future year's cash flow would have to be discounted at 12% per year. We will refer to this case as Proposal A. Exhibit 5.1 shows that if the payments are not discounted, you make $50 on the deal. However, if you discount the payments at your cost of money (12%), you lose almost $10. But what if you can borrow at 6%? Now this becomes quite a good deal (see Proposal B in Exhibit 5.1). The net present value of the deal at a 6% discount is $17.30.

Two points should sink in at this stage:

1. NPV is meaningless until the discount rate is specified.
2. A negative NPV does not mean a person is losing money (unless the discount rate is zero); it means that the person is not earning his own cost of money.

Once you understand the concepts of NPV and discount rate, you can understand another widely used financial concept: *internal rate of return* (IRR).[2] Internal rate-of-return is the discount rate at which the net present value of a stream of cash inflows and outflows equals zero. In plain English, IRR is the rate of return produced by the investment.

		Year 0	Year 1	Year 2	Year 3	(NPV)
Actual cash flow	DCF at 0%	($250.00)	$100.00	$100.00	$100.00	$50.00
Proposal A	DCF–12%	(250.00)	89.29	79.72	71.18	(9.82)
Proposal B	DCF–6%	(250.00)	94.34	89.00	83.96	17.30
NPV = O, Internal rate of return 9.7%		(250.00)	91.16	83.10	75.75	0.00

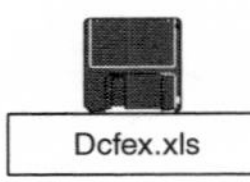

Exhibit 5.1 Discounted Cash Flow Example

In our example, we had these cash flows: one up-front negative outlay, followed over the succeeding three years by inflows of $100. IRR tells us the rate of return on this deal. Here, IRR is 9.7%. This tells us that if your cost of money were 9.7%, the transaction would be strictly breakeven for you. You would neither gain or lose. However, if you could borrow money for *less* than 9.7%, the proposal would be favorable. If your money cost were *more* than 9.7%, you should decline to play.

Note: The mathematics of IRR assumes that all cash flows received are not spent, but invested at the same rate of return as the IRR. For example, the $100 payment received at the end of years 1, 2, and 3 would have to be invested at 9.7%.

We can look at the proposition one more way. We mentioned borrowing money, and the cost of money. The DCF concept is relentless about exposing the true costs of money. For example, the NPV of Proposal B at a 6% discount rate is only $17, whereas there is a nominal profit in current dollars of $50. Where did the other $33 go? The answer is revealing and not as simple as it seems (see Exhibit 5.2).

Assume it was necessary to borrow money to finance this entire transaction. The simple interest of 6% would have to be paid on annual remaining negative balances of $250, $150 and $50 and a positive balance of $50 in year 4. The simple interest at 6% would be $15, $9 and $3 for each of the first three years, or $27 in total, offset by $3 of positive interest we could earn in year 4. We have accounted for $24.

Year	Raw Balance	Simple Interest	Cumulative Interest	Interest on Interest
1	($250.00)	($15.00)	($15.00)	($0.90)
2	(150.00)	(9.00)	(24.00)	(1.44)
3	(50.00)	(3.00)	(27.00)	(1.62)
4	50.00	3.00	(24.00)	(1.44)
		(24.00)		(5.40)
	Cash flow	50.00		
	Interest	29.40		
	Remainder	20.60		
	Discount factor (4 years)	0.8396		
	Present value	17.30		

Exhibit 5.2 Analysis of Proposal B

But we have forgotten the "interest on the interest." That has to be borrowed, too. Calculating that reveals another $5.40, bringing the nominal profit from the transaction down to $20.60. In fact, this is the amount one would clear at the end of the day with monies borrowed (and subsequently invested) at 6%.

But there is still more than $4 to be accounted for. The answer is that the deal terminates three years from now. The remainder of $20.60 discounted at 6% per year for 4 years is $17.30. Or, if you really want the $17.30 now, your friendly lender should in principle be happy to give it to you in exchange for a commitment to pay $20.60 four years from now. The $17.30 is *net* present value.

Economic Value Added

The popular term, *economic valued added* (EVA), is closely related to the concept of IRR. The term "economic profit" means the same thing. EVA represents the value added by business operations *above* the cost of money. When they say that an R&D project has contributed $10 million in economic value added to the corporation, they mean that the project has:

- Paid for all the cash outlays made on behalf of the project.
- Paid for the cost of those cash outlays at the company's cost of capital.
- Produced an additional $10 million for the corporation.

In effect, any project that produces economic value added has an internal rate of return that is higher than the company's cost of money. In Proposal B, the economic value added was $17.30. It is much less than the nominal $50 profit booked, and on which the Internal Revenue Service will levy taxes. The IRR of 9.7% exceeded the 6% cost of money by 3.7%.

Before moving on, note these three points:

1. Absent any element of risk, the discount rate is the cost of money, and is identical to the interest rate in simple debt financing.

2. The impact of discounting future rewards back to the present increases as (a) the discount rate becomes higher, and (b) the time span between the present and the receipt of the rewards increases.
3. NPVs and IRRs are readily calculated using spreadsheet software such as Lotus 123™ or Microsoft Excel™. These spreadsheets facilitate what is called *sensitivity analysis*—changing one or more assumptions (such as discount rate or price) to determine the effect on the final outcome, a technique explored in detail in Chapter 10. Using spreadsheets, an analyst can determine the range of assumptions that meets or fails to meet his or her financial return objectives.

The Cost of Money

The preceding material provides the background needed to calculate the *cost of money*—one of the key factors that the DCF method uses in project valuation. The cost of capital for a business is the return that its creditors and shareholders expect to earn. There is no single cost of money for a corporation, any more than there is for an individual borrower. The cost of money depends on both the marketplace for capital and by the lender's perception of the borrower. Interest rates and perceptions both fluctuate with time, as does the cost of money.

In determining the cost of money, it is tempting to look back to see what the corporation has paid for borrowed funds, particularly funds it is still using. For example, the firm issued 15-year bonds last year at a rate of 8%. Is this the firm's cost of funds today? Perhaps. But it may not reflect the firm's borrowing cost *for its next project*.

The cost of money should be determined at the margin, that is, on an *incremental basis.* The fact that the company may once have negotiated a terrific rate on some long-term bonds that are still outstanding does not mean that it could negotiate that rate today. The interest rate environment may have changed. The corporation's creditworthiness may have changed. It is the cost of money *today* that really counts. This cheap borrowing is already factored into the valuation now being placed on the company in the stock market. The appropriate rate is therefore not the average rate in the portfolio but the rate the corporation would pay if it went aborrowing today.

DEBT AND EQUITY

As discussed in Chapter 3, corporate treasurers normally target a ratio of debt to equity in their total capital structure, and will from time to time adjust the capital structure as costs of financial instruments change in the marketplace. Such day-to-day incremental decisions with regard to financing—short-term debt, long-term debt, equity issues—are not relevant to a particular project that may have a lifetime of many years. Hence, it is appropriate to use a *weighted average cost-of-capital (WACC)* under current conditions, where the weights relate to the broad mix of debt and equity. Because the cost of capital is based on market considerations, market values should be used for debt and equity in establishing these weights. For debt, this is usually a minor distinction. However, for equity, where company stock can trade at a substantial multiple to book value (often 2 to 10 times book), *using market values instead of book values makes an important difference!* And it usually raises the cost of capital.

In determining the cost of money for a corporation, one must factor in both debt and the equity components, and consider their respective percentages in the firm's capital structure. Because debt is tax deductible, the appropriate equation for the weighted average cost of capital (WACC) is

$$\text{WACC} = (\%\ \text{debt}) \times (\text{after-tax cost of debt}) + (\%\ \text{equity}) \times (\text{cost of equity}).$$

Recall the hypothetical company in Exhibit 3.8. It has total capital of $10 billion, of which $4 billion is debt. On a *book* basis, debt represents 40% of its total capital, and equity represents 60%. However, *on a market basis,* with a price-to-book ratio of 3.15, the market capitalization is $18.9 billion and the total capitalization of the company is $22.9 billion. The percentages of debt and equity in this capital structure are 17.5% and 82.5%.

Generally speaking, only long-term debt is used in the calculation of WACC. Short-term debt from lines of credit and the like, in parallel to the requirement to have some cash available, may be regarded as working capital.

The Cost of Debt

The cost of debt is the rate of return (interest rate or yield) that the corporation must offer creditors to induce them to provide funds. For the typical corporation, which has debts of different types and at different rates, cost of debt is a blend of many rates negotiated by the corporate treasury. Indeed, getting the right blend and the lowest rates is an important part of the treasurer's job. At any point in time, the lowest rate available may be debt secured by liquid assets such as inventories or receivables. Debt secured by other physical assets such as land or equipment may also have attractive rates. Unsecured debt, as a general obligation of the company, will be more costly to the firm depending on its seniority and the creditworthiness of the firm. In general, senior debt will carry a lower rate than junior debt because senior creditors are the first ones paid when a borrower finds itself in financial straits. The junior debt becomes progressively more expensive as the company becomes more highly leveraged (i.e., as total debt becomes a larger portion of total capital).

The treasurer is faced with an interesting dilemma in minimizing the cost of money. Up to a point, the cost of debt is cheaper than the cost of equity, so he or she may wish to raise the proportion of debt. However, this process encounters diminishing returns since debt holders demand higher and higher rates as the percentage of debt increases. The treasurer must also retain some additional borrowing capacity against a rainy day because the equity markets (i.e., the market for issuing new shares to the public) cannot be counted on as a source of capital.

Business history is strewn with the bones of overleveraged companies. Hence, the treasurer will be reluctant to leverage to a level at which the incremental cost of debt financing matches the cost of new equity. The range of choices is quite large: in a leveraged buyout, for example, debt may be more than 90% of total capital and will be lent at high rates; whereas in a conservatively managed corporation, debt may be 30% or less of total capital.

The Cost of Equity

The cost of debt is relatively straightforward, but evaluating the cost of equity is much less so, and there are several ways to go about it. The heart of the matter, however, is that the cost of equity to the corporation is identical to the returns expected by investors. If shares of other corporations *of comparable risk* offer better returns, investors will sell yours and buy those of others, causing the market price of your equity to drop. This process will continue until equilibrium is reached. In this sense, the cost of capital of the firm equals the opportunity cost of its shareholders.

A typical approach to calculating the cost of equity begins with this concept:

Cost of equity = Risk-free return + Risk premium.

The prospective contributor of equity capital begins with a "base rate" for letting the corporation have the use of his or her money (the risk-free return), then adds to that base rate appropriate compensation for the risk involved. The higher the perceived risk of the corporation, the higher the premium.

The *risk-free return* is easily understood and is generally pegged to the interest rate paid by U.S. Treasury bills—the nearest thing on planet Earth to a risk-free investment. This rate is quoted every morning in the newspaper. Let's say that it is 5.2%.

The basic idea behind a risk premium is simple enough. A typical stock, in the course of a year, may vary from its low to its high by 50% to 100%. Investors demand a premium for this volatility because they have no assurance that if they are forced to sell that their timing will be favorable. But what premium do they demand?

The total risk in a security can be measured in statistical terms, by noting how widely its value fluctuated about a mean. If the fluctuations are random, they will follow a bell curve or "normal distribution," whose width can be specified in terms of a *standard deviation*—a concept familiar to most scientists and to financial professionals. In a normal distribution, the value will fall within one standard deviation of the mean 68.3% of the time and within two standard deviations 95.5% of the time. Both the mean value and the standard deviation for any stock can be calculated from historical stock price data, and standard deviations as a percentage of stock price are published by many financial information services. This measure of risk is discussed in Chapter 12 relative to evaluating financial and technology options using the Black-Scholes formula.

However, the risk premium paid for equities does not depend simply on the volatility of individual stocks, as measured by their standard deviation, because investors in stocks can diversify away unique risk by buying a large number of stocks, if their objective is to minimize risk. They cannot, however, diversify away systematic risk—the risk to their portfolio if the stock market as a whole collapses. For this reason, it is extremely useful to divide the total risk in a stock into systematic and unsystematic (or unique) components.

To understand the power of diversification, consider a $1,000 bet based on the flip of coins. The payoff is double for heads, zero for tails. But I am given the choice of betting the amount on any number of coin flips I choose. If I choose one $1,000 flip, it is very risky—I can lose the entire $1,000. If I diversify over one thousand $1 flips, I only stand to win or lose a few dollars. The diversified portfolio is much less risky. There is no systematic risk here, only the unique risk in each coin flip, and I can diversify that away.

What is left, then, is the systematic risk built into the stock market as a whole. This is not the same for each stock or each company. Utilities, which pay regular dividends and may earn a regulated, automatic return on invested capital are far less vulnerable to a market decline than airlines, biotechnology stocks, or personal computer companies. In fact, the relationship between their price movements and that of the market as a whole can also be calculated statistically from historical price data.

The term that defines this relationship is called *beta*. In statistical terms, beta is proportional to the covariance of the stock's return to the market return.[3] By definition, beta for the market as a whole is set at 1. Thus, if a stock has a beta of 1.0, we know that its stock movements have been about the same as the market as a whole. If another stock has a beta of 1.2, we know that its price fluctuations have been 20% greater than the market as a whole: on average, moving 20% higher when the stock market rises, and dropping 20% more than the market during market downturns. Typically, utilities would have betas about 0.5, new technology stocks sometimes have betas of 2 or greater, and many traditional industrial companies fall into the 0.8 to 1.2 range.

Betas are calculated essentially by measuring the correlation of stock movements with respect to a benchmark index over a meaningful time frame[4] and are published by many financial information services. Calculated betas for individual stocks differ depending on the particular service and methodology used. Note also that if there were no correlation between the stock market as whole and individual stocks (just as there is no correlation between individual coin flips), there would be no systematic risk, and all risk could be eliminated with a sufficiently diversified portfolio.

The use of betas to value securities is an extension of the random walk theory[5], which presumes that stock movements (although demonstrably driven by events) can be efficiently modeled by statistical fluctuations, and follow what statisticians call a normal distribution. Investors react to events and use their business expertise and superior information in trading, but the random walk hypothesis implies that the actual stock price at any time in an efficient market is determined by a balance of equally knowledgeable bulls and bears, and the next move of the stock will be essentially random.

The Capital Asset Pricing Model (CAPM), used by financial scholars, takes the final step and concludes that the risk premium for an individual stock (or equity) should be directly proportional to its sensitivity to market risk[6], that is to its beta. ("Market," in this context, means the securities market, not the market for a company's products.) The risk premium for Company A is then defined as

$$\text{Market risk premium}_A = \text{Beta}_A \times \text{Risk premium for the average stock.}$$

Fortunately, the average rate of return for equities and the risk-free return have been tracked for decades, and the premium above treasury bills over the past 70 years has averaged 8.4%. And since betas for individual securities can be readily looked up in ValueLine and other places, the cost of equity for a publicly traded firm is not difficult to determine using CAPM.

Example
Company A, a utility, has a beta of 0.5. The risk-free rate is 5.2%; the market premium is 8.4%. Thus the cost of equity for Company A is

$$\begin{aligned}\text{Cost of equity}_A &= \text{Risk-free rate} + \text{Risk premium}_A \\ &= \text{Risk-free rate} + (\text{beta}_A \\ &\quad \times \text{market premium}) \\ &= 5.2\% + (0.5 \times 8.4\%) \\ &= 9.4\%.\end{aligned}$$

Although we use the CAPM model in this book, there are other approaches, both simpler and more complex. One rule of thumb is that the cost of equity is about double the prime rate. Another is that the cost of money for an average corporation is obtained by adding a risk premium of 5% to 6% to the going rate for long-term Treasury bonds.[7] Arbitrage Pricing Theory (APT) provides a more sophisticated macro-economic model. In addition to market risk, it assigns a series of other factor risks for issues that may differentiate individual stocks, such as foreign exchange risks, interest rates, and GNP growth. Although it may provide a superior fit to the historical data, statisticians and skeptics will recognize that in any least-squares analysis, the fit can only be improved by adding additional variables.

Neither CAPM nor APT takes into account factors that are *unique* to a company—new risks and opportunities. Both assume that the marketplace averages out these unique, or unsystematic, risks. They do so because an investor can eliminate virtually all unique risk through diversification. In fact, in direct analogy to flipping a "portfolio" of many coins, company-unique risk can be reduced substantially when the investor holds only 5 to 20 individual stocks in a portfolio.[8] Unique risk in a portfolio is essentially proportional to the inverse square root of the number of independent securities in the portfolio; that is, 9 stocks will have about ⅓ the unique risk of a single stock, and

100 stocks will have ⅒. The fact that investors successfully discount unique risk through diversification is extremely important to the thesis of this book because there is an obvious and quantitative analogy to the use of diversification to reduce risk in a portfolio of individual securities to the use of diversification in a portfolio of R&D projects. The empirical evidence for the validity of normal distributions of investment outcomes is strong: the Black-Scholes model for pricing options, based on normal distributions, is commercially proven, and it is impossible to verify that a statistically significant proportion of professional stock-pickers (such as money managers) can beat the S&P 500.

CALCULATING WACC

With this background, we can calculate WACC under the conditions shown in Exhibit 5.3 for our model corporation (assuming that the bulk of debt financing comes from corporate bonds). Using this information, we calculate the weighted average cost of capital for this hypothetical firm as:

$$\text{WACC} = (.175 \times .043) + (.825 \times .136) = 0.1196.$$

This can be rounded to 12%, the value we use throughout the book. This is the discount rate that we seek for use in discount cash flow analysis. Getting this figure right is extremely important since too low

Market Parameters	
Corporate bond rate	7.0%; pretax cost of debt
Corporate tax rate	38.0%
After-tax cost of debt	4.3%
U.S. Treasury bills	5.2%; the short-term risk-free rate
U.S. Treasury Bonds	6.0% the long-term risk-free rate
Average market risk premium	8.4%; relative to Treasury bills
Cost of equity	13.6%; Treasury bills + 8.4%
Corporate Parameters	
Beta	1.0
% Equity financing	82.5%; based on *market* value
% Debt financing	17.5%

Exhibit 5.3 WACC Example

a discount rate will make projects appear more attractive than they really are. Too high a discount rate, in contrast, will jeopardize the net present value of future cash flows, perhaps scuttling good R&D projects. (Note that if we had used the alternate approach of adding a 6% risk premium to a long-term Treasury bond rate of 6% we would have an equivalent result.)

Finally, for companies with several different lines of business, analysts may use different discount rates in valuing different assets. A gas transmission company owning chemical assets may value its pipeline assets using a lower cost of money than its chemical assets, since utilities tend to have low betas and are also able to borrow proportionately more of their total capital than is the case for cyclic industries.

Managing Earnings

Company executives are increasingly interested in what is politely called "avoiding earnings surprises" or, less politely, "managing earnings." Their motivation is partly to reduce the cost of equity capital for their shareholders. Earnings surprises increase volatility and thus increase beta. Higher betas result in higher cost of equity, higher perceived risk, and loss of shareholder value. This can be explained by both the CAPM model and by the tendency of the market to punish earnings surprises.

Earnings can be managed legally by accelerating or delaying customer shipments (usually with the agreement of the customer), by deferring discretionary expenses, by write-offs of inventory or fixed capital, and by adjustments of reserves for bad debts, environmental remediation, or product liability. For example, if a company regularly adds to a reserve, it can adjust the rate of addition downward by satisfying auditors that it is adequately reserved, or upward on the basis of "conservatism." Such adjustments will affect quarter-to-quarter earnings comparisons.

Of course, if managing earnings only means increasing earnings, the cupboard will soon be bare and management will have to face the music. The wise CEO will not wish to book all the gains of a great quarter and risk creating unrealistic future expectations for his company. He will quietly see that the cupboard is restocked. For these reasons, analysts look to the "quality of earnings."

Discount Factors

Having gone to some lengths to derive a discount rate, it is worthwhile to look at its implications. For reference, and for those who do not have a spreadsheet program at their fingertips, Exhibit 5.4 shows the discounted value of money at 12% over a 25-year period. The algorithm used to calculate the rate is to multiply each successive year by $1/(1 + M)$ where M is the cost of money (i.e., by 1.12 in this example). In the *n*th year, the discount factor is $\{1/(1 + M)\}^n$.

The severity of this discount rate is striking. At 12%, money loses half of its value in 6 years. At 10 years, it is worth 32% of its value, and at 20 years less than 12%. These numbers are extremely significant in the context of longer range research, since past experience tells us that many successful technologies took 10 years or more from concept to commercialization, and often many more years to reach their full market potential. Were the R&D expenditures that brought us these technologies bad investments? For that matter, can any long-term project withstand the rigors of DCF analysis? The answers lie ahead.

Year	Discount	Year	Discount
0	1.0000	13	0.2567
1	0.8929	14	0.2292
2	0.7972	15	0.2046
3	0.7118	16	0.1827
4	0.6355	17	0.1631
5	0.5674	18	0.1456
6	0.5066	19	0.1300
7	0.4523	20	0.1161
8	0.4039	21	0.1037
9	0.3606	22	0.0926
10	0.3220	23	0.0826
11	0.2875	24	0.0738
12	0.2567	25	0.0659

Exhibit 5.4 Discount Factors at 12%

Risk-Weighted Hurdle Rates for R&D

Financial analysts are very comfortable with idea of translating increasing risk into higher costs of money. Therefore, it is common for people to say that different discount rates should be assigned to different categories of investment. For example, one financial textbook cites these different discount rates:[9]

30% for speculative ventures
20% for new products
15% for expansion of existing business
10% for cost improvement of known technology

These numbers appear to be based on practical experience. There is an element of sound thinking in this approach, but also a major pitfall. To avoid the pitfall, it is imperative to distinguish between *hurdle rates* and *discount rates.*

Taken individually, high-risk projects should offer exceptional rewards. These can be captured in a calculation of internal rate of return. It is often reasonable to demand an IRR of 20% for a new product proposal. This represents a *hurdle rate,* or the minimum expected return that the corporation or investors require for undertaking a project. In effect, the hurdle rate is the "project cost of capital."[10] Indeed, a matrix of hurdle rates can be established for new proposed projects depending on their individual risk characteristics. In the context of an R&D portfolio, however, the unique risks of any individual project are reduced by the substantial diversification of the total portfolio. To take an extreme case, if experience shows that a drug company must synthesize 10,000 new compounds to create one commercially successful new drug, the synthesis of any single new molecule would seem to be an unacceptably high risk. No venture capitalist would invest in a company that proposes to synthesize 1 molecule or 10 molecules for drug screening. On the other hand, a major drug company may set out to synthesize 30,000 molecules using combinatorial chemistry to diversify its risks to an acceptable level, consistent with investor expectations that it will come up with a few hits every year. Thus, the hurdle rate can be appropriately reduced by the degree of diversification in the portfolio.

Using the same logic, a sufficiently diverse portfolio of high-risk/high-reward projects can exceed the expected value of a portfolio of moderate-risk/moderate-reward projects. This is a principle of venture capital investing.[11] However, even if industry experience has validated that 20% is an appropriate hurdle rate for screening a new product proposal, it is not reasonable to discount the entire project's costs or its rewards at 20% to establish *value*. As described in Chapter 2, well-managed R&D is in part a process of risk reduction. If risks are not being reduced with each progressive R&D expenditure, the project is likely to be unsuccessful and probably should be terminated. During the project's life, monies will be spent in stages of progressively lower risk, starting with conceptual research (with risk of failure at 90% or higher) and ending in advanced development (where the risks may be 15% or lower). And because development is much more expensive than laboratory research, most of the expenditure should occur at the less risky end!

In the reward stage, the project soon becomes an established, ongoing business. The appropriate discount rate for continuing investment in a successful product becomes the cost of capital. *It is conceptually incorrect to apply a uniform risk-weighted discount rate to a situation where risk is changing.*

Probability Weighting of NPVs

It follows that a superior approach to calculating net present value for the R&D portfolio is a probability weighted sum of the NPVs of the individual projects using the corporation's overall cost of money. A probability of failure should be assigned to each activity, and value should be established as the difference between the value of success times the probability of success less the cost of failure times the probability of failure, aggregated over the R&D portfolio. This powerful concept is worked out in detail in Chapter 12, in the discussion of *decision trees* and their applications.

As also discussed in Chapter 12, the use of probabilities enables two sources of value in sophisticated R&D management that are buried by overly aggressive discounting of risk. First, risk is substantially mitigated by the option of terminating projects early. Second, the rewards of R&D can be enhanced disproportionately by the existence of lower probability upside cases.

When Hurdle Rates Make Sense

Do hurdle rates based on experience make sense? Certainly a company needs to set a minimum hurdle rate for all its investments, since projects that only earn the cost of capital cannot create value. Setting the hurdle rate for investment at an achievable level somewhat above the cost of money will prevent financially weaker projects from competing for scarce resources. However, they make less sense when used as an initial screening tool to minimize risk, since risk is inherently correlated with economic profit. Moreover, they appear to be a single surrogate for four quite different risk-related variables:

1. The corporation's cost of money.
2. The degree of initial unique risk in the project.
3. The degree of portfolio diversification.
4. A blended average of the decreasing level of unique risk at each stage of the project.

As such, they are too crude to function in *valuation* of a technology portfolio.

Terminal Value

Terminal value has a number of aliases, including *residual value, horizon value,* and *continuing value*. By any name, the concept of *terminal value* is critical to the task of assigning a value to a technology or a business, using the DCF method. Indeed, excepting the cost of money, nothing is more critical. This is because most of the net present value of many projects, especially those with high revenue growth rates, is found in the terminal value.

We shall use the *free cash flow/growth in perpetuity* method for valuation purposes throughout the remainder of this book. Although it is quite sophisticated, it appears to be becoming the gold standard for financial analysts. Understanding it, its definitions, and mathematics, is worth the effort.

This section will discuss four approaches to determining terminal value. They differ in degree of conservatism. Conservatism in estimating value is considered a virtue by many. But bear in mind that being too conservative can lead to poor decisions and the destruction of value. Also remember that terminal values are calculated in an "out" year, say year 10 or year 20, where it no longer makes sense to project independent estimates for the individual factors (markets, margins, depreciation) that create value. Simply stated, detailed *pro forma* estimates based on specific customers or specific capital investments cannot be made 10 or 20 years far into the future and are generally replaced with a mathematical extrapolation or a business judgment regarding future value.

At the expense of some space, we shall do projections over a 20-year time horizon. Financial analysts, in valuing companies, typically use a time horizon in line with management's business plan (often 3 or 5 years), and calculate terminal values by the growth-in-perpetuity method for all subsequent years. While elegant, this can lead to significant undervaluation if high-value R&D projects are embedded in the business and have weak or negative short-term cash flows. So, "when in doubt, make a longer rather than shorter forecast."[12]

Terminal values must be discounted back to the present, and although still significant, they will be much smaller in present-day dollars owing to discounting over many years. Using Exhibit 5.4, a terminal value of $200 million in year 20 would be currently valued at about $24 million assuming a cost of money of 12%—a deep discount of 88%.

Liquidation Value

The most conservative method of dealing with terminal values is to assign them zero value. This approach assumes that the assets will no longer be productive at the end of a stipulated time, operations will cease, and the project will be liquidated. This can occur in a real business, for example when a mine is exhausted. In a technology business, it may also occur when a patent expires and would be an appropriate method for valuing a license agreement that terminates with patent expiration. A liquidation scenario contains the possibility of either a gain (e.g., from the sale of real estate or a trade name), or a loss (e.g., shutdown costs, environmental cleanup, severance pay, etc.), but those would normally be small compared with the value of an ongoing and profitable business. If these can be estimated, they should be used for terminal value.

As shown in Chapter 3, businesses may be valued and sold at a multiple of net income, of pretax operating earnings (EBIT), or of EBITDA. A phantom liquidation simply assumes the ongoing business can in principle be sold to a financial buyer in the horizon year at a price reflecting one or several of these valuation methods. The appropriate multiple or ratio will be based on actual transactions involving similar businesses. Anticipated growth rates in the horizon year are likely to be an important factor: other factors being equal, PE ratios are higher for more rapidly growing businesses. Typically, single-digit PE ratios are applied to companies that experience low or negative growth; large multiples (30+) can occur in aggressive growth situations; whereas a large band of companies with moderate growth cluster at multiples in the 12 to 25 range. Phantom liquidation can be based on DCF valuation, but that is equivalent to the growth-in-perpetuity approach to be discussed.

Terminal Value as a Perpetuity

A *perpetuity* is a security that pays out a fixed sum forever. Treating terminal value as a perpetuity is also the special case of the growth-in-perpetuity approach discussed next, where growth is assumed to be zero. This is a simplified, relatively conservative assumption that after a stipulated period the business has constant earnings in perpetuity. Its appeal is its simplicity. In mathematical terms, the value of a perpetuity is determined as follows:

$$\text{Value} = \text{Annual payment}/\text{Cost of money}.$$

Hence, an income stream of $1 million per year, forever, would be valued at $8.33 million when the cost of money is 12%.

Some people are bothered by the improbability that any business can survive in perpetuity. However, owing to the time value of money, most of the net present value would be earned in the earlier years, so the extension of time to infinity is not nearly as extreme an assumption as it may seem. For example, about 90% of the value of the perpetuity is paid back in the first 12 annual payments, if the payments are discounted at a 12% rate.

Perpetuities are useful as conceptual financial instruments, but in a competitive business you either grow or die. If liquidation is not foreseeable, the business will somehow manage to carry on—it will grow—and the paramount question will be "at what rate?"

The Growth-in-Perpetuity Method

The growth-in-perpetuity method is essentially a phantom liquidation based on direct valuation of extrapolated future *cash flow* in the horizon year. Its derivation is mathematically sophisticated, but directly addresses not only the issue of estimating future growth, but also the issue of whether that growth is profitable. Hence, it is a more reliable guide to valuation than PE ratios or multiples of EBIT or EBITDA. It appeals to financial analysts who are using it extensively, but suffers from the drawback of being less intuitive.

Fortunately, the results closely track more intuitive concepts such as measuring terminal value as a price/earnings ratio. And, because of its importance, we will discuss it in some depth.

The growth-in-perpetuity method provides simple and mathematically precise results given three assumptions:

1. The annual cash flow in the horizon year X.
2. The discount rate (or cost of money).
3. The rate of growth into perpetuity.

Perpetuity really means the time it takes a mathematical series to approach its limit. This will be a short time if the discount rate is considerably higher than the growth rate, but if a high growth rate is assumed (say 10% per year) coupled with a low cost of money (say 5%), the series will not converge at all. Which is to say in this happy but impossible circumstance, the owner would eventually become infinitely rich! As a practical matter, *the method should only be used when the growth rate is several percent less than the discount rate*. Unless this condition is met, mathematical convergence will be slow and unrealistically large values will be generated.

The mathematics are predicated on the assumption that in any given year, the cash flow will be larger than the cash flow in the previous year (P) by a percentage. Call this percent rate of growth g. That is,

$$P_{\text{next year}} = P_1 = P_0\ (1 + g) \text{ where g would be 0.06, for 6\% growth.}$$

However, the value of $P_{\text{next year}}$ must be discounted by the cost of money M, where M is 0.12 or 12% to arrive at present value.

$$PV_1 = P_1/(1 + M)$$

In the following year both factors are applied again:

$$P_2 = P_1\ (1 + g)/\ (1 + M) = P_0(1 + g)^2/\ (1 + M)^2.$$

To get the value out to "perpetuity" then one must multiply P_0 by a set of terms:

$$1 + (1 + g)/(1 + M) + (1 + g)^2/(1 + M)^2 + (1 + g)^3/(1 + M)^3 + \ldots.$$

Now let's define $X = (1 + g)/(1 + M)$ and make a simple algebraic substitution. We are saying P_0 must be multiplied by

$$X + X^2 + X^3 \ldots \text{ etc. out to infinity.}$$

Mathematicians call this a power series and know its sum to be simply $X/(1 - X)$. Therefore, it is straightforward to make a table (Exhibit 5.5) that shows the value of this factor for reasonable values of g at reasonable costs of money M.

In this table, the top line (zero growth) represents simply the perpetuity method previously discussed, and the now familiar ratio of value to annual cash flow at a 12% discount rate of 8.33 can be found in the top row. Again, the very large multiples near the diagonal are

Growth Rate	Cost of Money														
	1%	2%	3%	4%	5%	6%	7%	8%	9%	10%	11%	12%	13%	14%	15%
0%	100.0	50.00	33.33	25.00	20.00	16.67	14.29	12.50	11.11	10.00	9.09	8.33	7.69	7.14	6.67
1%		101.0	50.50	33.67	25.25	20.20	16.83	14.43	12.63	11.22	10.10	9.18	8.42	7.77	7.21
2%			102.0	51.00	34.00	25.50	20.40	17.00	14.57	12.75	11.33	10.20	9.27	8.50	7.85
3%				103.0	51.50	34.33	25.75	20.60	17.17	14.71	12.88	11.44	10.30	9.36	8.58
4%					104.0	52.00	34.67	26.00	20.80	17.33	14.86	13.00	11.56	10.40	9.45
5%						105.0	52.50	35.00	26.25	21.00	17.50	15.00	13.13	11.67	10.50
6%							106.0	53.00	35.33	26.50	21.20	17.67	15.14	13.25	11.78
7%								107.0	53.50	35.67	26.75	21.40	17.83	15.29	13.38
8%									108.0	54.00	36.00	27.00	21.60	18.00	15.43
9%										109.0	54.50	36.33	27.25	21.80	18.17
10%											110.0	55.00	36.67	27.50	22.00
11%												111.0	55.50	37.00	27.75
12%													112.0	56.00	37.33
13%														113.0	56.50
14%															114.0

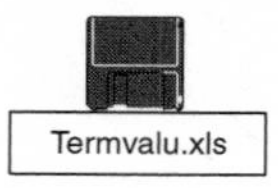

Exhibit 5.5 Terminal Value Factors

mathematically correct but not to be trusted in a business sense because of the mathematical convergence issue already discussed.

It is also illustrative to see that the terminal value changes about 10% for each 1% change in the cost of money in the range of interest.

DCF: An Extended Example

Now that all of the key elements of discounted cash flow have been introduced, we can apply them to a fairly typical situation. Exhibit 5.6 is based on a hypothetical project, which is described in detail in Chapter 9. We do not need the details just now, only the resulting income and free cash flow. The project begins with a substantial initial investment in year 1. Further net investments follow in years 2 through 6. Eventually, the project produces a growing stream of revenues and net cash flows in excess of investments in the business. Over time, as so often happens, the business matures, and the growing stream of revenues slackens. By year 20, the growth rate is about 5%. The net present value at 12% of the free cash flows through year 20 is $25.2 million and the internal rate of return is 21.3%, before any consideration of terminal value. An NPV can be calculated by multiplying the free cash flow in each year by the discounted value of money (Exhibit 5.4) and summing them from years 1 through 20. The strongly negative contributions of

Year	Growth	Revenues	Depreciation	EBIT	EBITDA	Net Income	Free Cash Flow
1	40.0%	$ 14.9	$ 1.0	$ 2.7	$ 3.7	$ 1.6	($10.9)
2	35.7	20.9	1.5	3.8	5.2	2.3	(1.7)
3	31.9	28.3	2.0	5.1	7.1	3.1	(1.7)
4	28.5	37.4	2.6	6.7	9.3	4.0	(1.6)
5	25.4	48.0	3.4	8.6	12.0	5.2	(1.1)
6	22.7	60.2	4.2	10.8	15.1	6.5	(0.4)
7	20.3	73.9	5.2	13.3	18.5	8.0	0.7
8	18.1	88.8	6.2	16.0	22.2	9.6	2.2
9	16.2	104.9	7.3	18.9	26.2	11.3	4.1
10	14.4	121.9	8.5	21.9	30.5	13.2	6.3
11	12.9	139.4	8.7	26.1	34.9	15.7	8.4
12	11.5	157.4	9.6	29.8	39.3	17.9	11.1
13	10.3	175.5	10.3	33.6	43.9	20.1	14.0
14	9.2	193.5	10.9	37.4	48.4	22.5	17.0
15	8.2	211.2	11.4	41.4	52.8	24.8	20.1
16	7.3	228.5	11.8	45.4	57.1	27.2	23.3
17	6.5	245.2	12.0	49.3	61.3	29.6	26.4
18	5.8	261.2	12.1	53.2	65.3	31.9	29.5
19	5.2	276.5	12.0	57.1	69.1	34.3	32.4
20	5.0	290.8	11.8	60.9	72.7	36.5	35.3
21[a]	5.0	305.4	11.6	64.7	76.3	38.8	37.2

NPV (at 12%) of free cash flow, years 1–20	$25.2
NPV (at 12%) of terminal value @ 15 × Year 21 free cash flow	$51.7
NPV assuming 5% growth in perpetuity	$76.9
NPV (at 12%) of terminal value @ 14.4 × net income	$51.7
NPV (at 12%) of terminal value @ 8.6 × EBIT	$51.5
NPV (at 12%) of terminal value @ 7.3 × EBITDA	$51.6

[a] year for calculating terminal value

Exhibit 5.6 Terminal Value as a Multiple of Free Cash Flow ($millions)

years 1 through 6 are finally outweighed by positive contributions in the later years.

But this calculation does not consider the terminal value of the business. How would we go about establishing a reasonable value for it?

If year 20 represents a state of maturity, and growth can continue indefinitely at 5%, Exhibit 5.5 tells us that the terminal value should be 15 times free cash flow in year 21 or $555.9 million. The present value of this sum is $51.7 million, discounting for 21 years at 12%. Therefore the net present value of the business is $25.2 million + $51.5 million = $76.9 million. More than ⅔ of the total value is in the terminal value.

Examining this conclusion in light of other valuation methods, the respective values for net income (earnings), EBIT, and EBITDA in year 21 are $38.8 million, $64.7 million and $76.3 million. As ratios to the $555.9 million terminal value derived from the growth-in-perpetuity approach, the implied PE ratio would be 14.4, not unreasonable for a modestly growing business. The implied multiples of EBIT and EBITDA would be 8.6 and 7.3 respectively. Should a real sale take place in year 21, the seller would emphasize the ratio that would seem to justify the best price, whereas the buyer would use the other three ratios as a reality check.

Again, we cannot overemphasize the importance of the selection of the terminal growth rate in a valuation exercise. Looking back at Exhibit 5.5, we see that the difference in multiplier (at a 12% discount) between a 3% growth rate and an 8% growth rate is 11.44 versus 27.00, or a factor of 2.4. When most of the value of the project is in the terminal value calculation, this choice is extremely critical to the result.

In selecting a growth rate in perpetuity, choose a rate modestly higher than the underlying rate of inflation. Also, since many market sectors grow at a constant ratio to the U.S. economy as a whole (which typically grows 2% to 3% per year), a growth rate characteristic of that market sector may be appropriate. This rate may be as high as 6% to 7% for a durable, high-quality business that will exhibit real growth over the long pull. If the business is expected to face serious long-term threats, much more conservative estimates are warranted. If the decision is a choice in investing in two alternatives where growth in perpetuity may be quite similar, the DCF method provides a fine apples-to-apples comparison regardless of the absolute values obtained for NPV.

Time Horizons

A perceptive reader may ask, "What is the rationale for assuming a 20-year time frame? What if we did the calculations for 15 years instead?" This is an excellent question because a reliable estimate of value should not be very sensitive to the choice of time frame. If we shorten the time horizon, in effect the growth-in-perpetuity phase begins in a year with higher growth rates—8.2% versus 5%. Since the earliest years are most heavily weighted in any estimates of terminal value, the growth rate should be adjusted upward if the "future" includes more years of the rapid growth phase.

The key to the value calculation is again the selection of the terminal growth rate. As an example, if we used for growth-in-perpetuity a rate of 6.3%, which is the numerical average for years 15 through 20, and use a multiplier of free cash flow of 18.4 (by interpolation in Exhibit 5.5), we obtain the results shown in Exhibit 5.7. (Also, the terminal value is now discounted to the present by 15 years rather than 20.)

The results for NPV are essentially equivalent to the 20-year case, but now over 90% of the total value of the project resides in the terminal value. The internal rate of return (IRR) is a bit higher, which suggests that if your object is to maximize returns you may wish to sell growth businesses as growth begins to slow.

If we shortened the time horizon still further, to year 10, terminal value would be the entire ball game, as it is in money-losing, high-tech start-ups. In this case, the NPV of the cash flows is still negative, so *more than 100%* of the total value would be in the terminal value part of the calculation. However, the growth rates in year 10 are still quite high in our model, raising issues of series convergence with regard to the growth-in-perpetuity valuation formula and making the value calculation too sensitive to the selection of the future growth rate. Use of the growth-in-perpetuity formula is premature and the longer time frames of 15 or 20 years should be selected.

Item	15 Years	20 Years
NPV free cash flow	$ 6.3	$25.2
Growth in perpetuity	6.3%	5.0%
NPV terminal value	$69.9	$51.8
NPV total	$76.2	$77.0
% NPV in terminal value	91.8%	67.3%
IRR (%)	28.9%	26.0%

Exhibit 5.7 Using a Shorter Time Horizon

The Pitfalls of Focusing on Cash Flows

In general, cash flow models are preferable to earnings models (EPS, EBIT, EBITDA) in the valuation process. The latter fail to reflect economic values, as discussed in Chapter 3, while the former take direct account of capital investment and, thus, flag the value-destroying process of growing a company through investments that do not return the cost of capital. However, the myopic use of cash flow models can also lead to poor decision making because profitable, rapidly growing businesses may have negative cash flow throughout the period of rapid growth. One could reach the erroneous conclusion that such a business has no value, especially if one is too literally into the "cash is king" school of thought. This error in reasoning is easily demonstrated in isolation (see below), but when negative cash flows from technology projects or growing businesses are embedded in an operating business with lower growth characteristics the error is much harder to detect. In these cases, cash flow oriented analysts are likely to calculate business value as *negative by difference*—the value of the business without the project is higher than the value of the business with the project. This error is bound to occur when the appropriate time horizon for the project is longer than the one being used for valuing the business.

The hallowed assumption of financial analysts that one can value the enterprise by valuing each of its parts simply does not hold if the time frames for the valuation differ. To see why not, consider a successful business growing at 25% per year, as in Exhibit 5.8, and earning a return on total capital of just over 20% over a planning horizon of 5 years. There is only one problem with this business—its free cash flow, line (10), is anemic. The positive cash flows from earnings, line (7), and depreciation, line (4), are almost offset by the increasing working capital, line (5), and fixed capital expenditures, required to grow the business at 25% per year.

Valuing the business based on the actual free cash flow is fraught with problems. The net present value of the tiny free cash flow in years 1–4 is only $32 million, line (16). And the terminal value of the

Pro Forma Financial Projection (M$) Line	Column / Comment	(1)	(2)	(3)	(4)	(5)	(6)	(7)
					Year			
		1	2	3	4	5	5	5
	Growth Rate	25%	25%	25%	25%	25%	0%	6%
(1) Revenues		$1,000	$1,250	$1,563	$1,953	$2,441	$1,953	$2,070
(2) Variable costs	60% of revenues	(600)	(750)	(938)	(1,172)	(1,465)	(1,172)	(1,242)
(3) Gross fixed capital	75% of revenues	750	938	1,172	1,465	1,831	1,465	1,553
(4) Depreciation	10% of fixed capital	(75)	(94)	(117)	(146)	(183)	(146)	(155)
(5) Working capital	30% of revenues	300	375	469	586	732	586	621
(6) Pretax profit	(1) + (2) + (4)	325	406	508	635	793	635	673
(7) Aftertax profit	60% of pretax profit	195	244	305	381	476	381	404
(8) Increase in fixed capital	Compare successive years	(188)	(234)	(293)	(366)	(458)	0	(88)
(9) Increase in working capital	Compare successive years	(75)	(94)	(117)	(146)	(183)	0	(35)
(10) Free cash flow	(7) – (4) + (8)+ (9)	8	9	12	15	18	527	436

Calculating Horizon Value (12% Discount Rate)	In Year 5	Present Value
(11) Actual free cash flow valued as perpetuity in year 5	$ 150	$ 85
(12) Free cash flow assuming 0% growth after year 4	4,393	2,493
(13) Free cash flow assuming 6% growth after year 4	7,692	4,366
(14) 20 × year 5 earnings assuming 6% growth	8,074	4,583
(15) 12 × year 5 EBIT assuming 6% growth	8,074	4,583
(16) NPV free cash flow years 1–4	NA	32

Exhibit 5.8 A Growth Business

free cash flow of $18 million in Year 5, if treated as a perpetuity, is only $150 million, which nets back to $85 million in the present. The NPV of the business would then be the sum of $32 million and $85 million or $117 million. This is clearly wrong, for if the business were to be turned into a perpetuity, it would no longer be eating cash to expand, and the free cash flow would be much higher. It is also wrong to value a business earning over $400 million per year at less than one times earnings. Using the growth-in-perpetuity approach for valuation is not an option here for the business reason that 25% per annum growth is not sustainable, and the mathematical reason that the formula is not applicable to situations where growth occurs at a rate higher than the cost of capital.

We can resolve the dilemma by assuming that management elects not to grow the business, and freezes revenues in Year 5 at the Year 4 level. *No more capital expenditures and no increase in working capital.* This case is shown in column (6). Year 5 earnings are down, but cash flow is up, from a tiny $18 million to a whopping $527 million. The valuation problem goes away. We now have created a no-growth perpetuity, the terminal value of which is 8.33 × $527 million or $4,393 million, which nets back to the present at $2,493 million.

Management clearly will not elect to freeze the growth of this wonderful business (absent external constraints or even better opportunities). Instead, it would forgo the opportunity to generate cash in favor of another year of 25% growth, at which time it could revisit the option to stop growth with even higher rewards. In fact, management would continue to grow the business *indefinitely* as long as it remains an outstanding performer. Any other course would destroy value.

But trees do not grow to the sky, and businesses cannot maintain high growth rates forever. The day will surely come when the 25% growth rate enjoyed by this business will decline to a sustainable but more mature growth rate—say at 6% per annum. Assume this happens in Year 5, a case shown in column (7). The cash flow for this case is nearly as high as for the zero growth case, and assuming growth in perpetuity of 6%, the horizon value is again 17.65 × the free cash flow or $7,692 million. Netting back to the present and adding in the small NPV of the cash flow from years 1–4, one obtains an NPV of $4,398 million.

A simpler approach to the issue would have been to focus on earnings *instead of free cash flows.* At a price earnings multiple of 20 × year 5 earnings, the horizon value is $8,074 million. The same $8,074 million value could be obtained as a ratio of 12 × earnings before interest and tax (EBIT), or pretax profit. The valuation is about the same as the free cash flow method, which implies a PE ratio of 19 instead of our assumption of 20.

Summary

It has often been said that discounted cash flow calculations will kill any good R&D project. The high discount given to distant profits works strongly against long-term projects, and a rapidly growing business may throw off very little cash. An intelligent approach to terminal value goes a long way toward solving these problems and corresponds to our historical knowledge that the value of good R&D projects often takes time to recognize, that high-growth businesses are desirable, and that quality companies seek both.

CHAPTER 6

R&D, Growth, and Shareholder Value

The concept of shareholder value contains good news and bad news for the R&D manager. If he or she is running a laboratory with a $50 million budget, eliminating the lab and capitalizing the "free" cash flow implied by its budget would create $500 million or more in shareholder value. That's quite a price to have on one's head! That's the bad news. No wonder executives elsewhere in the company are continually badgering the R&D manager, in effect, to "Justify yourself"—even in good times. In hard times, they simply axe the R&D budget.

The good news is that industrial R&D is far from dead, for reasons rooted, like the bad news, in shareholder value. R&D does not exist out of corporate benevolence, but because it has the power to create value. That value-creating power can be observed in the quantitative relationships between R&D, corporate growth, and price/earnings (PE) ratios.[1] If a $50 million per year laboratory is unproductive, it would be utterly rational to shut it down and claim the $500 million for the shareholders. The real issue explored in this chapter is *how* productive does that laboratory have to be?

To pursue this question, we need a corporate model. Real corporations, and the global economy in which they function, are enormously complex; they are driven by many cyclic factors and unpredictable one-time events. Underlying this turmoil, however, is the fact that most large global manufacturing businesses endure for relatively long periods, with differing and partially predictable degrees of success. Their long-term dynamics lend themselves to a steady-state model,[2] which is the basis for the calculations in this chapter. But to support these calculations, we need to look at the actual behavior of financial markets as a reality check on our conclusions.

What Is the Value of 1% in Added Growth?

Consider a hypothetical, healthy, corporation in a simplified, steady-state format. In this model, all key corporate parameters (revenues, income, capital, etc.) are assumed to grow at the same rate. Assume that this corporation is able to consistently and credibly gain a 20% after-tax return on its total capital, and seeks to maintain that return. Its growth will be controlled by the degree to which it chooses to reinvest its free cash flow. Here is why. Assume the company has sales of $5 billion and total invested capital of $3.5 billion. It also has net operating income of $700 million (20% of $3.5 billion). Its gross cash flow is then $700 million plus depreciation. To grow, this company must make net investments in fixed assets and working capital (see Exhibit 6.1). To grow its total capital at 5%, it must reinvest a *net* 5% of $3,500 million or $175 million in the first year. (The gross reinvestment will be $175 million plus depreciation.) Accordingly, its free cash flow,[3] the amount left over for other purposes and not

	Rate			Year		
	(%)	1	2	3	4	5
Capital base		$3,500	$3,675	$3,859	$4,052	$4,254
Operating income	20	700	735	772	810	851
% reinvested	25	175	184	193	203	213
Free cash flow		525	551	579	608	638
Growth	5					
Capital base		3,500	3,710	3,933	4,169	4,419
Operating income	20	700	742	787	834	884
% reinvested	30	210	223	236	250	265
Free cash flow		490	519	551	584	619
Growth	6					

Exhibit 6.1 Growth and Free Cash Flow (in $millions)

required to grow the business, is $525 million. The new investment is also assumed to earn 20% per year.

In Exhibit 6.1, the growth rate is the return on capital invested times the investment rate. For 5% growth, the investment rate must be 25% of operating income,[4] that is 20% × 25% = 5%. If the company wants to grow faster, it must reinvest more of its free cash flow. For example, to grow at 6%, its reinvestment rate must be raised to 30%, as shown in the second tier of the chart. The real-world issue is how to find investment opportunities that earn 20%—or any other rate above the cost of capital. Doing so is not easy in a competitive world, which is why some companies perform R&D.

An unfortunate and inevitable consequence of faster growth, however, is the reduction of free cash flow in the first few years of the model—which, taken by itself, obviously has a negative effect on a value calculation based on time-discounted sum of free cash flows. Over time, however, the growth in the rate of free cash flow always creates value, despite the initial penalty of starting from a smaller base (provided that such growth earns the cost of money). This can be seen in Exhibit 6.2, which shows the value of a business with total invested capital of $3,500 million at various growth rates and at various returns on invested capital. This business has a 12% cost of capital. The higher the growth rate, the more valuable the business; and the higher the rate of return on invested capital, the higher the dollar value of each incremental 1% of additional growth.

The case in which the business just earns its cost of capital (12%) is especially interesting. Its net present value is exactly equal to its invested capital, and no amount of growth will make it worth more. This growth is essentially profitless. If the rate of return is below 12%, growth actually destroys shareholder value, as many business executives who focus on earnings growth alone learn to their sorrow.

Exhibit 6.2 shows the intrinsic value of long-term sustainable growth rates. We know that in the real world it is difficult to sustain high rates of growth. There is also a problem in this exhibit owing to the large terminal values created by the growth-in-perpetuity formula as growth rates approach 12%. Though the result is mathematically correct, the results in the right-hand columns should be taken with a grain of salt. Nonetheless, the results are striking: an increase of 1%

Net Present Value Return on Invested Capital (%)	Growth Rate 5%	6%	7%	8%	9%	10%	11%
20.0	$7,500	$8,167	$9,100	$10,500	$12,833	$17,500	$31,500
17.5	6,250	6,709	7,350	8,313	9,916	13,125	22,750
15.0	5,000	5,250	5,600	6,125	7,000	8,750	14,000
12.0	3,500	3,500	3,500	3,500	3,500	3,500	3,500

Incremental Value of 1% Growth Return (%)	5 to 6%	6 to 7%	7 to 8%	8 to 9%	9 to 10%	10 to 11%
20.0	$667	$933	$1,400	$2,333	$4,667	$14,000
17.5	459	641	963	1,604	3,209	9,625
15.0	250	350	525	875	1,750	5,250
12.0	0	0	0	0	0	0

Incremental Value of 1% Growth Return (%)	5 to 6%	6 to 7%	7 to 8%	8 to 9%	9 to 10%	10 to 11%
20.0	8.9	11.4	15.4	22.2	36.4	80.0
17.5	7.3	9.6	13.1	19.3	32.4	73.3
15.0	5.0	6.7	9.4	14.3	25.0	60.0
12.0	0.0	0.0	0.0	0.0	0.0	0.0

Exhibit 6.2 The Value of 1% Long-Term Growth: DCF Calculation at Various Growth Rates (in $millions)

in the growth rate can be worth 5%–20% in value creation—in most cases more than 10%. For many industrial corporations, that 1% of added growth easily translates into $500 million to $1 billion.

Growth and Profitability—The Trade-Off

Business executives must constantly choose between growth and profitability when making resource allocation decisions:

- "Should we fund an R&D project that promises future growth at the expense of current income?"
- "Unit A is faster growing than Unit B, but Unit B is more profitable. To which should we allocate more of our scarce capital or R&D resources?"

R&D executives are more than a little interested in these choices, and in the trade-offs they force companies to make.

The analysis performed in creating Exhibit 6.2 allows us to determine the trade-offs of growth and profitability because we can compare value at a constant growth rate as a function of return. It turns out the relationship between value and rate of return is linear—the difference does not depend on the rate of return itself. The results for the same hypothetical company are summarized in Exhibit 6.3.

The values are of the same magnitude—a 1% change in return can affect value by $500 million to $1 billion (with the exception of growth rates approaching 12%). Roughly speaking, it is reasonable to trade about 1% of long-term return for 1% of long-term growth, *as long as the returns stay well above the cost of money*. At lower returns, growth does you little good, at least in financial terms.

A more precise calculation of the trade-off can be made for any particular corporate situation by calculating the slopes of the value curve for whatever point the business is at, using the methods we have just developed. For example, the data in the preceding two exhibits tell us the NPV for a business growing at 9% and returning 16% on capital is $8,167 million, while one growing at 6% and returning 19% on capital is worth $7,584 million. Both are of the same magnitude, but one does not want to trade three points of growth for three points of return.

In the world of high-growth businesses, the waters are muddied because a remarkable percentage of profits must be plowed back to sustain

Growth Rate (%)	Incremental Change in Value[(a)]
5	$ 500
6	583
7	700
8	875
9	1,167
10	1,750
11	3,500

[(a)] For 1% change in return on invested capital.

Exhibit 6.3 Growth Rates and Incremental Value Changes (in $millions)

the growth rate and to establish the strongest possible future competitive position. Therefore, the book return on capital can be low or even negative in the early stages of growth. And accounting profits may be anemic or negative. But these traditional metrics may mask the real wealth-generating potential of the business. High expense levels for R&D, market development, recruiting, advertising, and training are more investments than ongoing costs of business. If these costs were amortized, rather than expensed, the return on investment might well exceed the cost of capital. Businesspeople accept these investments primarily because of the high strategic value of being the first to stake out a position in a new market and also because new opportunities are likely to be uncovered in the R&D process.

The Business Life Cycle

No real business can grow at 20% or more indefinitely. Market saturation, competition, or hubris eventually causes the mightiest engines of growth to sputter. The typical business, in most cases, progresses through a life cycle of several sequential stages: (1) an initial period of exponential growth; (2) a period of moderate, linear rates; and, (3) a final period of stasis or decline. This pattern generally conforms to an "S-curve," a pattern discussed in Chapter 8.

In the early stage of growth, return on capital may be low or negative because of the need to constantly add staff and facilities to take advantage of future opportunities. In the later stages, profitability can be increased above 20% as the company gains economies of scale, exploits earlier investments, reduces overheads originally created to foster growth, and runs plants whose assets are largely depreciated. In the final stage, sales growth generally flattens or declines as the market matures. The final stage may take one of two courses. If the business has achieved a dominant position in its field it may become a "cash cow" and enjoy high profitability owing to its largely depreciated assets. Or it may become vulnerable to newer competing products, in which case profitability will diminish as competition over the now saturated market leads to price-cutting and margin erosion. In time, this business will have to be "harvested."

Exhibit 6.4 shows a hypothetical model through which a new business or product line might plausibly evolve while maintaining a magic number of 25 for return on capital plus growth through all stages of its life cycle.

As a practical rule of thumb, it seems likely that any technically based business for which the return on capital plus the growth rate (expressed as percentages) sum to the magic number of 25% or more is likely to be a good long-term investment. But there are caveats: all new investments must in time return at least the cost of capital; and, this rule of thumb will not work for long at low levels of profitability. Even so, it should be remembered that the low returns in the first few years of the S-curve will be on a much smaller capital base than the higher returns in later years.

The key to managing the product or business cycle is to *trade growth for profitability* as the product or business matures, while staying at or above 25. This conclusion is supported by the preceding analytical conclusion: in many cases, a percent of growth has about the same value as a percent of return. Although in principle one can move in the other direction by force-feeding growth, overinvesting in R&D and fixed assets, this strategy is at loggerheads with the law of diminishing returns. It is better to start up a new S-curve. Those businesses that manage to stay above a magic number of 20 are likely survivors, but as their magic number drops to the mid-teens, businesses start looking like losers.

	Return on Capital (%)	Growth (%)
Development	NA	NA
Year 1	−10	35
Year 2	0	25
Years 3–5	5	20
Year 6	10	15
Years 7–10	15	10
Years 11–20	20	5
Years 21–30	25	0

Exhibit 6.4 Profitability and Growth through a Hypothetical Business Life Cycle

Portfolio Balance Can Be the Key

From the point of view of a major corporation, a balanced portfolio of businesses (Exhibit 6.5) can solve many of the problems associated with growth and profitability. A balanced portfolio might include:

- Businesses that are growing slowly but throwing off lots of cash ("cash cows").
- Businesses enjoying healthy growth and a good capital return.
- High-growth businesses with negative cash flows that will sustain growth and shareholder value as they mature; these require external financial support from the cash cows.

Exhibit 6.5 describes such a portfolio. There, one very high-growth business (A) is supported by two highly profitable "cash cows" (B and C). The portfolio also includes two businesses that have both acceptable growth rates and acceptable returns (D and E). The portfolio as a whole has an attractive growth rate and abundant free cash flow. The attractiveness of the portfolio will be enhanced when and if the high-growth business, which is not yet adding value, begins to trade some of its growth for return.

One would expect all businesses to shift gradually toward the pattern of maturity. In fact, left to its own devices, any portfolio will eventually become a collection of moribund operations. Wise management anticipates this shift and keeps the portfolio vibrant by launching new

	Return on Capital (%)	Growth (%)
Business A	12	20
Business B	25	0
Business C	20	5
Business D	15	10
Business E	15	10
Average	15.4	9

Exhibit 6.5 A Balanced Portfolio Solves Many Problems

products and businesses at a rate that will keep overall portfolio growth and returns in balance.

R&D: Trading Current Profits for Growth

In the previous section, we discussed profitability and growth largely in strategic terms. We will now move to the level of the R&D budget, and attempt to link growth and shareholder value, while recognizing explicitly that there is no free lunch—the costs of R&D fall directly to the bottom line. We begin by viewing the trade-off between R&D expense and growth from the perspective of R&D productivity.

Assume that our hypothetical corporation, with sales of $5 billion, is currently growing at 6% and maintaining a targeted return of 20%. Exhibit 6.2 valued this case at $8,167 billion. Assume further that the *after-tax* R&D expense for this corporation is 3% of sales or $150 million. That level would correspond to about $250 million of pretax R&D spending, or 5% of sales, a number common in process industries with longer product life cycles and good growth opportunities.

Now assume that management, under pressure to deliver more growth, is considering whether to increase the *after-tax* R&D budget by $50 million, bringing it to 4% of sales. Based on experience, it calculates that over the long run this added expenditure will produce new products that will increase annual sales by 2% (initially $100 million). (Because we are using a steady-state model, we can ignore the time lag factor; it may take several years for the increased growth to manifest itself.) For this case, we are assuming a productivity factor of 2 for R&D spending, where the *productivity factor* here is defined as the *incremental amount of annual sales attributable to R&D divided by the annual after-tax R&D budget.*

This company has a hurdle rate of 20% for its investments; therefore, we also assume that the new products commercialized will, on average, meet this investment grade criterion.

Exhibit 6.6, coupled with calculations already made in Exhibits 6.1 and 6.2, shows the result. In Exhibit 6.2, we had shown that a 6% growth rate in our company implied a free cash flow of $490 million.

	Productivity Factor		
	1.0	2.0	3.0
Growth			
Before R&D increase	6%	6%	6%
After R&D increase	7%	8%	9%
Value			
Before R&D increase	$8,167	$8,167	$ 8,167
After R&D expense	$7,389	$7,389	$ 7,389
Value added by growth as %	11.4%	28.6%	57.1%
Value with additional growth	$8,233	$9,500	$11,611
Value gained/lost	$66	$1,333	$ 3,444
Value G/L as %	0.8%	16.3%	42.2%

Exhibit 6.6 R&D Productivity and Value (in $millions)

Management is proposing to reduce that by $50 million to $440 million, or 10.2%. Since value is directly proportional to free cash flow, the value for this case, $8,167 million (shown in the first line of Exhibit 6.2), will also be reduced by 10.2% to $7,389 million. However, we have yet to account for the value added by 2% additional growth, which is 28.6%. (That number can be derived from Exhibit 6.2 by comparing the values in line 1 for 6% growth ($8,167 million) and 8% growth ($10,500 million).) The new value of the company, after trading $50 million of free cash flow for 2% more growth, is $9,500 million, an increase of $1,333 million or 16.3%.

Exhibit 6.6 also tells us a great deal about the importance of R&D productivity. Following the identical logic, if the productivity factor were 1—a 1% growth in sales corresponds to a $50 million R&D increase—the value gained would be only 0.8%. A productivity factor of 1 is on the margin.

On the other hand, a tremendous spurt in productivity to 3 would give an enormous 42.4% jump in value.

This tells the CEO several important things:

1. A one-third increase in R&D spending in return for the *possibility* of 2% higher growth may be worth the risk.
2. If the productivity of the R&D lab falls below 1, any R&D spending is destroying value and the situation needs urgent attention.

3. Raising the productivity factor of the lab to 2.5 or 3 would create enormous value for the company by creating additional growth opportunities from within the *existing* R&D budget *and* by making additional investments in R&D even more effective at creating value.

The CEO is not the only person who gains from knowing the marginal benefit of R&D spending. The R&D director can also use this knowledge of R&D productivity to good effect. He or she sees many projects and project proposals, and is familiar with the odds of success. Techniques described in Chapter 9 can help this manager calculate the potential value of these projects, and rank-order them for risk and reward. But now, the R&D director has also identified the cutoff point, and knows that he or she must accept only those projects that are consistent with maintaining a productivity factor of 1 or more *at the margin*.

Productivity factors defined this way have the advantage of being readily tracked from corporate data. My own experience with the productivity factor metric in a specialty chemicals business was that a productivity factor of 1 was a "slam dunk," a 2 could usually be achieved and sometimes exceeded, and a productivity factor of 3 to 5 was characteristic of our *targets*, but extremely difficult to achieve in practice. In other words, in a competitive world, R&D resources determined largely by management's "gut feel" were being spent productively; we were not overinvesting past the point of diminishing returns. This is as it should be. Business is competitive at the margin, and any company that performs R&D consistently at target levels will quickly outperform its competition. The important lessons to be drawn are that (1) R&D productivity is critical to corporate value, and (2) the requisite degree of productivity to create new value can be quantified.

Can R&D's Contributions to Value Be Separated from the Other Functions?

This is usually a "hot button" in corporate discussions: the question often takes the form of whether R&D is improperly "taking credit" for the value created because all functions—sales, manufacturing, and marketing in particular—have contributed. In a purely financial sense, the answer is a clear "no" since we presume that without the R&D the *opportunity* to grow at a rate that earns 20% on capital would not have occurred. After that, the cost of capital and the expenses of the other functions are fully accounted for in the return. Once any opportunity to earn 20% is defined, it is a no-brainer for a company with a cost of capital of 12% to chase it. After the fact, it is financially irrelevant who contributed to it.

Nonetheless, the original innovative idea, and many of the subsequent ideas that give rise to a successful R&D project, often originate outside the lab; for example, from a customer through sales and marketing. The innovative contributions of the entire project team must be viewed as part of the R&D process, and the efforts of all contributors should be recognized.

Other Sources of Growth

We have demonstrated that even 1% of added profit growth can mean big gains in shareholder value, and R&D bears a major responsibility for sustaining this growth. There are, however, other

Source	R&D Input	Sustainability	Key Factors
Price	Low	Medium	Inflation Capacity/demand
Market growth	Low	High	GDP growth Geographical expansions
Market penetration	High	High	Substitutions Technology-driven products
Market share	High	High	Product performance and cost position
Manufacturing cost	High	Medium	Competitive pressure Cost of capital
Overhead reduction	Low	Low	Competitive pressure

Exhibit 6.7 Sources of Corporate Growth

sources of corporate value growth, and these must be understood and thoughtfully managed. Exhibit 6.7 identifies several.

Price

The first source of growth is price. In the long run, price increases are driven by inflation. In commodity and semicommodity businesses, they are also driven by the current relationship of supply to customer demand, which tends to follow a medium-term cycle. Currency relationships also affect pricing in global businesses. None of these factors relate significantly to value added. Unjustified price increases may result in lost market share. Therefore, from a corporate viewpoint, real value growth through price is seldom sustainable and is often illusory. It has virtually no R&D component.

Also, a growth goal of, say, 10% measured in current dollars, is much less challenging in a world of 6% inflation (4% real growth) than with 3 % inflation (7% real growth).

Market Growth

Market growth is a very important and highly sustainable source of corporate earnings growth. Mature businesses generally grow at rates comparable to demographic growth or the growth of a nation's gross domestic product (GDP). If GDP is growing at 2%, this contribution is significant. If your business is participating in strong economies growing at 5% to 10% per annum, it is *very* significant. In addition, many markets grow for long periods at significant multiples of GDP growth. In past decades, plastics and electronics shared this characteristic; today, the health care industry enjoys this type of growth.

Strategic positioning in the right markets affects growth projections, independently of R&D intensity.

Market Penetration

Market penetration is a sustainable and predictable source of growth, and has an important relationship to R&D activities. A classic example is rigid containers, where aluminum cans and plastic bottles have made steady inroads against steel cans and glass bottles. These substitutions are technology-driven and provide an intrinsic advantage to the attacker, who gains steadily in productivity and scale economies while moving up in the experience curve, whereas the defender must reduce margins and overheads to protect an entrenched position. The rate of penetration is related to the level of R&D support provided by the suppliers, but may be intrinsically limited by the customer's incentives to convert technology. Because of the latter factor, the completion of a cycle of substitution typically takes decades and can be enormously profitable. However, as the new technology begins to saturate the target market, the contribution to growth must inevitably drop.

Market Share

As competitors vie for position in a key market, some will win and some will lose. Participating in a growing market or market penetration scenario does a company little good if it cannot attain or hold market share. Attempts to capture or hold market share through pricing alone are seldom successful, as competitors simply match prices. Sustainable share gains are based on overall competitive superiority in all aspects of business performance: marketing, manufacturing, technical support, and innovation. R&D is a major contributor to business performance and, thus, to share performance.

Manufacturing Cost

Manufacturing cost can be an important component of growth strategy. Some outstanding companies have achieved their success by relentlessly attacking their manufacturing costs, often through quality programs and statistical process controls. Process R&D, both incremental and revolutionary, has also played a major role. Key driving forces for manufacturing cost reduction are competitive pressure (which can be quantified through benchmarking) and the need to use capital efficiently. However, the limits of a manufacturing cost strategy must be recognized, since only proprietary process improvements are likely to provide sustainable advantages. For example, the installation of new and better equipment reduces manufacturing cost, but competitors can quickly buy and install similar hardware. Quality programs may initially make major yield improvements, but are subject to the law of diminishing returns and are copied by competitors. In mature and efficient process industries, direct conversion costs are often negligible when compared with raw material, energy, and capital charges, which are less easily controlled. Nevertheless, manufacturing cost reduction is one of the most attractive sources of medium-term income growth.

Overhead Reduction

Overhead reduction is the final source of income growth. It is not R&D driven, and, if pursued aggressively, is not sustainable. Corporate bureaucracies inevitably grow in response to real and perceived problems and opportunities. Shrinking them is difficult and subject to limits. Like an overloaded boat, a corporation can only throw so many "not essentials" overboard.

Corporate reorganization from centralized to decentralized structures (or the reverse) can expose opportunities for overhead reduction and productivity gains. Despite the current urge to become "lean and mean," some corporate overhead must remain to provide direction and vision, ensure control and compliance, attract and retain good employees, and communicate with external stakeholders. In a well-managed company, overhead reduction may only be justifiable if there are offsetting productivity gains. Hence, the sustainability of overhead reduction is also subject to diminishing returns at the same time that some organizations find it essential for short-term survival.

To summarize, R&D in a manufacturing company is critical to only three of six identified sources of income growth, but it is concentrated in areas of sustainable competitive advantage.

Two Strategies for Corporate Growth

One view of corporate management holds that its real job is to allocate the discretionary resources of the company over the long term—that for the most part, short-term operating results in the current year are outside its control. These critical resources are capital, R&D, and its most talented employees. If so, it makes a great deal of difference what the strategy is. Consider two hypothetical strategies, as shown in Exhibit 6.8.

Strategy I

Company X plans to grow at 10%, and has the cash flow to finance it. It expects to gain 3% on price. Since it is already the dominant global producer, share gains are anticipated to be zero. Fortunately, this company is positioned for market growth, since its existing markets grow at 2% per year. It also has an exceptional opportunity for geographic expansion worth an additional 3%, but doing so will require heavy capital spending and developmental "overhead" expense. Therefore, it expects total market growth of 5%. It has a proven record of reducing its annual manufacturing costs by 1% per year.

The sum of Company X's growth opportunities now stands at 9%. Top management had these strategic choices for the last 1%:

- Reduce overhead by 1% of current earnings per year.
- Double manufacturing cost reductions (and further increase capital spending) by deploying a new generation of equipment.
- Redirect development programs to penetrate markets not currently served.

These choices differ in sustainability and risk. Company X's managers have chosen B, and are highly likely to achieve their growth goal primarily through a strategy of geographic expansion and strong capital programs.

	Price	Market Growth	Market[a] Penetration	Market Share[a]	Mfg.[a] Cost	Overhead Reduction	Subtotal R&D Intensive	Total
Company X	3%	5%	0	0	2%	0	2%	10%
Company Y	3	2	4	3	1	–1	8	12

[a] R&D Intensive

Exhibit 6.8 Two Growth Strategies

Strategy 2

Company Y is new product driven and, like Company X, has a goal of 10%+ growth. It also expects pricing gains of 3% owing to general inflation. It is committed to having 25% of its product portfolio represented by products launched during the past 5 years. Its experience is that these products alone grow at an average rate of 25% per year and, as such, contribute more than 6% per year to corporate growth, half of which is attributed to market share and the other half to market penetration.

Overall, Company Y's markets are growing modestly at 2% per year, but its older products still have the momentum to add 1% growth through market penetration with no loss of share. Manufacturing cost gains are 1% per year, but high developmental costs and an increasingly tough regulatory environment *add* 1% in overhead cost per year.

The Two Strategies Compared

Company Y grows at 12% per year and is highly reliant on R&D for 8 of those 12 percentage points. Company X, on the other hand is committed to its very heavy capital program and relies on R&D primarily in the form of process research to sustain its manufacturing cost position (2 of its 10 percentage points of growth); thus, it has a very different strategy from that of Company X.

Both strategies have been made to work, as have many others. The weakness of Strategy X is likely to become apparent if its customer base is attacked by innovative products at home while it pursues geographic expansion abroad. The weakness of Company Y may be its overreliance on innovation in markets that are growing fairly slowly.

Comments and Caveats

The world would be all too neat if a CEO could simply raise R&D spending and observe shareholder value respond directly. If this were the case, every CEO would immediately begin shoveling money into R&D. End of story. We know from experience that R&D spending in many cases produces no increase in shareholder value. In most other cases, a substantial time lag separates the decision to fund R&D and tangible bottom line results. In fact, the linkage between R&D spending and shareholder value is so long-term and so confounded by time and chaos that it has proven difficult to demonstrate by statistical means. Indeed, raising R&D expense with no short-term benefit assured may actually reduce shareholder value, depending on investor perceptions. The stock market in 1997 applauded when Boeing announced it was terminating a major development program for a large new aircraft. That program was perceived by investors as a drag on value.

Nevertheless, both experience and research indicate that sustained commitment to R&D pays off for investors. Bean, Guerard, and Stone have created an econometric model[5] to show that a firm's stock price is positively and significantly associated with R&D expenditures. They used this model to analyze a sample firm. Their calculations showed that this firm's R&D expenditures were only 36% of the level that would have optimized shareholder wealth.[6] In other words, management was underinvesting in R&D.

However, these authors failed to show that a successful stock market strategy could be created by investing in stocks whose share prices do not fully reflect their companies' past investments in R&D; in other words, in those stocks that would appear undervalued by an R&D measure. Part of the problem may be that the relationship between growth and shareholder value is itself statistically weak. In fact, Copeland et al. present data[7] showing that the expected correlation between PE ratios and growth rates of earnings per share is not impressive in actual practice. The reasons are well known. The principal reason is that growth alone does not create value: earnings growth earned by investments that return only the cost of money are value-neutral. Those that return less *than* the cost of money destroy value.

Many other factors skew actual PE ratios. If a solid company has a bad year and profits drop by 75%, as occurs in cyclical businesses, its PE ratio may be quite high in anticipation of full recovery in a year or two. But its profit growth is negative. At the top of the cycle, its PE ratio will drop in anticipation of the next trough, and the situation reverses itself.

Another factor affecting PE ratios is that certain industries tend to band together—PE ratios of Merck, Bristol-Myers, and Lilly may be in one band while Exxon, Mobil, and Texaco will be in another. At any given time, though, these individual companies will be growing at different rates. These bands are likely to be determined by analysts' perceptions of the long-term growth rates of the respective industries (e.g., the drug vs. the petroleum markets). There are also chaotic factors, inherently unpredictable events. The energy crises of the 1970s and the Clinton health care proposals put energy and pharmaceutical stocks on roller coaster rides. At other times, interest rates and the behavior of the central bank become the market's principal preoccupation. Internal management turmoil may also depress PE ratios, which recover smartly as these issues are resolved.

The stock market would be too simple if growth rates and earnings were the whole game, just as if R&D were the whole game. What is clear, however, is that as earnings grow, long-term shareholder value will increase proportionally. Equally clear is the tendency of PE ratios to revert to the historical mean after excursions to either the upside or downside. The job of the R&D director in this churning pot of influences is to keep an eye on long-term growth, which is something that he or she can, to a degree, control, and let investors worry about short- and medium-term market fluctuations—which no one can control.

Summary

Although R&D affects only a fraction of a corporate growth, this fraction greatly affects competitive advantage and shareholder value.

One percent in sustainable growth rate does not sound like much, and is difficult to measure. Nonetheless, the value of growth is much higher than is often assumed and is reflected in valuations for stocks that are generally consistent with the hypothetical models described here, indicating that analysts and investors understand the differences among corporations with differing growth prospects.

This chapter has shown quantitatively what we know intuitively, that the productivity of R&D is all-important in creating shareholder value. If the R&D director and other senior executives watch this parameter closely, the long-term rewards will be substantial.

CHAPTER 7

Strategy: Driving Value in the Competitive Arena

Corporate strategy is an intellectual and organizational battleground with business managers and technologists generally on one side and company bean counters on the other. The former are the corporation's gladiators, men and women who view strategy in terms of beating the competition. Beat the competition, they believe, and other corporate interests, including value creation, will take care of themselves. The financial analysts, on the other hand, view greater shareholder value as the main objective. They focus on the *value drivers* of the business such as returns on capital, turnover, margins, and revenues.

Strategic planning in the multidivisional corporation too often addresses the interests of the financial analysts. Financial measurements (which have the virtue of being objective) are stressed to the detriment of deep analysis of changing trends in technology and the marketplace (which are partly intuitive). This heavy emphasis on financial measures may occur because most strategic planners are financial professionals. They are more comfortable analyzing historical returns on capital, and their sources, than determining which of the firm's product lines or technological initiatives hold great promise or are threatened with obsolescence. Financial information is, after all, reassuringly factual, and "facts" give analysts many opportunities to apply their tools. Unfortunately, "there are no facts about the future,"[1] and the future is what strategy is all about. If strategy focuses on financial history, no matter how factual, it will direct the corporation into the future based on what appears in the rearview mirror. This is a sure formula for driving into the ditch. To the extent that technical and marketing executives are not important players in the strategy-creation process, there will be few early warnings of threats and opportunities on the road ahead.

Strategy attempts to focus the resources of the organization and the energy of its employees on a set of coordinated goals designed to create greater value for shareholders. Most experienced managers find that strategy is only as effective as the process through which it was developed. If strategy is developed by a handful of people with offices on the top floors of corporate headquarters, it is not likely to survive its first encounter with the mean streets of competition.

Effective business strategy requires the collaboration of both the gladiators and the bean counters. The financial analysts are correct in that the objective of the corporation is to create greater value for the shareholders—and they have identified the drivers through which this is accomplished. Those value drivers are like dashboard indicators on an automobile. They tell us how well the vehicle is performing, but do nothing to create performance itself. They suggest when we should be operating with a richer or thinner fuel mixture (high or low leverage), when we are carrying too much baggage (overhead), and when energy should be shifted from one system to another (resource allocation) to gain the optimum output (returns). But they do nothing to identify the curves and potholes in the road ahead. For that, strategy depends on the creativity and insights of technologists, product developers, marketers, and business managers. In fact, a competitor who knows that your strategy is largely financially driven can devise deadly counterstrategies, which is precisely what Japanese firms did to their U.S. competitors in auto and consumer electronics during the 1970s and 1980s. They learned to cut price to gain share in the United States, while using high margins earned in captive markets in Japan to finance market share gains. They then relied on the financial pressures they had placed on American CEOs to abandon now unprofitable product lines, allowing the Japanese attackers to expand their positions. This, and much other experience, tells us that effective strategy is not simply about value drivers. It requires an understanding of competitor strengths, weaknesses, and strategic options, and analysis of one's own capabilities and options. Together, an understanding of value drivers *and* the competitive environment give a corporation the potential to develop a powerful and effective strategic plan.

The Role of Value Drivers

There is nothing mysterious about value drivers. We dealt with them regularly in three previous chapters. They are lines in the income statement and the balance sheet, or the ratios between them. They include return on invested capital. This is always a value driver, although it is driven by a cascade of other value drivers, which may be identified in progressively greater levels of detail.

Gross margin is a driver for return on capital, and may itself be driven by productivity (of labor, purchasing leverage, or other activity).

Growth (in income) is another powerful value driver *if* returns are higher than the cost of capital. Chapter 6 indicated the degree to which a 1% change in growth rate, or a 1% change in return, can affect value. We learned that it is not the same in all circumstances. We could likewise perform the same type of analysis on a more detailed pro forma spreadsheet, using *any* of its components. For example we could examine the impact on value (as calculated from free cash flow) of a 1% decrease in working capital. This is referred to as *sensitivity* analysis. Different businesses have very different sensitivities and very different drivers.[2] Understanding them helps strategists to determine their priorities. For example, a project manager may need to decide whether to institute an R&D program to reduce fixed capital in a proposed plant, but the trade-off seems to be a year of delay in start-up. Quantitative analysis of the drivers should tell us whether cycle time or capital cost is a stronger value driver in this business. This information is necessary to make the decision, but it is not sufficient. A year's delay could significantly affect customer and competitor reactions in ways that cannot be fed into a simple pro forma model. If so, this raises deeper strategic issues that cannot be determined by value drivers alone. However, those decisions could still be assisted by quantitative analysis using *decision trees,* if the potential impacts on other value drivers such as margins and sales volumes can be determined and their likelihood estimated.

There are counterarguments to the use of quantitative value-based models in R&D decision making; indeed, a leading book on the link between R&D and corporate strategy argues, "The rigor implied by NPV or DCF considerations becomes not only meaningless but potentially harmful."[3] That potential harm must be recognized. Its source is found in low-quality assumptions that find their ways into spreadsheet analysis of alternative strategies. But where business data exists that allows a calculation of the relative impacts on value of alternative technical strategies, not using that data and considering its implications is difficult to justify.

What Is Strategic Planning?

The dictionary defines strategy as a "careful plan or method, especially for achieving an end." This definition is military in origin and implies a clear end point, usually winning the war. In business, however, there is seldom a single end point. This is where the confusion begins.

Absent an end point, timing is paramount. Some shareholders are long-term investors; others plan to sell when their shares reach a target price. Senior employees may view their retirement date as an end point, whereas junior employees view the company as a vehicle for long-term career-building.

A promising approach to resolving these issues is to define corporate strategy as the creation of sustainable competitive advantage. Although this approach will help to achieve agreement among stakeholders (and is also currently fashionable), it papers over inherent differences among the players. In exactly *what* business is sustainable competitive advantage to be created? And in what time frame? The term "sustainable competitive advantage" itself begs for a definition. Finally, in some of the more difficult and interesting business cases, the goal of sustainable advantage itself is unrealistic; for example, when the battle is already lost. In these cases, the real goal is to *maximize value* as one retreats.

Another source of confusion relates to priorities among ends and means. Some business authors see strategy as a statement of how objectives will be pursued and place their value on the *plan*. Others see more value in the strategic planning *process*, to leave scope for the reexamination of objectives, adjustment to changing conditions, and innovation. Yet others see more value in strategic *architecture*, defined as a "high level blueprint for the deployment of new functionalities, the acquisition of new competencies, or the migration of existing competencies, and the reconfiguring of the interface with customers."[4]

To a considerable extent, the balance among plan, process, and vision will depend on the predilections of the chief executive. Top-down CEOs may view themselves as the chief strategist and use a strategic planning department to assess resource availability and to project the probable results of his plan. Bottom-up leaders may request plans from each business unit and use the process to sort out the inevitable issues of credibility and of resource allocation.

Certainly, a caveat from military science applies to business as well: No battle plan ever survives its first contact with the enemy. Substitute the word *marketplace* for *enemy*, and you have it. The gladiators understand this.

Most senior executives have several years of strategic plans on their bookshelves. Very few seem relevant after as few as two years. What has happened?

Typically, a strategic plan is constructed of assumptions that are only probabilities, many based on simple linear extrapolations of past trends. Probability theory tells us that an outcome based on several highly probable assumptions may, itself, be improbable. For example, seven independent assumptions, each with a probability of 90%, have a collective probability of only 47%. It requires only three independent assumptions with a probability of 75% to produce a collective probability of 42%. Most business plans contain several major assumptions regarding pricing, sales volumes, product mix, costs, interest rates, and so on. They may also assume stable competitive conditions and the absence of individually improbable events (natural disasters, fires, mergers of competitors or customers, patents, lawsuits, etc.). As a result, it is not surprising that the shelf lives of many strategic plans are very short.

The Case for Strategic Planning

Despite its failings and pitfalls, there is considerable evidence that corporations that perform strategic planning do better in financial terms and enjoy greater marketplace performance over extended periods than do those in which the strategic planning function does not exist.[5] Several reasons account for these successes. The first is intellectual. Since every strategic plan rests on a set of assumptions, those who do strategic planning have an opportunity to challenge their assumptions and to uncover new threats and opportunities. Organizations that take planning seriously are less likely to run up against unpleasant surprises.

The second reason is organizational. Strategic planning requires executives to think deeply about the role of each corporate function and how those functions can be aligned to achieve greater power. When that alignment is reflected in strategy, each function understands the part it can plan in the larger organizational agenda. In the absence of corporate strategy, it's every department, function, and business unit for itself. Each unit then pursues its own strategy without examining the impact of that strategy on total resources, or on what it will be required of other functions in the corporation. For example, to meet its ultimate goal, a research program may require the creation of a dedicated sales force, the construction of a major plant, and a trip into the financial markets to raise additional capital. That project may not make sense if the corporation's capital resources during that time frame are already committed to a program of geographic expansion.

In the absence of a strategic plan (or a close-knit relationship between functions), a project like this can churn along and consume resources for several years before the rude awakening occurs, dashing the high hopes of researchers. Had strategic planning been a part of this corporation's culture, and had the researchers been active participants in it, this wasteful exercise might have been recognized at an early stage. Resources could have been redirected toward projects with lower capital requirements, such as new products that use the existing manufacturing base and sales force. Problems like this one have encouraged many companies to make strategic R&D management a core process.[6]

There is another side to this coin when resources are abundant. In a cyclic business, such as petroleum or chemicals, there are times when cash flow reaches very high levels. At these times, the corporation may lack investment opportunities superior to those available in the financial markets. Under these circumstances, the corporation faces a future of either diminishing returns or being a cash-rich takeover target. Hence, there may be a strategic need to create a pipeline of innovative products that provide opportunities for capital investment. The good news for technologists is that a well-developed portfolio of investment options may be available from a balanced and forward-looking R&D program.

The third reason for a strategic plan is external. Although the strategic plan itself is a highly confidential document, the CEO, CFO, and head of investor relations (at least for publicly traded companies) must communicate with shareholders and analysts regarding the company's future prospects. When they do, they are subject to knowledgeable questions regarding business conditions in the company's markets, and are asked to explain deviations in projections from historical trend lines. The strategic plan is an extremely useful document to craft credible answers.

Resource Allocation

"Resource allocation" is too often a euphemism for a fight about money. People and dreams are also involved. At the top, the CEO must protect the income statement and balance sheet. When calls for expanded R&D and market development programs hit the bottom line of the income statement, the CEO must ensure that the sum of these requests does not diminish the profit growth expected for the corporation. Or if it does, he or she must justify these investments, putting personal credibility on the line.

Calls for capital expenditures, such as for new manufacturing facilities, affect the capital budget and through it the long-term debt on the balance sheet. They will also affect the income statement through depreciation and start-up costs. Large projects with delayed profitability can have a serious effect on the corporation's financial position.

In addition to managing the cumulative effect on the bottom line, the CEO and staff must make resource allocation decisions between competing projects and product lines. These affect the competitive positions of these businesses and may in the worst cases impair their future value. Emotions can run high.

Beyond dollar-denominated resources, the corporation has a limited supply of talented managers, technologists, and salespeople. Deploying "stars" to staff the new initiatives leaves big gaps in other parts of the organization.

Finally, innovative research has a special position in the strategic planning process. New ideas start with low probabilities of success, and move to higher probabilities as money is spent to resolve the technical and market issues. Good strategic planners know that budgeting for success in each project is unrealistic.

Pitfalls in the Strategic Planning Process

A major pitfall in strategic planning is requiring unreasonable amounts of detailed preparation from business units. This can be onerous, costly, and unwarranted, given the brief shelf life of these plans. This problem is compounded when the planning function imposes a standard format that is substantially different from the information systems used to operate individual business units. One business unit may allocate insurance at the plant level using formulas based on historical costs and risks. Another may take it at the headquarters level and allocate it according to revenues, whereas yet another may choose to allocate it from headquarters according to capital employed. Strategic planners will want to compare apples to apples and give detailed instructions on their preferred method. The business unit accountants may have considerable work to do to fulfill these requests. As a practical matter, requests for seemingly unnecessary information create resentment and make strategic planning a target when the corporate mood turns to cost-cutting. If R&D staffers complain that their work always gets the axe when times are tough, they have the satisfaction of knowing that the corporate planners will put their necks on the block before they do.

The second pitfall is arrogance. Despite the abundant evidence that past plans bear little resemblance to actual outcomes, planners strive for ever more perfect planning systems. Many strategic planning departments evolve toward a purely financial focus, forgetting that real value is created only in operations and the marketplace, and that competition always occurs at the business unit level.

The final pitfall is vagueness. The key factors that create value and account for a company's competitive advantages can be lost in a search for overarching concepts, as exemplified in vision statements and strategic architecture. At their best, these can be powerful guides to the future.[7] But people who have participated in these exercises know that enormous effort is often spent—if not wasted—in a search for wording that will generate consensus among managers who have fundamental philosophic disagreements. These plans can create cynicism among the troops and may lead senior management to stray from reality.

Any of these pitfalls can hasten the demise of the strategic planning department.

Developing a Strategic Plan

This section outlines a systematic approach to developing a strategic plan from the viewpoint of R&D: initially R&D must be a participant in developing the strategic plan for the business as a whole, then it must develop its own R&D strategic plan that supports the overall business strategy. This approach is more "bottom-up" than "top-down," with top-level executive input coming only late in the following eight-step series:

1. Inventory core competencies and technologies.
2. Identify target markets.
3. Assess competitive position in the target markets.
4. Formulate a strategy proposal and a set of alternatives.
5. Identify gaps, resource requirements, and time frames.
6. Use formal or informal valuation methods to select among alternatives.
7. Adopt, modify, or reject.
8. Set targets and implement.

Step 1. Inventory Core Competencies and Technologies

A common approach to determining a viable strategy is to begin with an inventory of the organization's *core competencies.* A working definition of a core competence is *a significant capability that is equivalent to or better than the top tier competitors.*[8] By definition, core competencies require years of investment in technology, employee development and training, and often include the operation of unique physical assets. It is difficult for a competitor to match all of these. *Core technologies* are simply a subset, usually a very important subset, of a company's core competencies. Our emphasis in this book is naturally on core technologies.

There is a major potential pitfall in this step. As we inventory our core competencies, it is extremely important to avoid defining them too aggressively or allowing the apparent logic of words to formulate a strategy based on false assumptions. For example, a core competency in "coal mining" may not equate to a core competency in "mining."

Step 2. Identify Target Markets

In this step and those that follow, it is important to distinguish between the corporation and the business. All competition takes place at the business unit level. However, the corporation in general, and its R&D function in particular, may have competencies that it wants to deploy in new arenas.

The phrase "target markets" can be ambiguous; there are many criteria for market attractiveness,[9] some of them quite subjective, and they are worth reviewing as part of any long-term planning process. Characteristics of attractive markets are size, growth rate, and the existence of a large body of customers with unmet needs. Other market characteristics include the behavior pattern of existing competitors and the types of margins they are able to earn.

It is important that technological performance be a basis of competition and a critical component in meeting the customer's unmet needs. The long-term durability of the market and its potential to generate new opportunities are also important.

Consulting firms are famous for having developed two-dimensional charts that plot market attractiveness versus a parameter that is a surrogate for risk. The market attractiveness parameter is often represented by placing a circle indicating the market size in a position that somehow weights the other attractiveness factors. The risk parameter usually weighs the company's technical and business strengths, and the degree of novelty the target market poses. The position of the circle on the chart is then determined by its overall degree of attractiveness and risk. In one form or another, these simplistic charts can be useful in discarding many ideas, and selecting a few for more detailed analysis.

Market targeting assumes that the target markets are known, are attractive, and that the issue is how to succeed in them. If there is no way to be successful, the decision is easy enough! The obvious should not be overlooked: among the target markets are those in which the business currently participates, and among the company's strategic alternatives are ways to upgrade the value of the products that serve these markets.

Step 3. Assess Competitive Position in the Target Markets

Once the company's core competencies have been identified, we must determine how they stand up to those of competitors in target markets. This is an extremely important step and we shall spend some time on the key methods, especially as they relate to R&D.

The Informal Approach

The simplest and least expensive approach to assessing the competition is informal—mining the deep industry knowledge within the employee ranks. In general, the sales force and technical service groups have direct contact with the competition, and the research staff is aware of a competitor's technologies and patent portfolio. This knowledge is embedded in the minds of the more senior employees and can be systematically tapped and integrated.

The downside to this informal, internal approach is a bias in favor of the company. It is difficult for managers and employees to admit that they lag behind the competition. To do so implies that someone (yourself?) has not been doing the job. A company needs more objective and more systematic approaches to assessing its own competitive position. Three approaches that fill the bill are benchmarking, competitor intelligence, and technological assessment.

Benchmarking

Benchmarking is a popular and objective way to probe beyond what employees and managers currently understand about their competitiveness in target markets. Benchmarking is a process through which a company can compare its own business practices with practices found elsewhere. For strategic planning purposes, the most useful form of benchmarking is for a group of competitors to agree to share data. The sharing may be open, or it may be through a third party, where the identity of the data sources will be at least theoretically anonymous. The third party may be a trade organization in which the benchmarkers hold membership, or it may be a consulting firm with accepted expertise and integrity. For example, 10 competitors may share data on costs of a key raw material, on energy consumption in their plants, or on selling cost as a percentage of revenues (some of these items are technical, and others nontechnical). The third-party administrator will then aggregate the data and tell each participating company where it ranks in each category (first quartile, fourth quartile, etc.). This information can be critical in identifying performance gaps, where future threats may lie, and where a competitive advantage can be pressed.

Benchmarking is a powerful tool for process improvement, especially when practitioners look beyond the small circle of their direct competitors to "best practices" in other industries. The pioneering case of this was Xerox's benchmarking of order fulfillment operations in the early 1980s. Xerox recognized the shortcomings in its ability to pick, pack, and bill small orders of one or several replacement parts, and looked beyond the best of its competitors in the office equipment industry to an organization that had developed a superior approach to this same operation—the direct mail retailer L.L. Bean. A Xerox team traveled to Freeport, Maine, where it observed and analyzed the Bean order fulfillment operation in action. This team then adapted the Bean process and successfully transferred it to its own logistics and distribution unit, creating superior performance and a competitive advantage in that one aspect of its total business.[10]

If "best practices" drawn from another industry are truly applicable in your industry, you will innovate before the competition can. These cross-industry benchmarking exercises are often initiated by specialists, such as analytical laboratory managers, technical information specialists, or human resource managers. Such groups may assess the state of the art in, for example, mass spectroscopy, on-line databases, or compensation systems for scientists. The information is useful not only to assess one's current competitive position, but also to report it objectively to management, and to identify directions for improvement.

Benchmarking is a useful tool for continuous improvement but does not guarantee competitive advantage. The danger is that the performance target may be set too low—and the real leaders are already moving forward to a higher level of performance through innovation.

Competitor Intelligence

Competitor intelligence is a systematic effort to determine the actions, intentions, and capabilities of a company's adversaries and can be extremely valuable in formulating a successful strategy.

Competitive intelligence is generally pursued through formal programs that make use of specialized personnel and methods. These methods range from entirely ethical means of collecting and analyzing publicly available information to cloak-and-dagger stuff. Indeed, former employees of intelligence, security, or investigative agencies often develop second careers in this line of work.

The most useful competitive intelligence information is generally found in available literature. Most experts in this field claim that 90% to 95% of what you need to know about your competitors can be found in public sources—from their own literature and public statements, recruitment advertising, annual and financial reports, environmental permits and applications, patents, and technical publications.[11] Although some projects will be carefully hidden behind a wall of security, publicly traded companies have a legal obligation to report accurately to their shareholders, and are unlikely to employ deception regarding their strategic intent. Small companies dependent on raising new capital at frequent intervals typically disclose far too much information from a competitive viewpoint.

Employees are an important second source of competitive intelligence. Many come across articles or have conversations with competitors, or obtain information from customer sources. These bits and pieces of information may have little significance in isolation, but when combined with other information, they often reveal a great deal about the strategies and technological trajectories of competitors. To bring those bits and pieces together, companies need to systematically gather competitive information from the heads and personal files of employees and assemble them into a competitive intelligence database.

So much information can be obtained today that information overload is a real threat to efficient analysis. Someone with technical acumen—a "gatekeeper"—is needed to identify and extract the truly useful information. Every company employs technical specialists, and these employees should be given formal responsibility for assessing the competitive intelligence found in publicly available information in their fields of specialization. For example, if a company is in the petroleum catalyst business, one of its scientists should have explicit responsibility for patents in that field as they are issued. This may seem obvious, but it is often not done because of competing priorities. The gatekeeper approach ensures that experts do the analysis of competitive information.

Competitor intelligence is a two-way street, in the sense that while you are gathering information about your competitors, they may be doing the same to you. There is evidence aplenty that major companies in the Far East maintain large staffs for the sole purpose of obtaining information about Western markets and learning as much as possible about new technologies developed by their Western competitors. They use this information to define their own strategies for new product development and market entry. There is nothing wrong with this practice as long as ethical methods are employed, but the scale of their efforts seems far greater than that practiced by U.S. companies. There is also evidence that intelligence agencies of Eastern European countries and even some U.S. allies systematically engage in commercial espionage.

Commercial espionage is not only unethical, but also often wrong-headed, at least with respect to technical information. First, those who are engaged in it are frequently technically unprepared to separate the crown jewels from the trash. They must bring home whatever information they can gather and let their internal experts assess it. This can be expensive and time-consuming when the information is massive, and may be of limited value if some vital piece of information, like the key to a huge lock, is unavailable. Copying competitor products and processes can be even dumber, since it is likely to commit resources toward designs that the more innovative competitors are already abandoning for better ones. In fact, in dynamic industries, the copyist is doomed to always being behind the curve of progress. Like the inept duck hunter, the copyist aims directly at a moving target—at where it is at the moment—never hitting it.

The truly effective (and dangerous) sources of competitive intelligence are former employees, consultants with ties to the industry, or current employees who are too willing to talk about their business. When you really want a piece of critical information about a competitor, you can often get it faster from the "old boys' network" of your own employees than through any other source. This too is a double-edged sword and it is imperative to educate technical employees that proprietary information should not be disclosed to *any* outside party.

Technology Assessment

Technology assessment is a formal methodology for evaluating the external competitive environment with regard to technology. It goes well beyond R&D projects and capabilities and therefore should involve technically knowledgeable participants from other functions.

Many such processes can be devised—what is important is that a group of experts gathers to evaluate what is important and where they stand in the industry. At W.R. Grace, we developed such a process in which the technology base of each business unit was evaluated against its competition by an assessment committee comprised of senior business unit technical management, senior corporate technical management, and senior business general management. This brought together a broad group of perspectives for rating each business.

In the first step of the process, business unit technical managers identified technologies that they deemed were important or *potentially* important to the business. We adopted definitions based on concepts and vocabulary introduced by Arthur D. Little and widely used in the industrial research community—terms like "base technology" and "key technology." We defined a *base* technology as one currently available to all competitors and generally known in the industry. *Base* technologies can be purchased for a fair price. A *key* technology is a technology of current competitive interest. *Key* technologies can differentiate competitors and are generally the subject of current R&D. As they mature, these may become base technologies. *Emerging* technologies are of unproved value, but may significantly affect the future. They may become key technologies. Conventional technology management strategy advocates minimizing resources spent on base technologies, focusing on key technologies, and monitoring emerging technologies for opportunities and threats.

Just which technologies should be included in the list of technologies to be evaluated is arguable, as is its classification as base, key, or emerging. However, the discussion and analysis provoked by the question is at least salutary and often eye-opening. It should be handled carefully, because, as in real life, the selection of the agenda often determines the outcome.

Exhibit 7.1 is an example of one product line in a forest products division of a major corporation. It is hypothetical and oversimplified,

Business: Towel and Tissue

		Competitor			
	Firm A	**X**	**Y**	**Z**	**Other**
Base Technologies					
Forestry	6	9	7	5	
Pulp production	9	7	8	7	
Papermaking	8	8	7	8	
Converting	8	8	8	9	
Key Technologies					
Wet strength agents	7	7	9	7	
Embossing	9	7	7	7	
Recycled paper	4	3	3	9	
Emerging Technologies					
Genetic engineering	5				8[a]
Dry papermaking		7			
Anthraquinone pulp			6		8[b]

[a] Company B.
[b] Company C.

Exhibit 7.1 Competitor Assessment Matrix

but illustrates the output of an assessment process. In this case, a unit of our firm, Firm A, makes paper towels and tissue products, primarily for the consumer market. It shares that market with competitors X, Y, and Z. The pulp and paper industry is mature, and knowledge about managing woodlands and operating pulp mills is generic. Therefore they are included among base technologies. Papermaking and converting equipment are generally available from vendors and are classified as base as well. There are differences among the competitors, however, owing to the locations of their woodlands and the age of their equipment.

An important form of differentiation is that Competitor Z derives much of its fiber by deinking and cleaning recycled paper, and has a much lower investment in woodlands and pulp mills. Z has an advantage when recycled paper is readily available, but is disadvantaged at other times. It can expand capacity cheaply, since woodlands and pulp mills are not required. Z's deinking technology is proprietary and not available to Firm A. It is a key technology.

Competitor Y has better wet strength technology than the rest of the industry based on superb polymer chemistry capabilities. Y can make claims on television that its products are stronger, or produce sheets of equal strength with less pulp—another key technology.

Firm A has made a patented breakthrough in embossing and can use it to create an image of luxury while actually using less fiber in a roll.

The technology assessment team initially considered a long list of 15 emerging technologies, but considered three worth monitoring closely and worthy of inclusion in its report to senior management. Firm A has started some research on genetically engineered trees, which are targeted to grow 30% faster than wild strains. We are aware, however, that the real leader in genetically engineered trees is a biotechnology firm, Firm B, which has no commercial operations, but could license its technology to A, or to X, Y, or Z. Firm X has taken a fling at a revolutionary technology to produce towels from dried pulp and to gain wet strength from latex adhesives, but so far other product drawbacks seem to be preventing this threat from materializing. Firm Y has a pilot plant using an anthraquinone pulping process licensed from a chemical producer, Firm C. It is known that this process significantly increases pulp yield from wood, but the overall economics remains unclear.

Note the importance of the last column. Here, one can recognize competitive technology in the hands of an organization that is not a direct competitor. In some cases, these firms have the potential to become competitors; more commonly, they possess the capability of forming strategic alliances with others in your industry. This is particularly significant when a *key* technology is involved: for example, in catalytic zeolite technology, Mobil, a major oil company which did not directly manufacture catalysts, developed skills and a patent estate that surpassed most catalyst companies.

Some of the most formidable technological threats appear "out of left field." In these cases, a discovery made outside the circle of competing firms is capable of undermining the entire industry. Industrial history offers plenty of examples. Plastics undermined paper and metal packaging, semiconductor-based transistors replaced vacuum tube technology, and the Internet now threatens vast sectors of the publishing business. If a competitor finds and protects a successful paradigm by which a new technology can enter your business, the effects on your market share and profitability can be dramatic.

Step 4. Formulate a Strategy Proposal and a Set of Alternatives

After a hard look at one's core competencies, the competition, and the dynamics of the markets in which one can compete, it is time to propose a strategy. The business is already pursuing a strategy, if only implicitly. Second, a group of new ideas will have been created in Steps 1 through 3 by the very process of defining core competencies, assessing their relationship to possible target markets, and assessing their relative strength versus the competition in each of these markets.

If the current strategy has been very successful, one should think very carefully before changing it, but one should think about it, nevertheless. If it has been moderately successful, this strategy can be used as a starting point. But when a business is troubled, strategy formulation is more challenging: selection of the right strategy could reverse the downward trend, while the wrong choice or the continuance of the current strategy could lead to irreversible decline.

The preceding activities may now be presumed to have generated a list of strategic alternatives. Some typical alternatives involve greater focus on current value drivers versus greater diversification in search of the new opportunities in target markets. Implicit is the degree to which the business is managed for current profits or for future growth. It is important to consider time frames—should returns be optimized 2 years out, 5 years out, or 10 years out? Shorter time frames allow fewer strategies.

Is it time to consider geographic expansion? It is very attractive to spread research costs and other overheads over a global sales base rather than a national one—but not if one is unprepared for fierce competition in other people's markets!

At the corporate level, should the company be a one-product business, modestly diversified, or even highly diversified? While focusing on the core has been a popular theme in recent business literature, executives such as J. Peter Grace have saved their companies from certain extinction via diversification (from passenger liners into chemicals), and in so doing, created shareholder value. Diversification is the last resort of the buggy whip manufacturer. But diversification that does not earn more than the cost of capital is of no economic value, and can distract management from doing what it knows best. (In hindsight, Mr. Grace later overdiversified his company and had difficulty continuing to create value.) And the stock market generally does not reward corporate diversification since individual investors can diversify risk within their own portfolios. Indeed, it usually punishes diversification with a "holding company discount." (The subsequent sale of many of the W[illegible] Grace operations unlocked a great deal of hidden value.)

This is also the time to apply management principles to the alte[illegible]tives. Many executives abhor "commodities" because of their susceptibility to business cycles and the relative ease of entry to competitors, which can create long periods of oversupply. As they mature, products may transition from specialties to commodities, and when that transition begins to occur, these executives believe it wise to concentrate resources on more attractive markets.

Another example of a management principle is Jack Welch's well-known view that General Electric should be number one or number two in any business or get out. Does one accept this? There are powerful arguments for this formula. Even so, some firms do very well as number three or four, and tremendous value has been created by entrepreneurs such as Gordon Cain who have acquired supposedly "weak" business units from corporations that no longer wanted them.

Another common principle is an insistence on minimum levels of return and profit growth from each business. This approach is directly value-oriented. The assumption is that a business without a credible plan for meeting this goal (often a very tough goal!) is a drag on total corporate performance. If value cannot be created by an existing business, one must evaluate the alternatives: selling it and reinvesting all the proceeds, or milking it slowly for cash that can be reinvested in growth opportunities.

Step 5. Identify Gaps, Resource Requirements, and Time Frames

The next step is to assess the resources required for each viable strategic alternative. Although a real strategy may be multifaceted, consider the implications of adding a single new target market to the existing business. It is useful to look at the potential gaps between what we have and what we need to be successful:

- *Technology.* We have completed a technology assessment and know our strengths and weaknesses relative to the competition. Any serious weaknesses must be shored up by an R&D program, licensed technology, or a strategic partner. The first may result in risk and delay, but may pay off in long-term value. Both R&D and licensing have costs. A strategic partner implies some sharing of the rewards.
- *Capital.* The value of our proposed strategy will be directly related to the capital we invest and our economic profit on that investment. If that capital produces a return that exceeds the company's hurdle rate, the investment should be welcomed, but it also should be budgeted. Procuring capital equipment often involves lead time (e.g., a reaction vessel made of a special alloy may take many months to be fabricated and shipped).
- *Skills.* The members of the internal team charged with implementing this strategy must be found within the company or recruited externally. Consultants may also fill key skill gaps.

Recruitment is necessary if the company has a gap in key functions, such as marketing or research, but hiring involves risks and time delays. If the target market is new to the company, individuals with experience in that market may be recruited to speed market development activities. Consultants are most often used in handling specialized one-time tasks, such as initial market research, the gaining of regulatory approval, or design engineering.

- *Time.* The timeline for the project will depend on the time lines of each major activity and will obviously affect both the project's value and risks. Since some tasks cannot prudently be started until others are completed, PERT (Program Evaluation and Review Technique) and Gantt charts and other project management tools may be usefully employed to evaluate the full implications of the proposed strategy.

Adding up the resources required to make the strategy a success can quickly separate wishful thinking from reality. Too many strategic plans do not accurately reflect these resources, support too many projects, and address more market opportunities than the business can realistically handle.

Timing is equally important. A research project may take four years or more to complete. It may take two or more years to design, build, and start up a plant. The same is true for recruiting and training a dedicated sales force. The strategic plan must efficiently coordinate these diverse functions and activities, and provide the basis for establishing hard targets, including milestones and stagegates.

Step 6. Use Formal or Informal Valuation Methods to Select among Alternatives

Valuation techniques, and an understanding of the value drivers of the business, are helpful in this phase of strategy development. Specifically, the time frames and costs of various alternatives lend themselves to value-based analysis. This subject is discussed in detail in Chapter 10, which describes how a "quick-and-dirty" financial model of a project proposal can be constructed and used to identify the value drivers through sensitivity analysis.

In principle, a net present value can be calculated for each alternative proposal, making the most *financially attractive* candidate apparent. This quantitative approach to value, however, must be seen in context. Good numbers are usually not available when strategic planning addresses unfamiliar markets. A critical review of the implications of the calculations by seasoned managers is indispensable. Nevertheless, with these caveats, DCF comparison of strategic alternatives embodying comparable assumptions is an excellent decision-making aid.

STEP 7. ADOPT, MODIFY, OR REJECT

Once the previous steps have been completed, it is decision time. The facts are assembled. The costs and risks of the strategic alternatives have been calculated. The resources required have been weighed against resources on hand.

Decision making should be totally objective—but it is not entirely, since people are in the loop. Each business function has a wish list—a new plant, a promising research project, or a modern business information system. Commitments to customers, suppliers, employees, or communities may have to be honored. A decision that seems financially correct may have serious consequences if customers are unhappy or if disappointed key employees desert to the competition. The most aggressive marketers and visionary technologists may not be able to coexist with a strategy of maximizing short-term cash flow.

"Decisions, decisions, the terror of the jungle!"

When all factors are weighed, key decision makers must choose among the alternatives. Typically, they will not be entirely happy with any of the alternatives; they will choose the best and ask the strategy team to evaluate modifications. This request may take the team back to gathering additional information and incorporating it in Step 4. Once or twice through this recycle loop and the plan will be accepted.

STEP 8. SET TARGETS AND IMPLEMENT

Even with the strategy adopted, the job remains unfinished. Each function or unit must establish its targets and make work plans to reach them. Short-term targets, which are highly constrained by budgets and available human resources, must be linked to new long-term targets. For R&D, this may mean budgeting new projects, terminating projects that do not support the strategy, and reducing general targets to specific milestones, tasks, and specifications. R&D's counterparts in marketing, sales, engineering, and manufacturing must do the same.

Summary

Despite the risks inherent in the actual adoption of a strategy, there is great merit in illuminating the linkage between assumptions and projected results. This becomes most apparent after the quantitative elements of timing and resource allocation are factored into the plan. The planning process also allows us to discard strategies based on flawed assumptions, on rhetoric, or on numbers that fail to add up and match the available resources.

Few businesses are simple, and all must adjust to changes in the business environment, both gradual and revolutionary. Even so, the strategic plan will have a short lifetime. The test of its value will not be how closely it matched actual results, but whether it promoted sound decisions and avoided serious errors.

There are three key points about strategy making for technologists to absorb:

1. Participate in the process. Your input is unique and valuable.
2. Remember that value is only created in the marketplace. So, focus on the value drivers and on the competition.
3. Expect change. Even though you have provided technical vision to past plans, be prepared to adjust your programs to current realities.

CHAPTER 8

Marketing: The Top Line

"Yes, you've made a real technical breakthrough," the marketing director told the young researcher, "but we cannot find any commercial potential for it."

How often have you heard this around your shop? In my own experience, commercial failures of products based on new technology are about twice as common as technical failures. Commercial failures can generally be traced to a flawed understanding of the market and the competition. In *Winning at New Products,* Robert Cooper offers ample statistical and anecdotal evidence on this key point.[1] Inadequate market analysis is cited as the leading cause of failure (45%) and lack of effective marketing effort comes in third (25%), just behind product problems or defects (29%).

Failure cannot be entirely prevented because the marketplace may change unexpectedly between the time when a project is begun and when its output is ready for launch. The world is a complex place, and while good research and good planning can reduce the number of things we do not know, it cannot eliminate them entirely. However, a failure for any cause is still a failure, and casting the blame on marketing after the fact will not undo the damage.

To create value, technology must be coupled with a market. In one way or another, that coupling will be through products. In the case of a new or improved product, coupling is reflected directly in the revenue line of the income statement, which is the subject of this chapter. If the technology is embodied in a better process that creates cost or capital savings, its impact will still impact the revenue line, though indirectly, by giving the company's products a cost or quality advantage.

This chapter has two purposes. First, in the discussion of valuation, it reviews the principles, pitfalls, and complexities involved in estimating the markets for a new technology. Again, the valuation of that technology will be determined by the products in which it is embedded. We will particularly focus on a prime concern of researchers: the special problems associated with estimating markets for "new to the world" products. Second, the chapter addresses the role of the marketing function in creating value, and highlights pitfalls for unwary technologists who base market assumptions on preconceptions instead of data.

The chapter introduces some key concepts, with real-world examples that show why these concepts are critical to creating commercial technology. It begins with some broadly used, but often misunderstood, terms such as *product*. It moves to some key marketing concepts, notably *segmentation*. It then addresses the vital issue of market research and what researchers can and cannot do in predicting the future. It discusses *product life cycles*. Finally, it looks at the subject of *technological S-curves* and *technology forecasting*, an underestimated and powerful tool for predicting when certain technology-driven events might occur. All of these concepts provide the basis for decision making on the magnitude and timing of resources required for project success.

An R&D Perspective on Marketing

Marketing is the science of positioning products in terms of all of their attributes, including price and channels of distribution. Many confuse marketing with sales, which is a subset of marketing. Sales has the job of getting the order, servicing the order satisfactorily and, in some cases, seeing that invoices are paid. These are important functions, but they are not the same as marketing. An organization that lacks effective marketing can be referred to as "sales-driven."

Young researchers who join industry to make scientific discoveries soon discover that the majority of business issues they encounter have to do with marketing, which to them is both mysterious and a frequent source of frustration. Successful projects, their managers and senior colleagues tell them, are "market-driven"—a term rarely uttered in the Ph.D. programs from which many have just emerged. Technology-driven projects, they quickly understand, are seldom welcomed. The way to make an enduring mark, they learn, is to make or create something with important applications in the commercial marketplace.

Even when scientists recognize the imperatives of the marketplace, tensions between commercial and scientific values do not always go away, but linger as a source of organizational and philosophic conflict. The experienced industrial scientist recognizes that to leave an enduring mark in the corporation or in his or her discipline, scientific work must have impact in the commercial marketplace. This can occur only through the commercialization of a technology to which he or she has contributed.

To achieve these ends, the scientist needs to understand the marketplace, and must see the marketing function as a logical ally—as an extension of his or her own work—since both are pursuing the common goals of commercial success and corporate growth. In addition, the scientist learns that marketing people can be a source of interesting new projects and concepts. These are, after all, the people who rub shoulders with customers in the field, who hear about technical problems that beg for solutions, and who see how customers are adapting and reconfiguring the company's existing products in creative and unanticipated ways.[2] Like pollinating bees, good field marketers return to the corporate hive with new and exciting ideas for researchers. Even better, many unmet market needs will require a strong component of radical innovation—exciting stuff to researchers.

Marketing is perhaps best understood by working with it. Books about marketing abound, but most focus on consumer marketing, a highly efficient and organized discipline. However, industrial and government/institutional markets are at least as important to technological innovators. Here, marketing practices vary widely, and even within the broad industrial sector, differences are vast from business to business.

What Is a Product?

Marketing people think of "product" more broadly than do most other people, including the research community, who have a tendency to think about a product very narrowly in terms of its most obvious characteristics. Scientists are particularly vulnerable to thinking about a product primarily in terms of its technological content. But the less obvious characteristics may be the ones that create value.

Consider table salt. From the scientist's point of view, table salt is old hat. It is one of the simplest and most abundant chemicals, sodium chloride, in one of the simplest of crystalline forms, face-centered cubic. Its physical properties are well known.

But there is much more to salt, and marketing's eye is attuned to see those qualities.

Table salt itself must be in an appropriate grain size for salt shakers and must have good flow and antiblocking characteristics. Its color is very important, as is the level of impurities and additives (such as iodine). It must be free of toxic materials and pathogens. Though seemingly simple, these technical specifications can be demanding.

The package in which table salt is delivered is another critical aspect of the product. At a minimum, it must catch the customer's attention in the supermarket, be durable enough to stand the distribution cycle, serve well in the household, and be innocuous on disposal. More recently, tamperproof characteristics have been added to many packages. It is now common for packages of low-bulk materials to cost *more* than their contents, and to generate complaints about "overpackaging." However, these costs are supported in the marketplace by customer demands for safety, attractiveness, and durability—a powerful argument for good packaging.

Another aspect of the product is the degree to which it is defined by advertising, promotion, levels of customer assistance, and the distribution chain that brings it to the consumer. In the case of salt, the creation of a recognized brand, such as Morton Iodized Salt (with its "When it rains it pours" slogan and trademarked little girl holding the umbrella), is at least as important as the technical specifications of that product. The product carries with it implied *warranties* for its purity and safety, and these are backed by quality control functions, laboratory testing, and regulatory experts. The reputations of brands such as Morton or Diamond create formidable commercial barriers to market entry when technical barriers are low.

Surprisingly to some, *price* is also part of the product, as are the terms under which the product is sold. Premium pricing may be linked to an up-market package, or in the case of salt, a special source such as sea salt, or trace "beneficial" impurities. Terms of sale can be another important product attribute. Terms of sale can be very straightforward (cash) for retail table salt, but more complicated and of vital importance when the product is an automobile or an airplane.

The process through which a product is made can also be a product characteristic. Process can offer both economic and marketing advantages. A low-cost process should give its owner a pricing advantage over competition. However, a process that is more reliable, or dependent on raw materials that are relatively secure, may also provide a marketing advantage.

To restate the point, while sodium chloride is a chemical, it is not a product. It, or a substitute, cannot be delivered successfully to the marketplace with the scientist's knowledge alone. The scientist is simply part of a team of professionals that manages the entire spectrum of product characteristics.

More complex products, like pharmaceuticals, often originate in the chemical synthesis of a new molecule. Important product characteristics for pharmaceuticals are determined through a series of tests. The most important of these is efficacy in treating a particular disease, as determined initially through screens and, eventually, through preclinical (animal) and clinical (human) trials. Toxicity and side effects are almost as important as efficacy. The form of delivery and recommended dosage must also be determined. If the new drug successfully negotiates these hurdles, a process for commercial synthesis must be developed and validated as conforming to "good manufacturing practices." Regulatory approval, usually in the form of a label sanctioned by a government agency (FDA), follows. Thus, for a pharmaceutical, the process is *explicitly and legally* part of the product. The product is not only the molecule (which carries a "scientific" name, such as lovastatin), but the formulation, and the brand name Mevacor™. The cost of generating the information required to win FDA approval can be enormous: a total development process can last 13 years and cost $300–$500 million per commercially successful drug.

The Importance of Processes

Scientists and businesspeople often fail to consider the broad and often uncharted territory that separates a promising laboratory development and finished goods they expect to market to customers. This is the territory in which manufacturing and delivery processes must be developed. How effectively these are handled often spells the difference between market success and failure because process development is a key determinant of product quality, cost, and the ability of salespeople to fill orders in a timely way.

An effective process can provide tangible cost advantages. This is of paramount important in the case of commodity products for which features, specifications, and quality are expected to be equal. Here, the low-cost producer has a clear market advantage, assuming that delivery, service, and other nonproduct functions are equal. These assumptions, however, do not always hold, and producers who lack clear cost leadership are often able to capture a part of a commodity market through other means. Consider electricity, a product that would seem to be impossible to differentiate. A kilowatt is, after all, a kilowatt. The low-cost production process in many locations is based on coal. However, electricity can be generated by different processes—hydroelectric turbines, nuclear reactors and gas, oil, and coal-fired boilers among them. Each of these options represents, to the customer, a different set of risks including future fuel prices, accidents, regulatory changes, railroad strikes, and drought. Within each category, there also exist economically important distinctions between generating facilities in boiler design, pollution control equipment, and so forth. In a competitive marketplace, an industrial purchaser will consider all these process-related factors, which translate to value, in negotiating a long-term contract. Hence, the process is part of the broader definition of even a standardized product.

Scientists who lack industrial experience, and even some executives, may be unaware of the huge gap that exists between making something in the laboratory and a robust manufacturing process capable of producing commercial quantities. But there is much more to a process than a chemical reaction or a physical demonstration. The process involves many important factors:

- Material and heat balances of a system.
- Specifications of major equipment.
- The number of operators required to run the process, and the procedures they follow.
- Process control philosophy and software.
- Technical solutions to all regulatory requirements.
- Above all, demonstration of a reliable and consistent operation.

Scientists and developmental engineers have not demonstrated a process until there is sufficient data for design engineers to begin drawing plans for a plant. This is a big job requiring time and money; yet many decision makers pay too little attention to this requirement. They are then disappointed when they learn that a new and exciting laboratory technology may be a half-decade from commercialization.

The biggest problem in translating laboratory data into manufacturing processes is that some steps that are easy and cheap in the laboratory are expensive and complicated in a production plant. For example, when a lab chemist generates wastewater, he or she throws it down the sink (assuming that this broth is nonhazardous, this step is simple and easy). In the production plant, that same wastewater may need a large treatment operation and a system for recycling process water, all of which require engineering data and substantial capital. The individual steps used in the laboratory must also be linked into efficient, continuous processes by mechanical engineers, industrial engineers, and process control experts.

A new material or a new invention is not the same as a new product, and a physical phenomenon or chemical reaction is by no means a process in the industrial sense of these words.

Other related pitfalls lurk in logistics. A product shipped on conventional means of transport must be rugged enough to withstand breakage or leakage on the trip. Similar requirements apply to warehousing and storage, where exposure to high temperatures over extended periods, or to crushing pressures from the stacking of pallets or cartons to the ceiling, may put physical or chemical requirements on the package and its contents that are not needed for their end use. The wad of cotton in a bottle of aspirin is there mostly to get the tablets to the customer without breakage. Certain products may require specialized handling such as refrigerated, cryogenic, or controlled atmosphere transport or storage. All of these issues are real and have caused major problems when they were not considered early in the R&D process.

What Is a Market?

Researchers quickly learn that the word "market" means different things to different people. For example, if lab personnel are busily working on new technology for an electric vehicle battery, they are probably thinking in terms of the *potential market* for products based on this technology: one battery pack for each motor vehicle on the planet. As they talk about this development with one of the company's market managers, however, the marketing person is probably thinking in terms of the *actual market*, defined by the number of electric vehicles in service—a very small percentage of the total vehicle population. Most of them may be golf carts and forklifts! That percentage is growing and the term used to describe that growth is *market penetration*. The market penetration of electric vehicles could be described as being 0.5% today, with growth expected to reach 5% within 10 years. If this projection were correct, one would calculate the market for electric vehicles 10 years from now by:

1. Projecting the potential market (using the annual growth rate of all vehicles).
2. Applying the projected market penetration at that time.

The term "market" should also be qualified in terms of domestic and global markets. A product developed for domestic sales may have no sales potential outside that market, given its design or specifications. For example, a consumer electronics product developed for the U.S. market may not meet safety requirements or electrical standards in other jurisdictions. A classic case of this is to be found in the "Global" steam iron developed by Sunbeam Corporation and its offshore affiliate, Rowenta, during the early 1980s. Their researchers found that Sunbeam's current line of steam irons were too expensive for most non-U.S. households, did not have a feature set favored by most non-U.S. users, and were out of compliance with safety codes in many jurisdictions. To penetrate the global market for irons, the two companies had to develop an entirely new design using fewer, less expensive parts, a different set of features (easily adaptable to different regions), and safety features that would meet the requirements of most local regulations.[3]

A third important dimension to defining the market is by access. We distinguish between the total market and the *available market.* In a commodity such as styrene, the total market contains two sectors: the *captive* and *merchant* markets. This occurs because the major end use for styrene is in polystyrene resins, and many producers of polystyrene are partially or fully back-integrated to styrene. They obtain much of their styrene from their own plants, which are captive producers. The captive market is not available to a new producer and, in this case, is a very material percentage of the total market. However, some producers make styrene only, or make more styrene than they consume. This stream will be marketed to polystyrene producers who are net buyers of styrene. They constitute the merchant market. In deciding whether to expand a styrene plant, both the total and the merchant markets must be considered.

Likewise, in other types of products the available market may exclude large sections of geography. The available market for military rockets or high performance computers, for example, is restricted to allied or friendly countries. Policies of foreign governments may exclude certain classes of products as well, taking those countries out of the available market.

As always, judgment is required in defining available markets. Shades of gray are involved in these definitions: long-lasting supplier relationships or contracts may make what appears to be part of the available market virtually captive, and restrictive trade practices may de facto exclude countries that claim to be open to your product. Ask yourself if it is realistic to expect to sell a new vehicle in Japan, or to displace a can manufacturer who has put a plant next to a major brewery.

Tremendous errors can be made if a common definition of markets is not shared by all members of the cross-functional product team. Researchers often misunderstand the scope of markets, tending to look at them more broadly than do their counterparts in other functions. Sales personnel, on the other hand, may take an unduly narrow view of markets, and miss opportunities in what they regard as "marginal" technologies and accounts—an affliction known as "the myopia of the served market."[4]

Market Segmentation

A cardinal factor guiding the strategy for any research program must be the definition of the target market, which is best determined by *segmentation*. Market segmentation is one of the primary tools used by marketers. It is the process of reducing a large heterogeneous market into smaller, more homogeneous subsets, or segments. If we were developing a family of products for boat owners, it might be strategically important to segment the boat-owning population in terms of different characteristics; for example, large boats versus small boats; sailboats versus powerboats. Exhibit 8.1 describes this approach to segmenting the boat-owning market on these two characteristics. Based on our competencies, resources, and existing marketing channels, we would then decide to research and develop products targeted for one or more of these particular segments.

End-Use Segmentation

The first cut at segmenting a market is usually by the product's end use. Consider coatings or paints. There are several major markets for paints: architectural, automotive, and industrial, to name just a few. In addition, there are a host of specialized applications such as nail polish, dopes for model airplanes, and fabric waterproofing. Each of these end-use markets has its own characteristics.

	Large	Small
Sail	Large sailboats	Small sailboats
Power	Large powerboats	Small powerboats

Exhibit 8.1 The Boat Owner Market

	$Million	Percent
Construction	3,010	30
Transportation	1,750	18
Packaging	1,550	16
Nonrigid bonding	1,250	13
Rigid bonding	1,100	11
Tapes and labels	750	8
Consumer	490	5
Total	9,900	100

Exhibit 8.2 Segmentation by End Use (Adhesives)

Now consider a more detailed example, using the partially hypothetical case of a researcher who accidentally discovers a promising new adhesive. He quickly learns that the existing adhesive market in the United States is approximately $10 billion (Exhibit 8.2). He also learns that over 700 firms currently supply this market. Does he have any chance of capturing a significant portion of those revenues with his new technology? This is an important question because the size of the market target will in part determine how much R&D can or should be supported. Segmenting may also reveal the most attractive initial targets, which will drive the initial directions of the research program. Research on these promising target segments can provide clues as to the gross margins the company might expect to capture, and this will help its financial analysts to estimate the value of any new technologies or products developed through research.

As it turns out, the adhesive market is highly fragmented and, in general, quite mature. It has very few large targets, and many of the small ones are vigorously defended.

Assume that management commissions a market study, which divides this market into seven major segments. None is larger than $3 billion. However, each major segment contains smaller and quite distinct *subsegments,* which in turn contain still other subsegments. For example, nonrigid binding includes several significant smaller segments, including textiles, bookbinding, and footwear. Within textiles, there are important subsegments for carpet, textile lamination, nonwovens, and so on. The segmenting process may go down further before the list is reduced to a small group of potential end users (customers) with similar needs. Marketers, in fact, use the term *niche marketing* to describe a strategy whereby companies specialize in serving one or more very small segments or subsegments. Examples include Rolex, which produces watches for a tiny fraction of the timepiece market. In the field of music, *Strings* magazine caters only to players of violins; *Fiddler Magazine* addresses an even smaller niche of the string-player segment with material on traditional music played on the same instrument. Niche marketing is often an effective form of protection for small producers, since the total potential revenues of the niche are too small to attract larger competitors.

A careful market study would include estimates of market growth rates in each segment. For example, transportation may be growing at 7% annually owing to the replacement of welds by adhesives, while growth in more mature sectors may be limited to 3% to 4%.

Segmentation by Technology

Markets may also be segmented by technology. This is extremely important to the strategy development process, since the points of attack will be determined in part by the business's technology strengths. Exhibit 8.3 lists 12 common families of adhesives. Each represents a different technology. Typically, end-use segments are dominated by several, but not all, of these families. And each family includes a group of competing suppliers, some of whom may only employ one or two of the competing technologies in its product offerings.

For example, epoxy resins are an important class of adhesives in automotive and electrical applications, but are too costly for most construction applications. Three powerful chemical firms dominate the market for epoxy raw materials (Dow, Shell, and Ciba), but often

Acrylics	Rubber-based
Animal glues	Silicone
Casein	Starch-based
Cyanoacrylates	Styrene-Butadiene
Epoxies	Urethanes
Hot melts	Vinyl acetates

Exhibit 8.3 Segmentation by Technology (Adhesive Types)

sell indirectly through formulators and distributors, a pattern well suited for fragmented markets. Many customers cannot change technology type readily owing to the type of equipment (dispensers, robots, dryers, or pollution control devices) in their existing plants. Epoxies are usually two-component adhesives, so dispensing is complicated. Hence the customer equipment factor must be considered carefully in defining the available market for a new epoxy.

"Super Glue" is a new adhesive technology that used market segmentation to good effect. Cyanoacrylates—the "super glues"—represented a new class of adhesives. They were very strong and very fast-setting, but expensive. Their discoverer had to determine which market segments to attack first. Many could be quickly eliminated: construction, because of cost; and nonrigid binding, because of technical characteristics. However, niche positions could be, and were, exploited in automotive, electrical appliance, and consumer markets, especially where the fast-setting characteristics gave processing advantages over competing epoxies and other adhesive families. These targets, determined in part by technical considerations, inevitably created organizational issues during commercialization.

In consumer markets, specialized consumer marketing skills applicable to retail selling were required to define the product, package it in tubes, and get it into retail channels of distribution. The novelty of the product gave its manufacturers considerable latitude in performance specifications.

At the other extreme, technologically sophisticated industrial customers, such as automotive OEMs (original equipment manufacturers), would be much more demanding in terms of specifications and testing. They would want to ensure that the adhesive would perform on their products and equipment. In many cases, the substitution of a cyanoacrylate for existing adhesive technologies would require the replacement or modification of existing capital equipment. OEM's would be unwilling to make these costly changes in the absence of clear evidence of cost and/or performance improvements. Thus, product introduction to the industrial segment would take several more years than introduction to the consumer market. Understanding OEM needs, purchasing practices, and specifications were critical to realizing the promise of the new adhesives.

Geographic Segmentation

In addition to segmentation by end use and technology, segmentation by geography can be extremely critical because the competitors in one geographic market segment may be entirely different from those in others. (The same can be said of customer behavior and competitive dynamics.) Product performance requirements in some product classes also vary greatly by geography, as do requirements for technical support. For example, beef packaging patterns are vastly different in the United States and in Europe. In the United States, virtually the entire wholesale market is served by three large firms (IBP, Excell, and Montfort), which run industrial-strength operations. Their counterparts in Europe are generally small local abattoirs with inefficient equipment and labor practices. The small local butcher shop is still an important outlet in Europe but has been largely displaced by supermarkets in the United States. Advanced packaging systems, such as wholesale boxed beef, are required in the United States, but these and associated technologies are just beginning to penetrate the more fragmented and individualistic European market. Nevertheless, a global perspective from the outset can still create value.[5] This is apparent in our example: supermarkets are increasingly prevalent in Europe, and their need for standardization will inevitably drive the expansion of boxed beef technology among the European slaughterhouse operations.

Segments and Strategy

Strategy is the final factor in defining the target markets. There are important differences between niche and broad industry strategies, and in choosing one, a company must understand what the business sees as its long-term objective. Some companies view themselves as specialists and are comfortable with profitable niche positions. Others need to achieve larger market positions to survive, and use niches simply as beachheads for developing such positions. Japanese automakers used the latter strategy effectively in gaining a foothold in the lucrative U.S. market. They entered through the narrow, low-profit segment that served small car buyers. This segment did not particularly interest the "Big Three" U.S. producers, nor was it one in which they had significant products or technical capabilities. The Japanese entrants used the small car segment to develop U.S. distribution networks, gain experience with American buyers, and create confidence in the value of their products. Once these initial goals were achieved, they successfully moved up into the more competitive—but more profitable—large car and luxury car segments of the market.

An analysis of market segments in light of business strategy and capabilities should provide the critical information required to guide new product decisions.

Consumer Markets

Much of the marketing literature is oriented toward consumer marketing. This is a highly developed field, particularly in North America, and has a language and methodology all its own. One of its characteristics is the availability of abundant market data, much of which is generated from surveys of consumers and retailers by independent consulting firms such as Nielsen, which make this information broadly available to anyone willing to pay for it.

The effect of a product innovation can be fairly rapidly determined in the consumer area. Here we use the term innovation in its broadest sense. A change in the method of distribution, in price, promotion, or in packaging may have as great an impact as a change in the product's performance attributes. For example, a manufacturer of toilet paper may propose to reduce the basis weight of its product by two pounds per ream, give the consumer a 10% price break, and promote it as a discount brand. This strategy may succeed or fail depending on the soundness of the manufacturer's understanding of customer preferences and its anticipation of competitors' reactions. Data is collected in fairly short periods of time in the marketplace, often in a limited number of test cities. Shrewd competitors may attempt to disrupt the innovation by cutting prices or offering their own promotions in those test cities, thereby destroying the validity of the market data, and perhaps rattling the decision-making process of the innovator.

Whereas scientists may be tempted to think that price and product features or performance attributes would be the most powerful levers for gaining entry or share in the consumer market, marketers often look to innovations in customer service or distribution to gain a competitive edge. Perhaps one of the best examples can be found in the success of Dell Computer Corporation. Michael Dell's innovation was in his methods of direct mail distribution and product customization. While other PC makers were pushing prebuilt machines through the retail channel, Dell created a direct interface with the customer, giving the customer an opportunity to select features, options, and installed software over the phone or through the company's Web page. Within a week or so of the actual transaction, a uniquely designed PC would appear at the customer's door, ready to set up and operate. Dell's machines had no particular technical attributes to set them apart in the highly competitive PC market, but his innovations in distribution very quickly made Dell a leading seller of personal computers.

Industrial Markets

Industrial markets involve products or materials that are incorporated into the production of other goods which are eventually sold to consumers or other industrial producers. The key difference between an industrial market and a consumer market is that an industrial market does not comprise products or materials for final consumption. The final end user is somewhere downstream.

Whereas no consumer product can be sold without marketing support, many industrial products are sold without a marketing organization. Many industrial companies believe they have a marketing function, but confuse marketing with selling.

Industrial marketing, as one would expect, is driven by the types of products the company has to sell. These fall somewhere in a product type spectrum defined by *commodity products* at one end, and *specialty products* at the other. A *commodity product* is one that is available from several firms and is largely indistinguishable in terms of physical or performance attributes (e.g., electricity, coal, polyvinyl chloride, polyester fibers, and standard grade body stock for steel cans). These products are essentially interchangeable and are transparent to the customer's customer; they tend to be sold primarily on price and secondarily on quality, reliability of supply, and long-standing relationships.

They are commonly sold in large quantities and with long-term contracts. The amounts of money involved in these contracts are so large that the chief salesperson may be the CEO or the group vice-president. In such cases, there may be no marketing specialist and little day-to-day contact with the customer. As a result, a researcher will find it difficult to obtain formal information about potential product improvements or market trends through normal intercompany channels. However, consultant reports on industrial commodities are generally available and can be helpful, since they often highlight long-term trends that can help or hurt the commodity producer. Coal may be replaced by gas, vinyl by chlorine-free polymers, and steel by aluminum.

In the broad middle of the industrial product spectrum are what might be called *performance products,* with attributes that are unique to some degree and that can be differentiated from competing products. Industrial catalysts, machine tools, servers, automotive cushioning, aircraft engines, and paints are all examples. Industrial customers can adapt their operations to these products over a reasonable time. (They need to qualify the substitute products in their processes and to verify that there are no adverse downstream effects for their own customers.) They have the power to change suppliers, but at a cost.

For these customers, information regarding performance and technical specifications becomes extremely important. A firm selling performance products is well advised to have a marketing function capable of answering all appropriate questions about product attributes, although, as noted, many do not. And it is very important that researchers talk directly with their counterparts in the customer company to be sure that they have a full understanding of needs and specifications.

The farthest end of the industrial marketing spectrum is the field of *specialty* or *custom* products, which are sometimes bundled with technical services and sold intensively on a technical basis. Flavors, fragrances, industrial instruments, adhesives, and pharmaceutical intermediates are typical examples. Volumes are usually relatively small, and price is often a secondary issue. In these cases, the decision makers may not be in the executive suite, but on the plant floor or in the laboratories of the customer. Selling and technical efforts must be guided by the needs of these decision makers. In many cases, the product is customized to the individual buyer's requirements.

One indicator of industrial market category is the amount of sales generated per salesperson In true commodities, salespeople may generate sales upward of $10 million per annum, while in true specialties the salesperson may have sales of only a few *hundred* thousand dollars. Performance products generally fall in the middle range. This sales-per-salesperson parameter is a useful approach to acid-test the market in which one is proposing to compete and whether market entry is consistent with marketing core competencies.

One of the best ways for a research scientist to understand customer needs is to visit a customer plant and/or laboratory, and discuss current product shortcomings and improvement opportunities directly with the customer's operating and technical staff. This is not as straightforward as it sounds: although salespeople may see advantages in showing the customer the powerful R&D support behind their products, they also may be concerned that the dialogue may undermine their own credibility. The visit will go better when both the researcher and the salesman understand the value chain and are motivated to enhance it using a team approach.

Government Markets

Government markets include a number of levels and jurisdictions. In the United States, these include all federal, state, county, and municipal government units and their many agencies. Some private organizations such as universities and hospitals may have very similar characteristics from a marketing viewpoint, and these are sometimes lumped together as the *institutional market*. Foreign governments often dominate national economies, and can be a major source of opportunity.

The skills and behaviors required to succeed in government marketing are very different from those required in the industrial and consumer sectors, and making the transition from one to the other is often difficult. Governments often purchase through a process of competitive bidding and may be under political pressure to provide equal access to all bidders; sometimes, the bidding process deliberately favors some politically acceptable or preferred bidders. The cardinal maxim of industrial marketing—to be close to your customer—takes on a new meaning with government. Here, being close to the customer may be illegal! Whereas a dinner and drinks with an industrial customer is often de rigueur, regulations may specifically prohibit entertainment of government officials and purchasing agents. The government customer may also be sensitive to oversight bodies, official or self-appointed, to which decisions must be justified.

In the typical government bidding processes, potential suppliers are given product specifications, quantities, delivery dates, terms, and conditions. They are asked to submit their bids by a stated date. The governmental body is then required to accept the bid of the low-cost bidder (if that bidder is deemed capable of performing according to the terms of the bid). This open bidding process in some cases fosters commoditylike characteristics and low profit margins that make a government market far less attractive than industrial markets or consumer markets of the same size.

Politics being what it is, certain bidders sometimes obtain preferred or unequal access. Some overseas governments virtually require the use of well-connected agents or local partners who make a very limited operational contribution. Other overseas markets are simply not open to companies from some countries, either because of protectionist policies or political factors.

Military markets are unique in that secrecy provisions and, in particular, "need-to-know" requirements may make it extremely difficult to understand the specific application and performance specifications one is targeting. In these cases, the researcher may be forced to choose technical directions without accurate or even adequate market knowledge. New product developments often occur on a trial-and-error basis with a risk of technical or strategic failure that is much higher than in consumer and industrial products. Even more important, poor communications—because the technology provider has only a vague understanding of his customer's needs—may greatly prolong the time frame between concept and commercialization, thus penalizing the net present value of the reward and increasing the risk factor.

Once a company is established in a military project, however, and is behind the wall of secrecy, its position can be unusually secure and profitable. Its product attributes can be converted into military specifications that competitors must now meet, an enormous advantage referred to as being "spec-ed in." For example, an established military supplier may shrewdly manage to have specifications on strength and corrosion adopted by the military that can be met only with materials in which it holds a proprietary position. And it will have close working relationships with its governmental counterparts that are not accessible to competitors, who remain outside the need-to-know fence.

The skills and behaviors required to succeed in government marketing are very different from those required in the industrial and consumer sectors, and making the transition from one to the other is often difficult. With government, the time between concept and commercialization may be longer and the risks higher, but the rewards may be greater. There may also be opportunities to use government-subsidized R&D to target government needs, and then extend the technologies and products developed from those sources to the commercial sector. This strategy can be very attractive to small start-up companies. The reverse can also work in a company's favor: a proven industrial product such as a desktop computer can be redesigned and "hardened" to military specifications. Valuation methods can be useful in assessing the trade-offs between the available strategies.

Market Research

Market research falls into two categories: secondary and primary. Secondary research is available from trade associations and publications, from government sources, and from consultants. Multiclient studies, produced by consultants and paid for by subscribing companies, can be a high-quality source if purchased from expert vendors; however, competitors have access to essentially the same information. Primary research is derived from studies based on questionnaires, telephone surveys, customer interviews and visits, and focus groups. As might be expected, primary consumer market research can be expensive.

At first glance, good market research would appear to be an answer to many of the uncertainties surrounding new product development. Unfortunately, the strength of market research—both primary and secondary—is in measuring the *existing* market. Its weakness is in providing insights into the future. Customers simply cannot respond effectively to research questions about products with which they are unfamiliar. Market research can help companies and its R&D personnel find new ways to improve or create variations to existing products, but it is generally weak when investigating "new to the world" technologies and products. Technologists and product developers are often frustrated by market research's focus on the present. As Wayne Gretzky is reported to have said, the art of playing hockey is not to skate to where the puck is but to where the puck *will be*. Because in many fields 5 to 10 years of R&D are not unusual from product concept to initial commercialization, decision makers need all the insights they can get as to where that puck is going to be. And market research is often unable to provide those insights.

A classic example of the inability of market research to determine emerging markets can be found in the serious conflict between Thomas J. Watson, Sr., and his son Thomas J. Watson, Jr., over whether IBM should enter the computer business. At the time, only a few computers had been built, mostly for military purposes. Watson senior, the company founder and then-CEO, asked for market research on the need for these machines. A study was duly conducted. It foresaw worldwide demand for little more than a handful of computers. Given the enormous cost as well as limited reliability and performance of the early machines—and the few applications to which those machines were being directed—the research may have been quite on target. Though the senior Watson used the research to argue against any major initiatives in computers, his son ultimately prevailed, and that decision not only preserved the company but led to its best years.

One approach to breaking through the limitations of traditional market research is through the technique of *conjoint analysis*, a form of choice modeling.[6] Conjoint analysis is a relatively new data analysis technique used to determine how respondents rank their preferences for various product attributes, especially when those attributes are offered in different combinations. Traditional methods typically ask respondents to rate the importance of different features. For example, a hotel chain may ask respondents the following: "On a 1-to-10 scale, indicate the importance of each of the following features: price, parking, proximity to the airport, room size, room cleanliness, faxing service, etc." A potential hotel guest could easily assign a 9 or 10 to each of these attributes, providing no sense of how he or she would prioritize these features, given their relative costs. In this situation, conjoint analysis would help us see how respondents make trade-offs between the price of a room and proximity to the airport, and so forth.

Conjoint analysis begins with a list of attributes believed critical to the acceptance of the proposed product, and with an identification of the target population. Consider the following:

Example

A proposal to build an implantable medical device to supplement the pancreas had as its target markets Type I and Type II diabetics, who differed in the severity and nature of their diseases. Some of the factors believed to be critical to acceptance were degree of control of glucose levels, frequency of maintenance, safety, and price. Besides the two target populations, which differed substantially in size and level of interest in the device, it was recognized that both physicians and insurers would play a major role in shaping the future market. Hence, panels of both patients and physicians were used for the study. Participants were asked to assign quantitative values to their preference for the product overall, and for preferences for different levels of each critical attribute. The methodology was "forced choice"—each respondent had only so many chips to divide among the attributes. A sophisticated computer model then calculated a utility value for each attribute.

The output of the conjoint analysis identified those attributes that contributed most strongly and least strongly to overall preference. Since price was one attribute, a crude price elasticity curve (to be discussed) was one important result. Differences in responses between the two target market segments could also be identified, as well as rather interesting differences between the viewpoints of patients and their physicians. These data guided researchers toward a supposedly optimal product configuration and gave them important signals on the sensitivity of acceptance to each of the product attributes. It also raised the confidence of decision makers that certain issues would not be showstoppers.

Pricing

As discussed, pricing is a key attribute of a product, and of course, a key responsibility of marketing and general management.[7]

Pricing decisions in most industrial markets tend to be based on two principles which, to a degree, conflict with each other. Principle 1 is that you should never give up market share on price. In effect this means that if a competitor undercuts your price, you should always meet it. The basis of competition will then be some other value factor—performance, delivery, service, or terms. Conversely, "buying" market share through price is seldom a viable strategy. This widespread form of competitive behavior is based on the assumption that competition will follow the same principle and creates the secondary consequence that lost market share is extremely expensive to buy back—it must be won on factors other than price.

Principle 2 is that a reasonable gross margin (selling price minus direct manufacturing cost) must be maintained for the business to be sound and sustainable. What represents a "reasonable" gross margin will vary with the business, but it is always well understood, through experience and intuition, by those in it. It can also be calculated from DCF business models, as shown in Chapters 9 and 10.

Pricing Strategies

Cost-Plus Pricing

The simplest pricing strategy is cost-plus. Here, the producer figures the costs and adds in enough to obtain a desired level of profit. This has been a method of pricing for some regulated monopolies and some defense contractor agreements. However, in the real world, where supply and demand rule, cost-plus is merely an executive's dream.

Reinvestment Pricing

Some industrial firms promote the concept of *reinvestment pricing*—a current market price that will support the full costs of a new plant built at the prevailing cost of money. Reinvestment pricing, they claim, ensures that new capacity will be added as sales and the market grow. This is wishful thinking and may not be supportable in the marketplace, particularly if lower priced substitute products are available, or if a competitor has a lower cost structure. Reinvestment pricing, in fact, runs against the grain of another business axiom, "Old (depreciated) plant will beat new plant every time." It is the nature of cyclic markets that when capacity runs short, pricing reaches reinvestment levels and beyond, but can later cycle to ruinous levels where only the low-cost producer has positive cash flow because capacity is in excess. The reasons are obvious—when the reinvestment price level for a product has been achieved, any producer can earn an economic profit and will be inclined to reinvest if the price seems sustainable. If the market price then reaches a level where several producers find investment attractive, overcapacity will inevitably result. At the bottom of the cycle, if cash costs for most producers are negative, many will shut down plants. Prices will start to rise, and the plants will come back on stream, most likely led by those with the lowest cash costs.

A counter strategy is to keep prices just below the reinvestment level to ensure that competitors do not build new plants under one's own pricing umbrella. An "umbrella" may be created by a prevailing market price set by the pricing leader in the industry. If a new entrant can make an attractive return at that level, and if the pricing leader does not react (a big "if"), the new entrant's profits will be protected by the leader's umbrella. Smart competitors leave no umbrellas. Dow Chemical produced a polymer called polyol from propylene oxide and sold *both* products to industrial customers. The end use was primarily urethane foam. Dow's basic pricing strategy was to keep the price difference between the urethanes and propylene oxide at just a level where no one would be tempted to buy propylene oxide at market price to self-produce polyol.

The margins needed by producers vary greatly, depending on product type. In specialty businesses, where the typical salesperson produces only a few hundred thousand dollars per year of sales, products must be sold at very high gross margins to offset high selling and customer service expenses. Commodity products, on the other hand, generally sell at low margins, and producers focus on *contribution* to total profits in pricing decisions.

Price Skimming

Skimming is yet another pricing strategy and is used widely by technology-based companies as they introduce new products. The price skimmer sets the initial price high and "skims" off the market segment that is eager enough for the new product to be relatively indifferent to price. Once that generally small segment has been satisfied, the producer lowers the price, increases production, and rolls the product out to a new and larger market segment.

Examples of price skimming are readily seen in consumer electronics and computer markets when the products in question are unique, new to the world, and without clear competition. In some cases, it is the only feasible strategy since the innovator company has not developed the plant capacity to move directly to larger market segments. And for companies that have spent millions developing the new products, it is a great way to quickly recoup the R&D outlays. Ultimately, however, price skimming invites competition and gives would-be rivals time to design and roll out their own products. The strategic choices for managers are:

1. Slowly work through the high-tier segments, lowering the price as you go.
2. Begin with a low price that will attract the broad market.

Strategy 1 will maximize profits in the short run, and give the company time to develop both the market for its innovation, product variations suitable for other market segments, and the experience it needs to reduce production costs significantly. The risk is that competitors will rush to join the party and capture market share that would otherwise go to the product innovator.

Strategy 2 may give the innovator a larger and perhaps enduring command of market share. This, in effect, is how Microsoft became the dominant purveyor of PC operating systems. Margins will be smaller, but greater volume may result in larger total profits. Competitors will be less attracted to the business at the lower margins. The risk is that the larger market will never develop, and the company could be stuck with plant overcapacity.

Racing down the Experience Curve

Closely related to skimming is the strategy of the *experience (or learning) curve*. The associated pricing concept is "pricing ahead on the experience curve," and was made famous by the Boston Consulting Group (BCG) in the early 1980s. Today that pricing strategy remains controversial, while the experience curve concept itself is widely accepted.

The experience curve concept is derived from the empirical observation that the more a firm produces of a product, the more it learns; and the more it learns, the more it is able to reduce cost through improved product design, improved production processes, and increased labor skill.[8] The empirical relationship between experience and cost turns out to be logarithmic and industry dependent. For example, in one industry, one may find that if cumulative volume produced is doubled, costs are reduced by 15%. In another industry, doubling cumulative production may reduce unit costs 40%. In technological terms, these observed gains might well come from cumulative R&D, increasing economies of scale, and purchasing power.

The strategic implication of the experience curve is that the first producer to jump in and start producing and learning will have a permanent cost advantage over latecomers. The BCG concept was that one should not base price on current costs, but on future costs as predicted on the experience curve. The benefits were the ability to gain market share on price, and to discourage competitor reinvestment by dropping their returns below the cost of capital. One could expect to recoup the initial costs through increasing market share and move down the experience curve even faster, leading to an unassailable position. Current margins would suffer, but there would be a pot of gold at the end of the rainbow.

The strategy was closely associated with an initially successful effort by several Japanese firms to capture the global DRAM memory chip market. They achieved their market share objective. But not all the consequences were foreseen. DRAMs became commodities, prices plummeted, and considerable value was captured by customers. But DRAMs, unlike microprocessors, never became as profitable as had been forecast. Later, competitors in Korea and elsewhere in Asia (often emulating the Japanese) entered the DRAM market with the advantage of much lower labor costs and built large plants with excellent economics of scale. Their entry served to maintain competition and to keep pricing low. The experience curve failed to protect the Japanese from this outcome and their pot of gold has yet to materialize.

An analysis of destructive pricing and the experience curve for the commodity plastic polyvinyl chloride has been detailed by Peter

Spitz.[9] It is a fine case study involving pricing dynamics, back integration and economics of scale.

While pricing strategy makes interesting table talk for managers, the effective arbiter of success is the marketplace in which the Law of Supply and Demand holds sway.

Price Elasticity

Economists use the term *price elasticity of demand* to describe the effect of price changes on product demand, or sales. It is a fundamental principle of economics that the lower one's price the more units of a product one will sell, and vice versa. (There may be an exception in superprestigious goods, such as Swiss watches.) So companies would expect to sell more units at a lower price, and sell less at a higher price. The price elasticity of demand helps companies to estimate how demand will be affected by price changes using the parameter:

$$\frac{\text{Percent change in quantity demanded.}}{\text{Percent change in price}}$$

If demand drops by a smaller percentage than the percentage increase in price, then demand is said to be "inelastic" (i.e., not highly responsive to price change). In the short term, some goods, such as gasoline and other forms of energy, are highly inelastic. People will continue to buy almost as much in the face of price increases. In the long term, however, they will find substitutes or install energy-saving methods that will reduce their demand for energy at the higher price.

Most goods, particularly nonessential luxury goods, are highly price-elastic. Raise the price of chicken by 25%, for example, and demand for chicken will drop like a rock as people turn to substitutes (pork, fish, beef, and vegetarian items).

An understanding of the price elasticity of demand can lead to a market entry strategy in which one first targets high-priced low-volume markets in which price is relatively inelastic. This "skimming" strategy requires a minimum initial capital investment and delays the need to develop economies of scale. As discussed in Chapter 4, this approach can lower the break-even point and initial risk. For example, pesticides introduced initially on high-value products such as those grown via horticulture in a greenhouse can stand a higher selling price than those acceptable in medium-value crops such as fruits and vegetables. These in turn have higher unit value than row crops such as corn and cotton. So a market entry strategy would be to initiate the business by selling at a premium price in horticulture, then move to medium-price crops such as fruits and vegetables, and ultimately address the row crops, which will demand very competitive pricing.

Delivering Value

Businesses exist to add value: to raw materials through manufacturing; to the customer's experience through services; and so forth. Researchers add value to the extent that they improve performance or lower costs through new or enhanced products, processes, and services. Managers find it useful to determine the value added in every step of the business process *from the viewpoint of the customer,* the person who ultimately pays the bill. Indeed, to understand the customer, the value question must often be extended to the viewpoint of the *customer's customer,* and so on down the chain. To the researcher, this extended value question produces important answers.

Typically, applied research adds value by increasing the performance of an existing product. Also quite typically, it costs more to produce the new and improved product. (If it costs less, the situation is a no-brainer.) Just how that added value must be shared with customers is an important aspect of marketing that researchers should understand, as shown in the following:

> *Example*
>
> A new plastic produces film that is 50% tougher by some accepted physical measurement, but costs 2 cents a pound more to produce. Assume the current grade of plastic is priced at 48 cents per pound and costs 32 cents a pound to produce, and no new capital is required to make the new grade. The researcher determines that customers can reduce their use of plastic by one-third by "down-gauging" from 3 mils to 2 mils and obtain film of equivalent toughness. Therefore, he believes that resin can be priced for as much as 72 cents per pound. He calculates the value-added of his

> contribution, as 24 cents a pound (72¢ – 48¢) less 2 cents of additional production cost or 22 cents. Unfortunately for the innovator, it is highly unlikely that his firm will be able to capture the entire added value. The customer will not change to the new plastic unless he can realize some benefit for himself.

In practice, added value is always allocated, either explicitly or implicitly, between innovator and customer. In the preceding example, it might work out as 11 cents each. If the customer has to share value with his customer, the innovator's share may be even less. In many cases, precedents guide the sharing formula. In the agricultural chemicals operations of W.R. Grace, the supplier could expect to retain only one-quarter to one-third of the per acre benefit that might accrue to a farmer from increased crop yield.

To appreciate why value added gets shared, look at the value proposition from the customer's viewpoint:

> The customer must be sure that the new product will run on his machinery at speeds and yields at least equal to the past. This will require costly trials. He will have the problem of working old film out of inventory and may have to sell some at a discount. Most importantly, he must be sure *his* customer will accept the down-gauged film and ascertain whether she will demand a lower price for it (she will try and this will take time). He must therefore attempt to analyze the value proposition from the viewpoint of his own customer—hence the concept of a value chain. In the end, the customer will weigh his share of the value against these potential costs, and determine whether his net gain offsets the perceived risks.

Capital costs must be part of value analysis. An innovation that renders part of a customer's existing plant obsolete has to offer cost-savings or performance enhancements at a much higher level than an innovation that fits seamlessly into a customer's existing process. The customer will also resist putting in incremental capital to adopt the innovation. Part of research's job is to try to design improvements that represent "drop-in" products at the customer level.

But capital considerations need not be entirely negative:

> If the new down-gauged product reduces the volume of waste, the customer may be able to buy a smaller waste incinerator. If a 50

> thousand foot roll takes up less space, expenditures on warehousing may be deferred or eliminated. And if the customer is about to invest in an entirely new plant, he may want to ensure that it is based on state-of-the-art technology to avoid costly retrofits.

Thus, knowing your customer's capital situation and strategies is an important part of value chain analysis.

Distribution Channels and R&D

An important early decision with regard to a new product is whether it will be sold through an existing direct sales force, a new dedicated sales force, or through a distribution network. Each of these channels, and there are many variations, has its own advantages and disadvantages.

A product sold through a direct sales force employed and controlled by the company has the advantage of being able to roll out a new product very quickly. As a servant of the company, this direct sales force can also be enlisted to do a lot of market development, even when a new product is in its final stages of development. And as a direct conduit between customer and company, it is a source of valuable market information, customer complaints and suggestions for improvement, and so forth. The potential disadvantage is the need to reeducate the sales force and to convince its members that the new product is an important improvement and not a distraction from the bread-and-butter products that now produce their commissions and bonuses. A product innovation is a major event and requires an extensive investment in training—both of the sales force and of the customers. Under these circumstances, incremental innovations that require extensive customer training may be something to avoid. Perhaps a new product should be introduced every three or five years, often enough to convince the customer that your firm is adding value through constantly improving technology, but not every few months. On the other hand, in areas where technology is fast moving, the salespeople may need a flow of new concepts to maintain customer interest and actually create suction for new ideas and innovative products.

The distinction between these two extremes is very much in the province of marketing and is extremely important to commercial outcomes.

Products that are new to the world have special problems: customers may not understand them or they may be used in unanticipated ways. New technologies, new software, new materials, new instruments, all require education to succeed. It is always best to avoid this problem by designing products that are simple to use (in today's phrase, "user-friendly"). Put another way, the KISS axiom, "Keep it simple, stupid," is a fine guide. A one-component adhesive system is always better than a two-component system, and three may be too much for customer acceptance no matter how well the final product performs. If the option to simplify is not realistic, then one may need a dedicated sales force.

The use of a dedicated sales force for a new product implies an entirely different strategy. Because the dedicated sales force has nothing else to sell, the producer is assured that their full attention will be given to the new product. In some cases, the impact of having a dedicated sales staff is impressive. Consider Saturn automobiles. When Saturn launched its first car models, it did not sell them through the dealership network of its parent company, General Motors. Those dealers were used to traditional selling methods, and incented to push bigger cars with more options. Haggling over price was part of their game. Understanding that its small, fewer frills cars would get little attention from the GM dealer network, Saturn set up its own, dedicated network of dealerships, educated its salespeople intensely about its small number of models, eliminated the traditional commission sales system, and went to a no-haggling pricing format. The result was good sales and the highest level of customer satisfaction with the dealer system.

For technical products, a dedicated sales staff generally requires extensive training and technical support. Creating it is expensive, time-consuming, and risky. In many ways, like R&D, its creation is an intangible capital investment.

The third option is to sell through a distribution network of distributors, agents, manufacturers' representatives, and other independent sellers. This reduces administrative overhead for the producer, but also reduces margins and separates it from the customer. Distributors cannot be relied on for giving good customer feedback.

Disposal of waste products and other environmental issues are also affected by channels of distribution. Sales forces can arrange for the return or disposal of used containers, whereas distributors may not. Manufacturers of toner cartridges sold through distribution now include mailers for return of the packages for recycling. This general subject is called *product stewardship* and is discussed further in Chapter 15.

Market Share

Market share is the percentage of total sales captured by one or another companies competing in the same market. For example, Intel's Pentium microprocessors had close to a 90% share of the new personal computer market in the late 1990s. In the U.S. auto market, Japanese producers collectively have a market share of about 30%. As an objective measure, market share is often used to assess the competitiveness of individual firms and their products, and how that competitiveness is changing over time.

Though the competitive instincts of business managers drive them to seek greater market share, there is no theoretical benefit in being the market share leader other than simple bragging rights. This is because the determination of market share does not consider profitability. Your company, for example, could surely increase its market share if those pesky financial people would allow the marketing folk to price your products at below the cost of production! You would go out of business just as surely.

While there is no theoretical benefit in being a market share leader as long as profitability is not part of the measurement, experience indicates that market leaders nevertheless enjoy certain benefits. These include:

- Economies of scale.
- The benefits of the experience curve.
- Higher returns on assets.
- The spread of certain expenses, such as advertising, over a larger number of units.

These advantages, in turn, allow the number one player to control pricing and use it to increase its lead over the competition.

By definition, not everyone can be Number 1. So what are firms operating in markets dominated by one or two firms to do? One strategy is to get out of their unfavorable positions and seek out markets or segments in which they can be Number 1. In other words, find a pond in which they can be the biggest fish. Market segmentation can offer a path, since a company can be Number 1 in some market segments and territories while being Number 2 or 3 overall. Market share of "passenger cars and light trucks" is not the same as market share for pickups or sports utility vehicles. Superb performance in a market segment can be the beachhead for attacking other segments.

Whether one is a major or minor player, strategic focus on dominating and protecting a technical field or market niche, large or small, can drive value. Bristol-Myers holds a powerful position within the segment of anticancer drugs. As such, it can be expected to have superior skills in testing and identifying promising new drug candidates in the cancer field. It will have an extensive network of contacts with the oncologists that shape the market and with the academic researchers who are defining the evolution of new cancer therapies. Other players are well entrenched in much more specialized niches. They may lack the technology resources of the pharmaceutical giants, but can compete with recognized brand names and focus on protecting that niche.

Technological Performance and Product Life Cycle

Over the years, analysts have observed that the technical performance of an innovation follows an *S-curve* when plotted against time. At first, performance is minimal and makes slow progress because the technology is poorly understood, and funding may be difficult to obtain. This has been referred to as the incubation stage. Gradually, however, progress is achieved and credibility grows, and advances in technical performance begin to accelerate. More technical effort is made, and the innovation begins to attract funding. Competitors set goals involving annual percentage gains over past performance (as we are seeing today in microprocessor speed). As this happens, performance gains become exponential.

Eventually, progress begins to level off as the law of diminishing returns sets in. Finally, the technology begins to approach a *performance limit* as technological opportunities become more difficult to identify, and gains become increasingly expensive. R&D investment in further advances also begins to drop off.

The technological S-curve is not a universal phenomenon, but an empirical observation supported by numerous case histories. Knowing the general pattern is helpful, but to assess financial impact, one must look beyond the general pattern to the specifics of how long a particular technology may remain in incubation, how rapidly it will advance once it is established, and where it will find its limits (see Exhibit 8.4).

Some authors[10] prefer to draw S-curves for performance versus effort (instead of time). Thus, effort may be a surrogate for cumulative R&D spending. In these cases, the front and back ends of the S-curve, where annual R&D effort is at a lower level, will shorten.

Technological Discontinuities

Technologies that reach their limits often set off a search for new technologies that can break through those limits. Hence, when propeller-driven commercial airliners advanced to a level of performance (speed) beyond which further advances were not cost-effective, aviation engineers and scientists did the work that led to turboprops, and then to jet propulsion. The transitions between these technologies represented technological *discontinuities.* We can observe a similar set of discontinuities in many varied fields, including the following:[11]

- *Word processing.* From mechanical typewriters to electric typewriters to computer-based word processors.
- *Imaging.* From daguerreotype to glass plate photography to celluloid roll film to digital imaging.

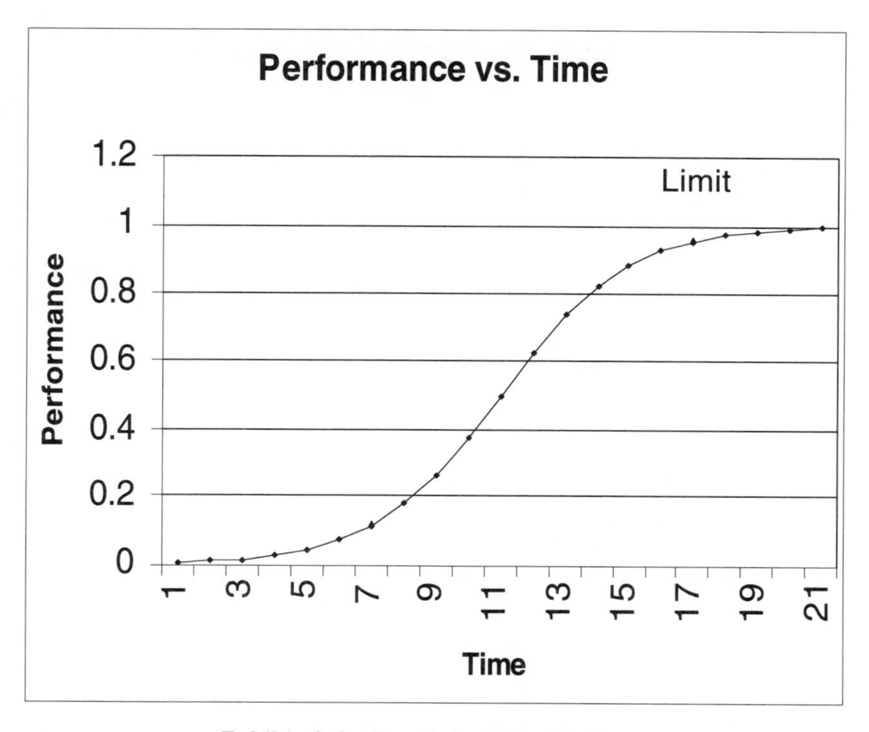

Exhibit 8.4 The Technological S-Curve

- *Refrigeration.* From harvested ice to machine-made ice to electromechanical refrigeration.
- *Electronics.* Vacuum tubes to transistors to microprocessors.

Companies that find themselves trapped in obsolete technologies are in very serious trouble (the business implications are discussed in this chapter). One way to avoid this predicament is to maintain research on technologies that afford an opportunity to change the basis of competition. As one technology starts to approach its mature phase, it is time to get on a new S-curve before it is too late. Translated to project terms, the concept is to at all times have a *pacing project*. This is corporate policy at 3M, one of the most consistently innovative U.S. companies.[12] Discontinuities must be anticipated and managed.

Product Life Cycle

Product life cycles are closely linked with technology S-curves, and generally follow their own S-curves. The two, however, are not the same. First, product life cycles are plotted in terms of sales volume or revenues on the X-axis, not performance. Second, technology performance capabilities do not decline in the terminal stage, but product sales often do. The four typical stages of the product life cycle, and their characteristics are:

I. Incubation Stage
   Search for credibility.
   Unpredictable duration.
   Industry-specific characteristics.

II. Rapid Growth
   Growth at expense of older products and technologies; faster than end-use market.
   Opportunity for geographic expansion.
   Required net investment.

III. Mature Growth
   Growth at some multiple of GNP (e.g., plastics at 2X).
   Not applicable to short product cycles.

IV. Maturity
   Gradual decline.
   Harvesting of overheads previously built to support growth, including R&D.

The incubation stage exists between the point of conceptualization and the time of full commercial introduction. This can be quite long in some industries and extremely short in others. And it can be quite predictable in some industries such as the auto industry, and unpredictable in others such as computers. In the rapid growth stage, the product's sales grow exponentially, often at the expense of other products and technologies. Geographic expansion may help fuel this growth. This stage of the product life cycle often requires a tremendous amount of staff support (overhead) and new investment to sustain it (Exhibit 8.5). Without that support, the hoped for growth rates won't be obtained, and future profits will be more limited.

The "mature growth" stage can last quite a long time. Here, revenue growth is maintained, but at decreasing rates. Growth is often described in relation to the economy as a whole, for example at "2 times the rate of GNP." Revenues may decline, slowly or precipitously, as the product enters the mature stage. Precipitous revenue decline is likely if competitors are attacking from the comfort of a new technological S-curve: as examples, electronic typewriters savaged their manual competitors; the sales of dot matrix computer printers went into a deep dive as laser and ink jet printers entered the market. The only hope for profitability for a product stuck in the mature stage is to reduce overheads, such as market development and engineering.

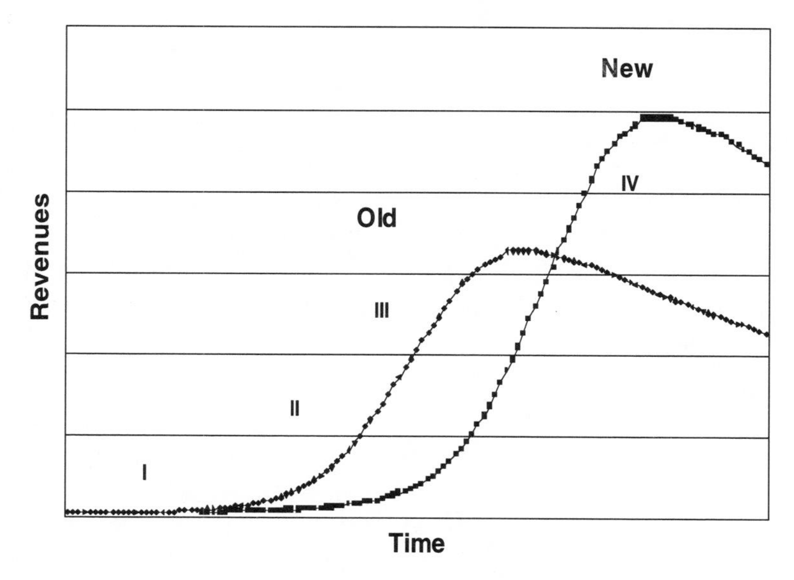

Exhibit 8.5 Product Life Cycles

These were growth drivers in an earlier stage, but are no longer required and simply burden the now-mature business. The byword becomes to reduce costs.

Richard Foster has made a powerful point in describing the advantages of the attacker in this situation when the attacker is in the rapid growth stage of his own product cycle. As the attacker grows, he has the advantages of increasing economies of scale, is rapidly gaining cumulative experience and reducing costs. This high growth rate should also translate into high PE ratios, increasing the ease of raising equity capital at lower rates. In the meantime, the defender sees a loss of market share and its attendant disadvantages; he will weaken his marketing and technical capabilities in the name of cost reduction, and the cost of his capital will begin to rise as his growth rate inexorably drops.[13]

Technology Forecasting

Many of us know from experience the severe forecasting limitations of conventional marketing research. Technology forecasting offers a different form of crystal ball that is highly relevant for researchers and is generally underutilized.[14] Absent discontinuities and given adequate data, technology is relatively easy to forecast, since the rate of progress is linked to the ongoing level of R&D efforts and the productivity of R&D. The latter slowly declines, but one can correct for that effect. Furthermore, the substitution of one product for another is closely linked with the substitution of one technology with another, and, absent new discontinuities, follows the pattern of the S-curve (see Exhibit 8.6).

The model for this substitution is called the Fisher-Pry equation,[15]

$$f = \tfrac{1}{2}\,\{1 + \tanh\,[\alpha(t - t_0)]\},$$

where f is the fraction of takeover in time t, α is ½ of the initial annual exponential takeover rate, and t_0 is the year in which $f = \tfrac{1}{2}$. Note we are now talking units of product, not technological performance—the stuff from which financial models can be built.

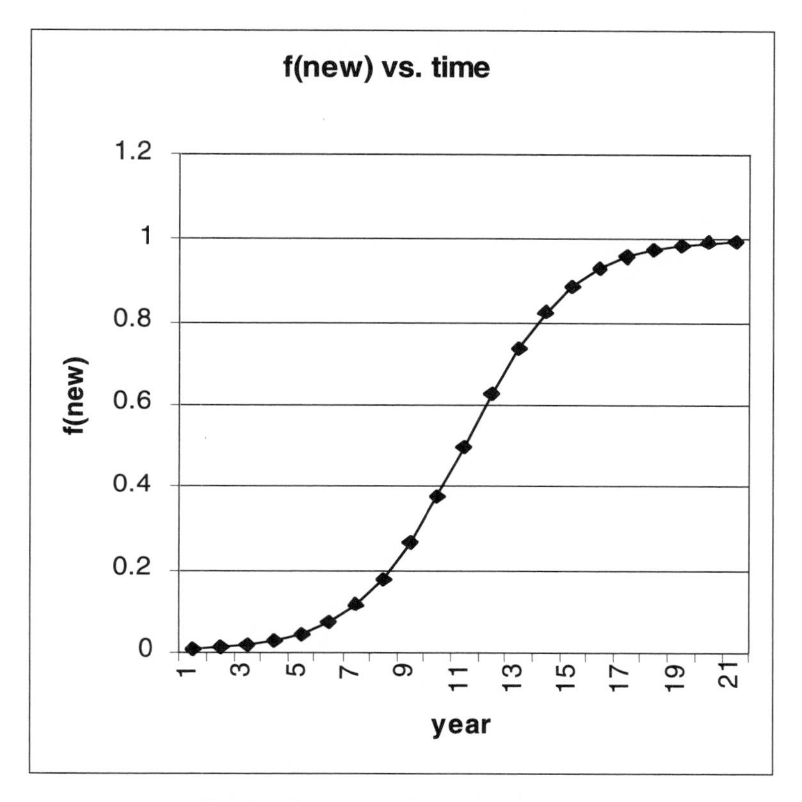

Exhibit 8.6 Ideal Substitution Curve

There is really no need to use the mathematical formula to predict the future. The Fisher-Pry model can be reduced to a very usable linear plot where the log of the ratio of new to old product is plotted on the Y-axis and time on the X-axis. For example, if steel beverage cans are 60% of the market, and aluminum is 40%, the ratio of the fraction of new {f(new) = 0.4} to the fraction of old {f(old) = 0.6} is 0.667. This plot can predict quite accurately when 50% (or any other fraction) of the market will be converted to the new technology, and therefore in what time frame R&D must be completed for meaningful participation in the substitution.

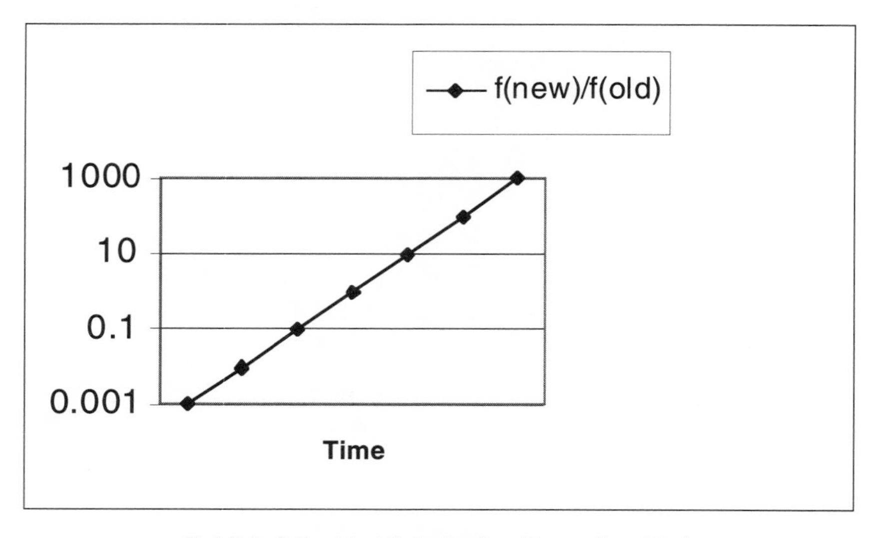

Exhibit 8.7 Ideal Substitution Curve (Log Plot)

Substitutions, however, assume that the older technologies are truly disadvantaged, and not temporarily impaired by current market conditions (Exhibit 8.7). Examples would include the substitution of steam locomotives by diesel, steel beverage cans by aluminum, and vinyl disks by audiotape cassettes. Predictability would be less sure for coal by gas, gasoline engines by electric, and so on.

Creating a Revenue Model

We now return to the core purpose of this chapter—to create a revenue model for the purposes of valuing technology. We have thoroughly discussed how to go about it. If the target market is familiar, the marketing function can define the goals. If it is less familiar, market research tools can be employed. And if we are in the fuzzy arena of radical innovation and technological discontinuity, a combination of technology forecasting and market segmentation may provide predictive power. Regardless of the methods used, the following steps can be used to build a base case revenue model:

1. *Estimate target revenues in a medium-term time horizon from time of commercial launch.* The time frame selected should be consistent with your firm's internal planning horizon. In our example, we use 5 years. This is the most critical step. The revenue estimate should meet two criteria: first, it must be doable at the customer level. If at all possible, identify individual customers who can adopt the product within 5 years from time of launch. The marketing organization should look at industry experience in rolling out comparable products as a guide to what might happen.
2. *The target must be doable at the organizational level.* Sufficient market development and support must be budgeted to ensure reaching the target. This exercise will create the all-important medium-term trajectory for product rollout.
3. *Look for natural limits to the size of the market or your own market share.* The market for shoe adhesives is limited by the number of shoes sold annually. This is, of course, a moving target. Competitor reactions or the desire of the marketplace for multiple sources of supply may cap your market share.
4. *Determine the nature of the end game.* Some products have natural life spans. These have been estimated as low as 2 years for software, 3–4 years for automobiles and computers, 6–7 years for chemicals and aircraft, 10 years for pharmaceuticals and biotech products, 12 years for communications systems, and over 20 years for timber products and transportation systems.[16] However, the devil is in the details: product lives of patented drugs can end suddenly, whereas other products can grow indefinitely as a result of continuous improvement, or newly discovered uses. In the former case, the long-term growth rate should be estimated based on industry experience. If the latter is expected to be the case, the R&D budget underlying continuous improvement should be included in the financials. If not, the business model should not be charged for R&D targeted at next-generation products.
5. *Assume that growth rates will decline, both because you will begin to saturate the available market and because your initial technical advantage will erode.* This is the S-curve at work.

Consider this example, which makes the following assumptions:

- Projected sales in year 5 of $48 million.
- A long-term target of 30% of a $360 million market growing at twice GDP. If GDP growth is 2.5% per annum, market growth will be 5.0% (the mature growth level is 5%). This market is estimated at $360 million in year 1 and therefore $960 million in year 20.
- Assume with continuous improvement the product life cycle is long, and growth will continue at 5% indefinitely. (We now know it makes little difference whether indefinite means 10 years past the horizon year, 20 years, or 100. Such an assumption might be valid for a new plastic such as polycarbonate or a new packaging format such as a PET bottle.)
- We assume that in the rapid growth phase, growth in revenues slows at a constant rate of 12% per annum over the forecast period, until the 5% per annum mature growth level is reached (see Exhibit 8.8).

This pattern of growth is generally consistent with the historical experience of W.R. Grace's research division over a period of 30 years.

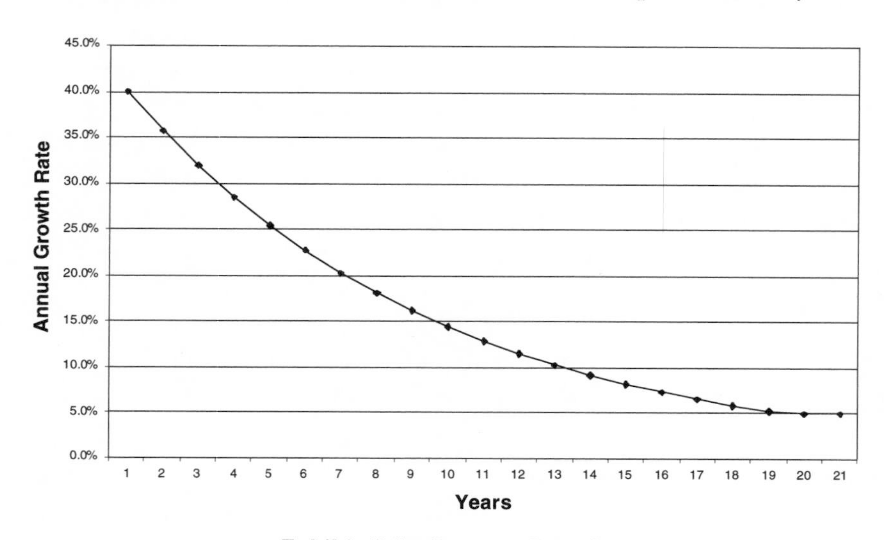

Exhibit 8.8 Revenue Growth

That unit showed an average growth rate of approximately 25% per year over the first 10 years of product life (new products), and a 7% per annum growth rate thereafter (established products).

The results of these assumptions[17] are charted in terms of revenue dollars as Exhibits 8.8 and 8.9 respectively. We use these results in the next chapter.

The most surprising feature of this growth model is the surprising linearity of the absolute growth per year in dollars. This is a result of rapid growth *rates* off a smaller base in the early years, and lower growth off a large base in the later years, and arguably is an artifact of our model. However, it is paralleled by real-world experience with long-lived products.

Anecdotally speaking, most new product lines in the chemicals, plastics, packaging, and health care businesses have continued to grow seemingly indefinitely once established. This would appear to speak both for the persistence and ingenuity of those associated with the business in achieving continual improvement, and the durability of customer needs. Of course, there have been exceptions. Grace's business in bead format catalysts for catalytic converters on automobiles was eliminated by a superior technology, ceramic monoliths, following a classic substitution curve. However, management was astute

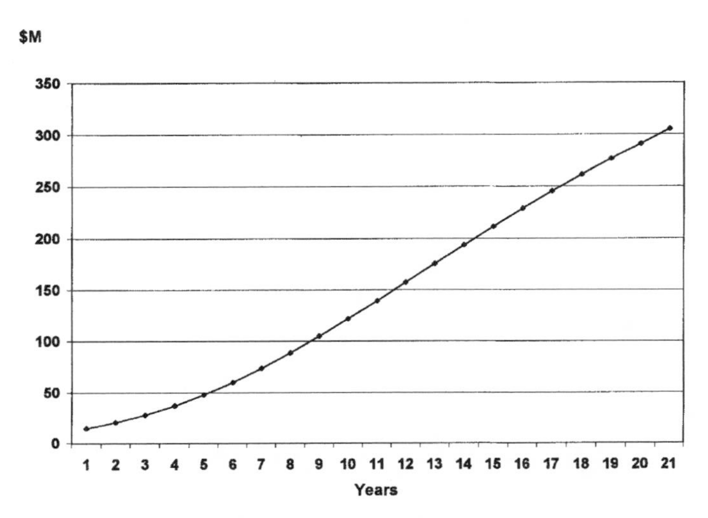

Exhibit 8.9 Revenue Projection

enough to achieve an adequate gross margin so that the project added value over its product life cycle. Airframes appear to be remarkably durable. Boeing's 707 and 727 have been discontinued, but the 747 has lived on for decades in a variety of updated reincarnations.

Summary

This chapter has introduced key marketing concepts through which the top line of a business plan can be developed, and has identified issues that researchers must consider. The results are used to generate a projection of revenues (using unit sales volume, price, and gross margin). However, marketing assumptions also impact other costs through decisions on channels of distribution and technical support philosophy (which may show up as factory overheads, R&D support, or selling costs). They also affect capital requirements and depreciation as choices are made to begin cautiously with a niche or skimming strategy (using existing plant or toll manufacturers wherever possible), or boldly with a broad-based product launch taking full advantage of economies of scale. If the decision is yet to be made, these and other alternatives can be tested using pro forma business models described in Chapter 9.

CHAPTER 9

Building a Pro Forma DCF Model

This chapter is the heart of the book. Previous chapters have built up a basic understanding of the financial, strategic, and marketing information that scientists and engineers need to create a pro forma model for a technical project. A pro forma model is one that incorporates forecasted or estimated values for a future activity within the same form used to represent current or past values, such as a balance sheet or income statement. Subsequent chapters use the pro forma model to introduce more complex valuation techniques—sensitivity analysis, decision trees, Monte Carlo methods, options and portfolio dynamics, and techniques for quick and convenient project screening.

Here we go through the mechanics of building a moderately detailed pro forma spreadsheet. Finally, we introduce a case in which valuation using pro forma models helped guide a series of R&D decisions in a way that would not have been possible using nonquantitative methods.

Purposes of the Model

Before beginning any pro forma exercise, we should ask, "What are we trying to find?" The answer may affect *how* we conduct the pro forma. If the purpose is *internal decision support*, that purpose will be satisfied if:

- The project can be determined to be either financially sound or unsound.
- Its financial attractiveness can be ranked against other projects.

This approach, which we shall use in our example, fits well in ongoing businesses, especially if all projects being prioritized have comparable assumptions—the most common situation in industry. It warrants *conservative* assumptions (i.e., 90% probability), since management may rely on the projections being met. If the project is strategically attractive, and is being proposed for other than financial reasons, it may still be helpful to run the financials to determine the minimum circumstances under which the project is also financially attractive.

However, if the purpose of valuation is a *financial transaction* (e.g., the purchase, sale, or licensing of technology), or if it is the decision to pursue a long shot, more aggressive assumptions, perhaps representing a 50% probability of achievement, may be more appropriate. As discussed in Chapter 12, a disproportionate share of value is likely to reside[1] in upside cases and it is important to include them in the calculation for a transaction. Decisions about the downside cases, those that produce little value, are less important.

Getting the Assumptions Right

Assumptions are at the heart of any pro forma model, and if these aren't reasonable, little else matters. The reason is in part mathematical. Long time frames, for example, magnify errors that originate in our assumptions. Just a 2% error in the assumption of annual price changes will have enormous effects on the profitability of a project over 10 years.

The creator of the model must be prepared to defend his or her assumptions. If the assumptions differ significantly from those of other experts, the creator's credibility will be suspect. It is, in fact, wise to have the major components of the model, including assumptions, *evaluated and calculated by recognized experts.* This point is essential if the model is to be used for significant resource allocation decisions (i.e., for supporting a major R&D program or capital expenditure).

Of course, the creation of a credible model is time-consuming and costly, and enlisting busy experts to check out its elements and assumptions may be premature. "Quick-and-dirty models," described in the next chapter, can be useful tools for early screening of projects, and for categorizing them as nonstarters, marginal, promising, or likely winners. However, the simplifying assumptions used in quick-and-dirty models make them inappropriate for high-level decision making.

Who, then, are the recognized experts? To whom should we turn when we seek a sanity check on our models and their assumptions?

Projected sales or revenues should be estimated if at all possible by people with field marketing experience. The marketing staff of the division that will commercialize your product should be able to estimate volume, pricing, and growth for a product based on their past experience (to the degree that the project fits their past experience).

If the product is new to the company, market research by internal staff or consultants may be required. This will be costly, but so are decisions based on inadequate information and false assumptions. Under some circumstances, cost-based pricing may be appropriate. For example, if the products in a business traditionally command gross margins of 50%, a price that is twice the "cost of goods sold" may be reasonable. If a better price can be achieved owing to perceived value to the customer, it will represent an upside case.

When we get to the cost line, the engineering staff of the company can usually supply numbers in which we can have a high degree of confidence. Engineers can calculate raw material costs, utilities, direct labor, and the other technical components of cost. They can also estimate capital requirements. Good capital estimates allow us to estimate depreciation expense and to determine the project's return on investment. In the early stages of a project, that capital estimate is likely to be done from a process flow diagram; but in the end, an increasingly greater degree of confidence will be obtained when detailed engineering has been completed and actual construction bids are received. Preliminary capital estimates typically have a reliability of 25% to 35%; this improves in stages to 10% to 15% as detailed engineering is completed and as actual bids are received from contractors and vendors.

Engineering calculations are usually critical to a project's prospects in the planning stage. Management needs to be certain that the engineers are not overly conservative in their cost estimates since many good projects are killed by high capital estimates. At the same time, underestimating capital costs will diminish a project's credibility if each successive level of engineering detail creates a cost escalation.

Finally, one cannot overemphasize the sensitivity of any income model to the *relative* rates at which costs and prices escalate. This can happen inadvertently if marketing is asked for the price projections and manufacturing or purchasing is asked to supply cost projections on raw materials, labor, utilities, and so on. In this case, just asking the experts may not work. Each set of experts is working with reasonable assumptions about the future, but if they have not coordinated their projections, their assumptions about the future will not be consistent, and the differences between them will grow in each successive year. Keeping gross margins—the spread between revenues and cost—in a reasonable range, based on business history and the existence of competition, is a good way to avoid this pitfall. Conversely, gross margins that are escalating or diminishing with time *for no intended reason* may indicate a planning error of this type.

Finally, pro forma analysis must be done in a way that is consistent with corporate accounting principles and corporate culture. The financial community within the company expects serious analysis to be conducted in the language it understands and respects. This language includes net present value, cost of capital, cash flows, and internal rates of return. If the research department's presentation of its case and its estimate of profitability are not articulated within the framework of financial language, it will surely (even if unjustly) be scorned.

Building a Pro Forma Model

We are now ready to begin construction of a pro forma model for a new technical project. The purpose is to calculate the net present value of the project. This significant undertaking is much easier with the use of spreadsheet software, such as Lotus 123 and Excel. Since the "cells" of these spreadsheets are dynamically linked, any change in an individual cell or key parameter, such as gross margin or growth rate, is automatically reflected on the bottom line. Once the basic model has been built, a thousand refinements can readily be explored. Endless "what if's" are possible.

The following example is designed so that the calculations are transparent to the reader and reasonably simple. A more complete and realistic model would include many additional adjustments (such as site-related costs) and more sophisticated accounting (such as using different depreciation schedules for different classes of capital items). These refinements, however, would needlessly complicate this demonstration.

> Researchers at Polymers & Materials, Inc. (PMI) have invented and patented "polyarothene," a new plastic that can scavenge oxygen from gases and liquids with which it is in contact. It appears processable and usable in the interior of a variety of packages where the shelf life of the contents is adversely affected by oxygen. However, considerable development will be required to create a process to manufacture it economically and to teach customers how to apply it to containers and films. Initial contact has been made with a beer company that plans to line beverage can lids with this material and create a new brand stressing assured freshness.

Since values in the pro forma model are represented in dollars, should we use current dollars or constant dollars to factor out inflation? My preference is to use current dollars, for these reasons:

- Inflation is already factored into the cost of money.
- Inflation has many other effects besides causing prices and costs to rise.
- Most corporate plans are written in current dollars.

The R&D investment by PMI that must precede the commercialization of polyarothene is not yet included in the financials (we look at those costs in Chapter 11). Nor are any financial assets or charges (e.g., interest) included since we are attempting to value the polyarothene project, not the company in which it is embedded.

Our spreadsheet is divided into three sections, each representing one of the three financial statements used by an operating entity: the income statement, the balance sheet, and an analysis of cash flow. These are linked, and those links and their implications will be discussed in detail.

The Income Statement

Our first task is to develop a pro forma income statement. Here we must estimate sales revenues going out a number of years, and subtract from these our estimates for cost of sales, factory overhead, depreciation, taxes, and so forth until we arrive at each year's estimated net income.

The top line of the income statement—revenues—drives just about everything else, and so it is important to get the best possible estimate for annual revenues. Generally, the revenue line is based on a marketing study that estimates sales volume in units (say pounds of polyarothene) and unit prices (say $1.00 per lb.) as a year-by-year buildup. We skip this step for polyarothene to simplify our demonstration. If several different products were included in the business, each set of volumes, selling prices, and revenues would be shown in separate columns, and then totaled.

Alternatively, management might simply set a revenue objective: The CEO of an automobile company might declare "We *will* sell 100,000 minivans at $17,000 per van in year 5!" and the organization would then set out to define the resources required to achieve that target.

The terms "net sales" and "net revenues" are frequently encountered. "Net" usually means net of returns and net of freight paid by the supplier. The distinction between sales and revenues is that revenues may include royalties, rents, services, and sources of income other than sales. For example, PMI may receive revenues from polyarothene sales in the United States and Europe and royalties from a Japanese licensee. In the polyarothene example (Exhibit 9.1), management has agreed to a target of $48 million in sales by year 5, as discussed in Chapter 8.

Year	Revenues ($M)	Year	Revenues ($M)
1	$ 14.9	11	$139.4
2	20.9	12	157.4
3	28.3	13	175.5
4	37.4	14	193.5
5	48.0	15	211.2
6	60.2	16	228.5
7	73.9	17	245.2
8	88.8	18	261.2
9	104.9	19	276.5
10	121.9	20	290.8
		21	305.4

Exhibit 9.1 Revenue Projection for Plyarothene

With these revenue estimates in hand, we can begin to build our spreadsheet, beginning with Exhibit 9.2, the Income Statement for the polyarothene business. The parameters determining gross margin, selling costs, taxes and other factors are stated explicitly in the boxes. In addition to Columns (1) and (2), the Income Statement consists of Columns (3) through (15) discussed in the following sections.

The formulas used for other calculations are detailed in Exhibit 9.3.

Cost of Sales (3) and Gross Margin (4)

Gross margin represents revenues less the *cost of goods sold,* sometimes called confusingly "cost of sales." Typically, the costs of goods sold includes raw materials, utilities, and operating labor. These costs are sometimes referred to as direct manufacturing cost (DMC). Gross margin attempts to capture variable costs at the plant level, that is, it includes in costs only direct inputs into the manufacture of a product.

Gross margin may vary from as little as 20% of revenues in commodity businesses with few nonmanufacturing costs, to 85% or more in businesses that are heavily sales or promotion dependent, such as luxury perfumes, and may reach the vicinity of 95% for proprietary drugs where huge development costs must be amortized in the selling

Year	(1)	(2)	(3)	(4)	(5)	(6)	(7)	(8)	(9)	(10)	(11)	(12)	(13)	(14)	(15)
1	0.400	14.9	7.5	7.5	1.2	1.0	5.2	1.5	0.7	0.3	2.7	3.7	1.1	**1.6**	10.8%
2	0.357	20.9	10.4	10.4	1.7	1.5	7.3	2.1	1.0	0.4	3.8	5.2	1.5	**2.3**	10.8
3	0.319	28.3	14.2	14.2	2.3	2.0	9.9	2.8	1.4	0.6	5.1	7.1	2.0	**3.1**	10.8
4	0.285	37.4	18.7	18.7	3.0	2.6	13.1	3.7	1.9	0.7	6.7	9.3	2.7	**4.0**	10.8
5	0.254	**48.0**	24.0	24.0	3.8	3.4	16.8	4.8	2.4	1.0	8.6	12.0	3.5	**5.2**	10.8
6	0.227	60.2	30.1	30.1	4.8	4.2	21.1	6.0	3.0	1.2	10.8	15.1	4.3	**6.5**	10.8
7	0.203	73.9	36.9	36.9	5.9	5.2	25.9	7.4	3.7	1.5	13.3	18.5	5.3	**8.0**	10.8
8	0.181	88.8	44.4	44.4	7.1	6.2	31.1	8.9	4.4	1.8	16.0	22.2	6.4	**9.6**	10.8
9	0.162	104.9	52.5	52.5	8.4	7.3	36.7	10.5	5.2	2.1	18.9	26.2	7.6	**11.3**	10.8
10	0.144	121.9	60.9	60.9	9.7	8.5	42.7	12.2	6.1	2.4	21.9	30.5	8.8	**13.2**	10.8
11	0.129	139.4	69.7	69.7	11.2	8.7	49.8	13.9	7.0	2.8	26.1	34.9	10.5	**15.7**	11.2
12	0.115	157.4	78.7	78.7	12.6	9.6	56.5	15.7	7.9	3.1	29.8	39.3	11.9	**17.9**	11.4
13	0.103	175.5	87.7	87.7	14.0	10.3	63.4	17.5	8.8	3.5	33.6	43.9	13.4	**20.1**	11.5
14	0.092	193.5	96.8	96.8	15.5	10.9	70.3	19.4	9.7	3.9	37.4	48.4	15.0	**22.5**	11.6
15	0.082	211.2	105.6	105.6	16.9	11.4	77.3	21.1	10.6	4.2	41.4	52.8	16.6	**24.8**	11.8
16	0.073	228.5	114.3	114.3	18.3	11.8	84.2	22.9	11.4	4.6	45.4	57.1	18.1	**27.2**	11.9
17	0.065	245.2	122.6	122.6	19.6	12.0	91.0	24.5	12.3	4.9	49.3	61.3	19.7	**29.6**	12.1
18	0.058	261.2	130.6	130.6	20.9	12.1	97.7	26.1	13.1	5.2	53.2	65.3	21.3	**31.9**	12.2
19	0.052	276.5	138.2	138.2	22.1	12.0	104.1	27.6	13.8	5.5	57.1	69.1	22.8	**34.3**	12.4
20	0.050	290.8	145.4	145.4	23.3	11.8	110.3	29.1	14.5	5.8	60.9	72.7	24.4	**36.5**	12.6
21	0.050	305.4	152.7	152.7	24.4	11.6	116.6	30.5	15.3	6.1	64.7	76.3	25.9	**38.8**	12.7

Key Parameters:
Gross Margin = 50% of Sales
Factory Overhead = 8% of Sales
Gross Fixed Assets = 70% of Sales
Depreciation 10-year straight line
Selling Cost = 10% of Sales
G&A = 5% of Sales
R&D = 2% of Sales
Tax = 40% of Pretax Income

Column Head Key:
(1) = Growth Rate
(2) = Revenues
(3) = Cost of Goods Sold
(4) = Gross Margin
(5) = Factory Overhead
(6) = Depreciation
(7) = Gross Profit
(8) = Selling Cost
(9) = G&A
(10) = R&D
(11) = EBIT/Pretax Profit
(12) = EBITDA
(13) = Taxes
(14) = After-Tax Net Income
(15) = As % of Sales

Exhibit 9.2 Pro Forma Income Statement for Polyarothene Business ($M)

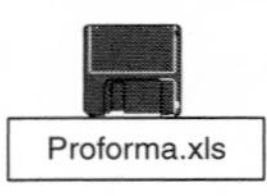

price. In our example of polyarothene, we have chosen to use a percentage gross margin of 50%, which implies a 50% cost of sales.

The concept of gross margin is useful in assessing incremental profitability, since, assuming that all other costs are fixed, an incremental sales dollar will return an amount equal to the gross margin to the bottom line. Also, if our gross margin is lower than that of competitors, our manufacturing costs may be out of line, in which case raw material costs, labor rates, and so on must be addressed to achieve cost-competitiveness. An adverse trend in gross margin is a signal that competition is gaining on us, even if we are maintaining market share (Exhibit 9.3).

Income Statement
  Cost of Sales = Sales × (100 – GM%)
  Gross Margin = Sales × (GM%)
  Factory Overhead = Sales × FO%
  Depreciation: 10-year Straight Line on Capital Expenditures (column 27)
  Gross Profit = col(4) – col(5) – col(6)
  Selling Cost = Sales × %S
  G&A = Sales × G&A%
  R&D = Sales × R&D%
  Pretax Profit = col(7) – col(8) – col(9) – col(10)
  Taxes = Average Tax Rate × Pretax Profit
  After-Tax Net Income = col(11) – col(13)
  Net Income/Sales % = col(14)÷ col(2)
Balance Sheet
  Inventories = Sales × Inv% (days/365)
  Accounts Receivable = Sales × AR% (days/365)
  Accounts Payable = Sales × AP% (days/365)
  Working Capital = col(15) + col(16) – col(17)
  Gross Fixed Assets: Cumulative Sum of Capital Expenditures {col(27)}
  Accumulated Depreciation: Cumulative Depreciation {col(26)}
  Net Fixed Assets = col(20) – col(21)
  Total Capital Employed = col(19) + col(22)
  % Rtn on TCE = col(14) ÷ col(23)
Cash Flow
  Capital Expenditures = (Current Year Sales – Previous Year Sales) × FA%, where FA% represents gross fixed assets as % of sales
  Incremental Working Capital = Current Year Working Capital – Previous Year Working Capital {see col(18)}
  Operating Cash = col(24) + col(25) – col(26) – col(27) = Free cash flow
  Operating Cash Including Terminal Value: Assume business is sold for cash after horizon year at appropriate multiple of operating cash using perpetuity formula.

Exhibit 9.3 Formulas for Pro Forma Financial Statements

Factory Overhead (5)

Factory overhead generally consists of all the fixed costs associated with running a manufacturing facility: supervision, plant engineering, plant level accounting functions, quality control, property taxes, and so on. Our example has used a factory overhead (FO) equal to 8% of sales. This is a fairly generous number, and reflects that a new product may need extra attention on the factory floor. Within a given industry, factory overhead as a percentage of sales is a good measure of operational efficiency, and is generally lower in larger and more modern plants.

In the short run, factory overheads are regarded as fixed costs, since one would not adjust them upward or downward because of a short-term fluctuation in orders. In the longer term, however, requirements for machinery, space, and other supposedly "fixed" overheads also grow, although one would expect them to grow more slowly than sales growth. Ours is a long-run model, but we have not built in any such productivity increases.

DEPRECIATION (6)

Depreciation is a noncash expense that reduces the value of long-term assets due to use or obsolescence. This value links directly to the fixed assets on the balance sheet (columns 20–22). In practice, different classes of items (buildings, computers, equipment, vehicles, etc.) are depreciated according to different schedules, reflecting their expected useful lives. These numbers are available from plant accounting, and should be used in preparing capital authorization documents. This level of detail is premature for research stage decision support.

Our example assumes that the entire plant is on a 10-year depreciation schedule, typical for heavy plant equipment. As fixed assets are added each year, additional depreciation is added for each of the next 10 years. After the 10th year, no further depreciation is taken for the first year's investment (for the mechanics of depreciation, see Exhibit 9.4).

Two points about depreciation are worth noting. First, unlike the other expense items shown in our model, depreciation is a noncash expense, hence the business will throw off cash equal to its net income *plus* depreciation. Second, tax authorities often permit a faster schedule of depreciation than what is shown here. This practice, called "accelerated depreciation," reduces income taxes as long as the "tax rate" of depreciation is higher than the "book rate," and thus improves short-term cash flow. As discussed in Chapter 3, management practice is to report earnings using the book rate and pay taxes according to the tax rate while creating a liability for the taxes that have been deferred. As long as a company continues a healthy rate of capital expenditure, this deferred tax liability will grow with it. In principle, the cash flow from deferred taxes could be added to our model, and would improve returns slightly. For simplicity, however, we ignore it here. Depreciation is an important link between the income statement, balance sheet, and cash flow statement.

GROSS PROFIT (7)

Gross profit is simply gross margin less factory overhead and depreciation. This is the plant-level profit with fixed costs included. However, there is more overhead in an operating business than occurs at the plant level, and we next consider these items to calculate *operating profit.*

SELLING COST (8), G&A (9), R&D (10)

Selling cost refers to the expenses of the sales force and would include such items as advertising and promotion. G&A is a term for *general and administration expenses.* These may arise at the business unit, division, or corporate level and should be identified appropriately. G&A typically includes the salaries of corporate officers, the general manager and staff, and legal and environmental expenses. These should be examined carefully since some corporate costs (e.g., large legal and environmental liabilities from unrelated operations) should not be allocated against individual businesses. This point is important in valuation, since if the project, viewed as a minienterprise, were sold, these headquarters charges would not go with it. Some corporate reports combine *selling* and general and administrative expense into a single item labeled SG&A when headquarters overheads are hard to explain!

Research and development in this particular example is taken to be the R&D expense required to support this product and the manufacturing processes that produce it. This expense item should not include additional R&D expenditures aimed at creating new products and product lines for the future. However, many companies do allocate such costs to existing products on the assumption that renewal is just a cost of business. Although this is a convenient way to account for the many projects that fail, it can then be double counting to charge each project or business for its up-front R&D expenses. This matter of business philosophy may significantly affect employee behavior and, on the margin, go/no go decisions. From the point of view of project valuation, I favor not charging renewal costs here.

EBIT/Pretax Profit (11)

This number is the gross profit less allocations for selling expense, G&A, and R&D. The main distinction between the terms operating profit and pretax profit is corporate interest payments, which arises from corporate debt. At the project level, they are one and the same. If one is evaluating a business or a project, one need not include interest—it is part of the cost of capital. However, one-time start-up expenses are sometimes taken at the pretax profit level to make operating profits look healthier to analysts.

Pretax profit before interest is also called EBIT (Earnings before Interest and Tax). Again, at the project level, it is the same thing.

EBITDA (12)

Like EBIT, EBITDA (Earnings before Interest, Taxes, Depreciation, and Amortization) is a term largely used by financial analysts and investment bankers for rule-of-thumb valuations based on the pretax cash thrown off by a business. It is not the same as free cash flow, because both the capital required to sustain and grow the business, and the taxes, are excluded. However, it is all important in valuing a leveraged buyout (LBO), since taxes will be largely offset by interest payments, and short-term growth will be deferred until profitability is restored (it seldom makes financial sense to grow a business that is not earning the cost of capital).

Taxes (13)

Taxes in our example are assumed to be 40%. Most of it is the federal corporate income tax. Actual rates will vary from year to year according to patterns of investment, state and foreign tax rates, write-offs, and many other circumstances. These year-to-year variations are detrimental to evaluating projects and cash flow; an average tax rate blessed by the CFO and reflecting only the taxes attributable to EBIT should be used for valuing technology.

After-Tax Net Income (14)

The after-tax net income is the *bottom line* for the polyarothene business. Getting to this number is the object of the income statement exercise.

Return on Sales (15)

The next and final column (15) in our pro forma income statement indicates net income as a percentage of sales. This is the project's profit margin or *return on sales* (ROS)—one of the better measures of the relative profitability of companies *within* an industry. It can usually be calculated from data included in the annual report. (Gross margin data is often a closely held commercial secret.) However, comparisons of income as a percentage of sales *across* industries are seldom useful and financial investors will look more to return on capital as a measure of profitability. Use ROS as a reality check as you construct your own pro forma, and look at how the competition performs with regard to this metric.

The Balance Sheet

An established business creates a balance sheet to give managers, investors, and regulators a point-in-time picture of the entity's assets and how they are financed with liabilities and shareholders' equity. We do something similar here for the technical project or new product line, creating a pro forma balance sheet that estimates assets at different times in the future. The goal here is to calculate the level of capital needed to support the revenues projected on the pro forma income statement. We are not concerned here with how those assets will be

Year	(16)	(17)	(18)	(19)	(20)	(21)	(22)	(23)	(24)
1	2.5	1.9	1.2	3.1	10.4	1.0	9.4	12.5	12.9%
2	3.5	2.6	1.7	4.3	14.6	2.5	12.1	16.5	13.7
3	4.7	3.5	2.4	5.9	19.8	4.5	15.3	21.2	14.4
4	6.2	4.7	3.1	7.8	26.2	7.1	19.1	26.8	15.0
5	8.0	6.0	4.0	10.0	33.6	10.5	23.1	33.1	15.6
6	10.0	7.5	5.0	12.5	42.1	14.7	27.5	40.0	16.3
7	12.3	9.2	6.2	15.4	51.7	19.8	31.9	47.2	16.9
8	14.8	11.1	7.4	18.5	62.2	26.1	36.1	54.6	17.6
9	17.5	13.1	8.7	21.9	73.4	33.4	40.0	61.9	18.3
10	20.3	15.2	10.2	25.4	85.3	41.9	43.4	68.7	19.1
11	23.2	17.4	11.6	29.0	97.6	50.7	46.9	76.0	20.6
12	26.2	19.7	13.1	32.8	110.2	60.2	50.0	82.8	21.6
13	29.2	21.9	14.6	36.6	122.8	70.5	52.3	88.9	22.7
14	32.3	24.2	16.1	40.3	135.5	81.4	54.0	94.3	23.8
15	35.2	26.4	17.6	44.0	147.9	92.9	55.0	99.0	25.1
16	38.1	28.6	19.0	47.6	160.0	104.7	55.3	102.9	26.4
17	40.9	30.7	20.4	51.1	171.7	116.7	55.0	106.1	27.9
18	43.5	32.7	21.8	54.4	182.9	128.7	54.1	108.6	29.4
19	46.1	34.6	23.0	57.6	193.5	140.7	52.8	110.4	31.0
20	48.5	36.4	24.2	60.6	203.6	152.6	51.0	111.6	32.7
21	50.9	38.2	25.4	63.6	213.8	164.2	49.6	113.2	34.3

Inventories–60 days of sales
Accounts Receivable–45 days
Accounts Payable–30 days

Column Head Key:
(16) = Inventories
(17) = Accounts Receivable
(18) = Accounts Payable
(19) = Working Capital
(20) = Gross Fixed Assets
(21) = Accumulated Depreciation
(22) = Net Fixed Assets
(23) = Total Capital Employed
(24) = % Return on TCE

Exhibit 9.4 Balance Sheet for Polyarothene Business ($M)

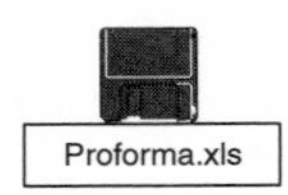

financed, and so there are no debt or equity sections in the pro forma balance sheet for polyarothene.

Exhibit 9.4 continues the pro forma model with a balance sheet for our example project. Columns 16 through 24 indicate the total capital employed in the project, according to standard balance sheet categories.

Like the pro forma income statement model presented earlier, this balance sheet is "hardwired" with formulas that define the assumed relationships between different categories. These are shown in the second tier of Exhibit 9.3.

INVENTORIES (16)

Inventories in a typical manufacturing business are between 30 and 60 days of sales. These include raw materials, work-in-progress inventory (intermediate products stored in warehouses, tanks, etc.), and finished products. Considerable variation can occur. A plant virtually dedicated to a single nearby customer (such as a can plant adjacent to a brewery) can minimize finished product inventories, while an export-oriented commodity business will require extra days of inventory owing to the time the inventory spends on ship and clearing ports. To be conservative, we have used 60 days of sales. The model spreadsheet is therefore based on Inv% = 60/365.

ACCOUNTS RECEIVABLE (17) AND ACCOUNTS PAYABLE (18)

Accounts receivable represents amounts owed by customers for completed transactions that remain unpaid. Typical payment terms are 30 days, but owing to late payers and modified terms, we have used 45 days of sales in the example (AR% = 45/365).

Accounts payable are monies owed by the project to its vendors (e.g., for raw materials, electricity, transportation services). A common policy is to pay bills within 30 days. (AP% = 30/365). Accrued payroll (wages earned by, but not yet paid to, employees may be included in this category).

WORKING CAPITAL (19)

Accounts receivable and accounts payable are, in effect, monies loaned to or borrowed from customers and vendors respectively. The term *working capital* is used to describe the sum of inventories and accounts receivable less accounts payable. As the business grows, so does its need for working capital. Working capital must be financed just as surely as additions to manufacturing capacity. If the business or project grows rapidly and at a moderate (but attractive) level of profitability, it is unlikely that this financing can be provided from operating cash flow; instead, the business must borrow money or raise capital by issuing stock. Fortunately, accounts receivable and inventories are generally excellent collateral, making it possible to obtain some financing from lenders on quite reasonable terms, thus mitigating the high cost of using equity capital for this purpose.

Gross Fixed Assets (20)

Gross fixed assets are the cost of plant, land, and equipment prior to depreciation. This number represents the cumulative investment in hard assets in the business. In the example, we have assumed the gross fixed assets are maintained continuously at 70% of annual sales (FA% = 0.7). This is an important choice, and has the benefit of simplifying the calculations.

Businesses are often characterized by their capital intensity. This is typically measured in terms of a ratio of annual sales to assets, or as a *turnover ratio*. There are several kinds[2] of turnover ratios, based on fixed assets, total assets, inventory, accounts receivable, and so on. These tend to be useful guides to the profitability of a business; for example, inventory turnover is an extremely important measure in operating a supermarket. More importantly, these ratios tend to be characteristic of a business. Ratios of fixed assets to sales (the reciprocal of a turnover ratio) vary between 0.4 and 2 in general chemical operations, with mixing and formulating operations at the low end, and petrochemicals at the high end. The term "light industry" is usually applied to industries with low ratios, and "heavy industry" to high ratios. Capital intensity is an enormously important parameter in establishing business profitability in manufacturing.

As discussed in Chapter 4, fixed assets are never added continuously, but stepwise, as a result of capacity expansion projects in the plant. The reason is simple: Plant equipment, such as production lines and reaction vessels, generally comes in standard sizes. The choice of an addition is determined by the time at which new capacity is needed and by the economics of scale achieved by adding larger units, which will be partially offset by likely overcapacity in the plant, when the unit comes on stream. Hence, our "smooth" ramp-up of capacity is unrealistic for an actual manufacturing situation, as discussed in Chapter 4 ("Matching Capacity to Demand"). In effect, we can be accused of improving our return on capital by eliminating idle plant capacity. However, this benefit can readily be eliminated by adding an additional capital charge for average unused "spare capacity." This simplification is probably better than guessing at the complex timing and site-specific issues involved in the actual capacity expansion decisions that will face the business in the years ahead.

Accumulated Depreciation (21)

Accumulated depreciation represents the sum of the annual depreciation taken against the gross fixed assets for every year of operation. For example, accumulated depreciation in year 3 represents the sum of the depreciation in years 1, 2, and 3.

Net Fixed Assets (22) and Total Capital Employed (23)

Net fixed assets are gross fixed assets minus accumulated depreciation, and represent the book value of the plant. Total capital employed (TCE) is the sum of working capital and net fixed assets. It is the book value of the business.

Return on TCE (24)

Return on capital, also known as *return on investment* (ROI) or *return on invested capital* (ROIC), is a key index of profitability in that it tells us how well a project and its managers are doing with the capital at their disposal. It is calculated by dividing net after-tax income by total capital employed.[3] This number is watched more carefully than ROS by investors since it represents the total profitability of the project or of the firm and can be compared with alternative investments in other companies. A low number (less than the industry average cost of capital) in this column would normally be viewed as a failing report card. It can also be a warning about future problems, since a business that is growing, but whose average return on TCE is dropping (but still acceptable), may not be earning the cost of capital on its newer capital projects.

But in the last analysis, return on TCE is an accounting measure, not a value measure. For example, an uneconomic return on new capital projects, which destroys value, would be obvious using DCF analysis. However, such projects might be approved in a company whose business philosophy is only to maintain an acceptable return on TCE in each of its businesses.

The financial statements for polyarothene reflect constantly increasing returns on TCE. This is typical in pro forma financial statements. But such accounting returns may create a false sense of security since they are solely the result of operating out of increasingly depreciated facilities as those facilities age and business growth slows. The real issue is not book return but whether the business is continuing to add value, which can best be determined by DCF methods. To quote a current text "Payback is a bad rule. Average return on book is probably worse. It ignores the opportunity cost of money and is not based on the cash flows of the project."[4]

The Cash Flow Statement

As used at the level of the company, the cash flow statement indicates the sources of cash to the company and the various uses to which that cash was put over a particular accounting period. The term cash flow itself refers to the amount of cash generated or used by a business or activity over that accounting period. Our pro forma cash flow statement—shown here as Exhibit 9.5 (Columns 25–30)—does the same, projecting what these values will be given certain assumptions. In the end, it nets out the cash generated and consumed, giving us an estimate of the actual cash generated by the project. We can then apply our discounted cash flow tools to these cash flow estimates to determine the net present value (NPV) and internal rate of return (IRR) of the project.

Year	(25)	(26)	(27)	(28)	(29)	(30)
1	1.6	1.0	10.4	3.1	(10.9)	(10.9)
2	2.3	1.5	4.2	1.2	(1.7)	(1.7)
3	3.1	2.0	5.2	1.6	(1.7)	(1.7)
4	4.0	2.6	6.3	1.9	(1.6)	(1.6)
5	5.2	3.4	7.4	2.2	(1.1)	(1.1)
6	6.5	4.2	8.5	2.5	(0.4)	(0.4)
7	8.0	5.2	9.6	2.8	0.7	0.7
8	9.6	6.2	10.5	3.1	2.2	2.2
9	11.3	7.3	11.3	3.3	4.1	4.1
10	13.2	8.5	11.9	3.5	6.3	6.3
11	15.7	8.7	12.3	3.7	8.4	8.4
12	17.9	9.6	12.6	3.7	11.1	11.1
13	20.1	10.3	12.7	3.8	14.0	14.0
14	22.5	10.9	12.6	3.8	17.0	17.0
15	24.8	11.4	12.4	3.7	20.1	20.1
16	27.2	11.8	12.1	3.6	23.3	23.3
17	29.6	12.0	11.7	3.5	26.4	26.4
18	31.9	12.1	11.2	3.3	29.5	29.5
19	34.3	12.0	10.7	3.2	32.4	32.4
20	36.5	11.8	10.1	3.0	35.3	35.3
21	38.8	11.6	10.2	3.0	37.2	558.7

Column Head Key:
(25) = After-Tax Net Income
(26) = Depreciation
(27) = Capital Expenditures
(28) = Increase in Working Capital
(29) = Operating Cash Flow
(30) = Cash Flow + Terminal Value

Exhibit 9.5 Cash Flow Statement for Pro Forma Polyarothene Project ($M)

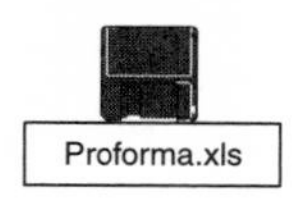

Like other pro forma statements presented earlier, this one is driven by embedded formulas that define the assumed relationships between different categories (see Exhibit 9.3).

After-Tax Net Income (25) and Depreciation (26)

Our figures for after-tax net income and depreciation are picked up directly from the pro forma income statement (Exhibit 9.2). Depreciation, of course, is not a cash flow item and must be added back to net income for cash flow analysis.

Capital Expenditures (27) and Increase in Working Capital (28)

In a profitable operation, after-tax net income and depreciation are positive numbers and create positive cash flow. However, the business will need to make capital expenditures to maintain existing facilities, which are depreciating in a real as well as a financial sense. These capital expenditures represent a use of cash. In a growing business, additional capital expenditures are required for capacity increases. Similarly, in a growing business, increased working capital is needed to support a larger sales base, and these annual increases represent a consumption of cash, often a significant one.

Operating Cash Flow (29)

The operating cash flow, or *free cash flow* (FCF), available to the owners of the business is the sum of operating cash flow from net income and depreciation, less funds required for capital expenditures and increased working capital.[5] When this number is negative, the project requires outside financing. If the number is positive, the profits can be returned to the corporate treasury and used for other purposes.

Free Cash Flow plus Terminal Value (30)

Column 30 is identical to Column 29 except for the addition of terminal value in the horizon year, in this case, year 21. Terminal value is an estimate of the amount for which the business or project, including its various assets, could be sold or liquidated. Chapter 5 explained some of the methods used for determining terminal value. Here we used the growth-in-perpetuity formula, assuming 5% long-term growth, to calculate terminal value.

We used the same data in Chapter 5 to calculate a net present value (based on a 12% discount rate) of $25.2M without terminal value, and $76.9 million with it (see Exhibit 5.6). The IRR is 26.0%, including terminal value. This superior result occurs even though operating cash flow is negative for 6 years! Why? Year 7 is the year when growth drops from 22.7% to 20.3%. This represents the point at which cash flow finally turns positive for this set of business parameters. Although prior earnings growth has been outstanding, the cash thrown off before then has been more than absorbed by growing requirements for fixed and working capital. Positive cash flow would have been achieved earlier had growth slowed sooner.

A Case Study

The following real case involves the use of pro forma models. Some details have been simplified for purposes of demonstration.

Several years ago, I accepted a position as director of a research laboratory that had recently undertaken a major research program. The concept behind the program seemed sound: combine a strong raw material position in a class of specialty chemicals (hydantoins) and biotechnology (taking advantage of newly discovered enzymes to produce amino acids) to create a new product line. The target end use was in nutritional supplements for health care. This product line would not be new to the world, but would be new to the company. The organic chemicals business unit was solidly behind the program and the idea was technically sound, but required a large R&D investment on the process side of the business. The program also appeared to be strategically sound, since the only serious producers of amino acids were overseas companies, such as Ajinomoto and Kyowa Hakko, which used traditional fermentation processes.

The U.S. market was dominated by three major customers—Baxter, McGaw, and Abbott—and these appeared to be eager for a reliable domestic supplier.

We had excellent information about prices, quantities, and market growth, and knew (and respected) the competition. We had also enlisted a powerful partner—a leading biotechnology firm, Genentech—which at the time had an expressed strategic interest in exploiting biotechnology in the chemical industry.

My initial reaction to this situation was that the R&D and capital investment to produce a reasonably complete line of about 20 essential amino acids might be too costly. Also, the terms negotiated with the biotech partner seemed rich by chemical industry standards (although the royalties were reasonable by pharmaceutical standards). This was not good news for the internal champions or the partner, and raising objections on intuition alone had serious downsides. Enter pro forma analysis.

Pro Forma Results

Our excellent information on the market and raw materials, coupled with our engineers' ability to conceptualize the manufacturing process flows, gave us the ingredients we needed to conduct a realistic analysis. We knew very little yet about yields and reaction times, but we could still make calculations on cost and capital, assuming that we substantially met *targeted* levels. Once the spreadsheet was built, we could later calculate ranges on these numbers to determine what level of technical performance would be needed to meet the test of financial soundness.

The results of our pro forma analysis were very disturbing: the target markets did not have sufficient scale to support either the capital or R&D investment for the product line. In other words, the initial project was not financially attractive.

However, the game was not yet over. What if we could focus on those few amino acids that had the market potential to afford economics of scale? Two amino acids, lysine and phenylalanine, did in fact have this potential. Lysine enjoyed a large market as an animal feed nutrient. But it was rapidly becoming a commodity product causing us to reject it. Phenylalanine, on the other hand, was enjoying rapidly growing demand as an ingredient in the newly patented sweetener, aspartame. Its sole producer, Searle, had made aspartame into a billion-dollar business. With a superior cost position, our calculations showed that we could enter this market. However, there were two important drawbacks: (1) we would have to compete with fermentation processes, and (2) we would be dealing (at least initially) with a single customer—always a dangerous proposition. The perils of the business were clearly visible in the case of Genex, a current producer of phenylalanine that was being badly burned. Would the same happen to us?

DCF analysis of a pro forma of the phenylalanine project showed that our partnership arrangement was a business killer: more than 100% of the project's NPV went to the partner via royalties and benchmark payments. Neither party had considered this possibility when the agreement was first formulated. Obviously, we wouldn't invest a dime in the project under this condition, and our partner agreed to reduce royalties. (This episode alone demonstrates the power of pro forma analysis in licensing negotiations.) Later, the partner backed out, owing to its belief (probably correct) that it had better opportunities in pharmaceuticals than in specialty chemicals.

Even without the partner, our work on the phenylalanine process progressed to the point that we believed we could, assuming economies of scale, produce phenylalanine at half the cost of the fermentation processes used by the competitors. The situation was now reversed—the production economics were good, but the market extremely risky.

Endgame

There remained one other possibility—produce aspartame itself. The timing was reasonable: Nutrasweet's use patent was due to expire within four years, just the length of time we would need to build a plant and enter the market. Research had also found a new coupling enzyme capable of efficiently converting our phenylalanine, purchased aspartic acid, and methanol to aspartame. It would bypass any aspartame process patents held by Nutrasweet and provide further economic advantage.

Another benefit of the aspartame alternative was that it would free us from the perilous single-customer situation. We would be free to sell to any end-user of the artificial sweetener, including the two largest users of aspartame, Coca-Cola and Pepsi-Cola.

Based on this analysis, a decision was made to continue the research program *provided* that take-or-pay contracts could be negotiated with key customers, who were verbally expressing their need for a second source and their concerns about Nutrasweet's monopoly position.

In the end, however, no such contract could be negotiated. As much as they wanted a second and lower-cost source of aspartame, the major soft drink makers were too afraid of alienating their key supplier and putting their thriving lines of diet drinks at risk. After all, they were generally able to pass along any aspartame price increases to consumers. In other words, value chain analysis indicted there was no value at that time in a less costly source. The "reasonable assumption" that the world needed a cheaper source of aspartame seemed wrong, even if the indicated economic profit was huge.

Senior management made the decision to pull the plug on the project, although some senior executives saw value in keeping the *option* open to enter this market later. A portion of the project's expenditures was eventually recovered when the patented coupling enzyme was licensed for use in a medical application, an outcome never anticipated.

Summary

Pro forma models were of pivotal importance at several decision points in the case just described. In the absence of that analysis, the outcome might have been quite different. Lacking rigorous pro forma modeling, decisions might have turned on other, less objective, factors: qualitative arguments, organization power, and plain wishful thinking. Experience shows that decisions made on these bases are usually regretted later.

Technology, by itself, was never an important consideration in any of these decisions. By themselves, each of many successful process developments could have been taken as a signal to continue. But even the best technology would not have improved the unfavorable risk-reward ratio in the medical nutrition business, nor would it have made up for the fatally unfavorable terms of the partnership agreement. In this sense, technology development and financial analysis must travel together.

This chapter has united two extremely important topics developed earlier in this book: discounted cash flow analysis and financial statements. It has demonstrated how market research and informed estimates can be used to create a picture of a multiyear future for a given technological or product development, viewing it as a business-within-the-business. Using multiyear operating cash flows and the terminal value of this business, the tools of discounted cash flow and internal rate of return can be effectively applied. Together, these give decision makers a powerful grip on an otherwise confusing and uncertain situation.

CHAPTER 10

Shortcuts and Market-Based Approaches to Value

Let's take a step back and review where we are and where we are heading. Chapter 9 (on pro forma analysis) was heavy on accounting terminology, spreadsheets, and alternative ways of valuing a far-off future. For the financial analyst, these are all useful approaches to measuring the value of a project or new business. Scientists, engineers, or managers faced with proposing or justifying a project, or prioritizing many alternative projects, are not interested in creating a plan that will stand financial scrutiny at the top of the corporation. They need a simple set of tools. They might even appreciate a shortcut method that would make it possible to simply read net present values and internal rates of return right off a chart.

The scientist may also recognize one of the obvious problems about pro forma models: the hard numbers needed to build them are only available after a great deal of the research has already been done and the project is well down the road—a classic "chicken and egg" problem. It is difficult to start drawing a process flow diagram for a plant, based on early laboratory indications that suggest something new and exciting has been discovered. It is even more difficult to project prices for new products that offer performance characteristics never before encountered. And consulting with market research, engineering, and finance for expert advice in the early phase of a project may seem premature or naïve.

So why have we gone through the pro forma exercise first? Why didn't we start with something simpler? The reason is that the pro forma is a foundation. Having built it, we can then develop simpler models based on the same methodology and vocabulary. Make the *right* simplifying assumptions and the calculations become much easier. As always, there is no escaping the correlation of the quality of the results with the quality of the assumptions.

Sensitivity Analysis

Sensitivity analysis is a tool for anticipating how a change in one parameter of a complex system will affect the system as a whole. For example, sensitivity analysis helps answer a question like this: "What would happen to NPV if market conditions forced us to sell our new product at $50 per unit instead of the forecasted $60?"

A system typically responds very strongly to some parameters and not others. For example, a 20% change in anticipated product price, or in the amount or capital required, may result in a major change in anticipated net present value. On the other hand, an equal percentage change in factory overhead or R&D expense might have very little impact on NPV. We have already used the sensitivity concept in Chapter 6, where we showed how the value of a corporation can change as a function of return on assets and of growth rates.

In this chapter, we look at the sensitivity of NPV and IRR (as calculated from the pro forma data developed in Chapter 9) to changes in two critical parameters, gross margin and fixed assets as a percent of sales (see Exhibit 10.1).

In a well-constructed spreadsheet, sensitivity analysis is simple: change one key parameter and obtain a new spreadsheet and a new bottom line. The data is useful in another sense: when the system is basically linear over the chosen range, one can interpolate the effect of smaller changes. In other words, if a 20% decrease in manufacturing cost will increase profits by 30%, then a 10% decrease in costs will increase profits by about 15%. Sensitivity analysis and interpolation are the key tools in constructing "quick-and-dirty" models. Exhibit 10.2 graphs net present value for the same data. The linearity of

Net Present Value (NPV) (M$)

	Fixed Assets as Percent of Sales		
Gross Margin (%)	40%	70%	100%
35	$ 6.0	$(19.4)	$(44.9)
40	38.1	12.7	(12.8)
45	70.2	44.8	19.3
50	102.3	76.9	51.4
55	134.4	109.0	83.5
60	166.5	141.1	115.6

Internal Rate of Return (IRR) (%)

	Fixed Assets as Percent of Sales		
Gross Margin (%)	40%	70%	100%
35	14.0	6.7	NA
40	22.7	14.8	9.6
45	30.5	20.7	15.1
50	38.7	26.0	19.7
55	48.2	31.3	23.7
60	59.4	36.8	27.7

Exhibit 10.1 Sensitivity to Gross Margin and Capital Intensity

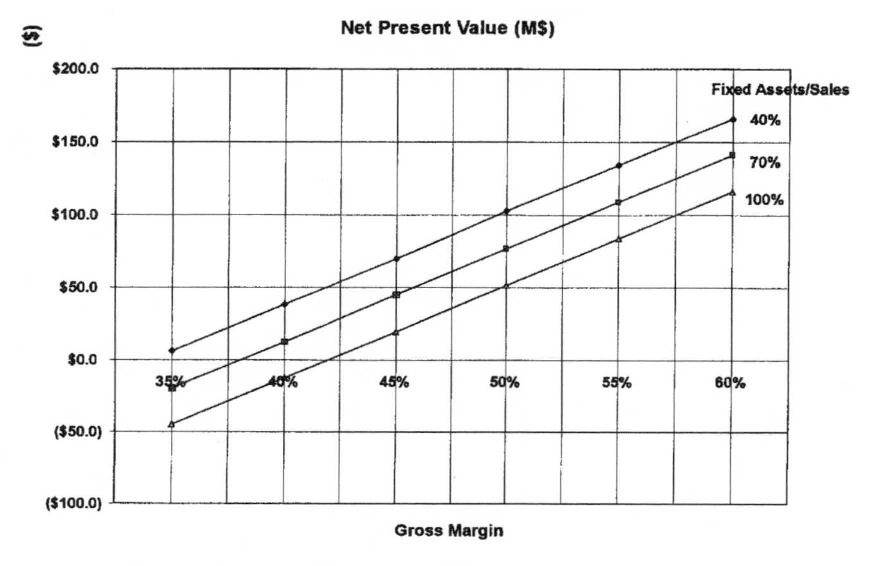

Exhibit 10.2 Sensitivity to Gross Margin and Capital Intensity

the data over a broad range is obvious. Here, gross margin is a surrogate for price, and fixed capital-to-sales is a proxy for capital intensity.

The overall range of values covered in the chart is very high, and some of the individual numbers, all upside cases, are very attractive. But things are not quite as good as they seem: the cost of the research program that will lead to these fine results has to be factored into the financials. So must the time discount for the deferral of the rewards. For example, if commercialization were to begin two years from now, these NPV values would have to be discounted by $1/(1.12)^2$. These important considerations will be addressed in Chapter 11.

It is also clear that there are some nonstarters in the upper-left-hand corner of Exhibit 10.1. A project requiring a fixed capital investment equivalent to annual sales cannot survive with a 35% gross margin. Exhibit 10.1 allows one to determine exactly the gross margin required to support a given level of capital investment—it is the point at which the curves cross the line for NPV = 0. If one wishes to use a hurdle rate for IRR of say 20%, the bottom tier of Exhibit 10.1 will allow us to pick those combinations of conditions that meet the 20% criterion. A project with a gross margin forecast of 40% and estimated capital intensity of 70% would earn the cost of capital but not meet this goal.

We explore some other key sensitivities in Exhibit 10.3; for example, the effect of a two-year delay and the corresponding effect of accelerating the project by two years. The sensitivity of the project to timing is quite significant. Other parameters, such as working capital and factory overhead have a lesser effect on project economics.

			NPV 12% ($M)	
Item	**Range**	**Low**	**Base**	**High**
Gross margin	± 5%	44.8	76.9	109.0
Fixed capital	± 30%	51.4	76.9	102.3
Price	± 10%	57.7	76.9	96.0
Timing	± 2 years	61.3	76.9	96.4
Sales volume	± 20%	61.5	76.9	92.2
Selling cost	± 5%	63.3	76.9	90.5
Factory overhead	± 3%	68.7	76.9	85.1
Working capital	± 30%	75.5	76.9	78.2

Exhibit 10.3 Key Sensitivities

The reader is invited to explore other business assumptions and scenarios, using the software and instructions packaged with this book.

Tornado Diagrams

Since each important parameter represents an uncertainty with some probabilistic range of outcomes, decision makers need to test each parameter to see which have the greatest potential impacts on NPV. Some use a "tornado chart" like the one shown in Exhibit 10.4, based on the polyarothene case, to make the sensitivities more visually clear. Each bar in this particular chart represents the estimated range of 90% of all probable outcomes for each uncertain parameter. The potential effect on NPV can be read directly.

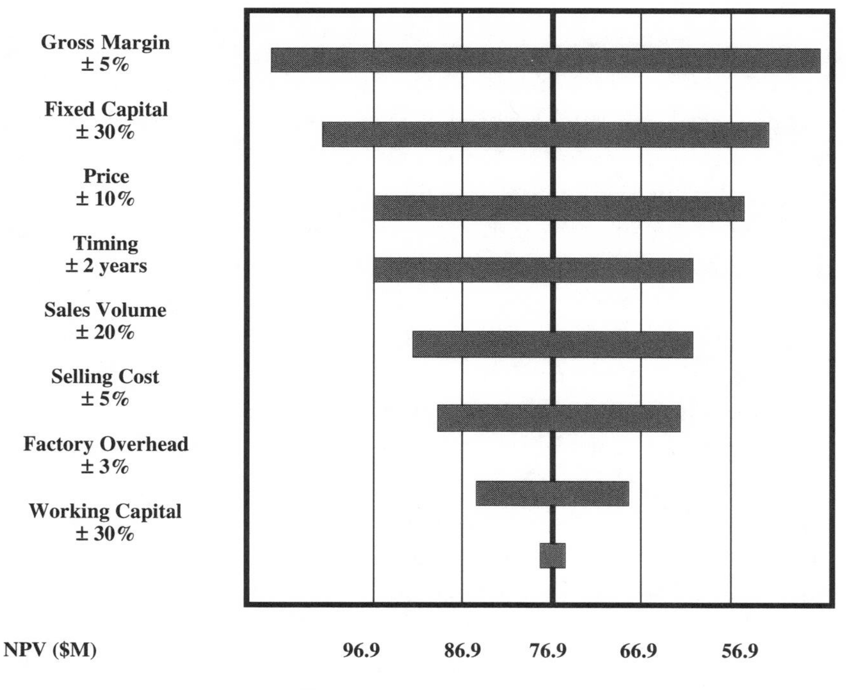

Exhibit 10.4 Tornado Diagram

By arraying these ranges in descending order (thus the tornado shape), decision makers can readily see where the potential big swings in NPV lurk. The parameters with the wide ranges can generally be narrowed if more study is given to them, thus reducing the range of uncertainty.

In the absence of sensitivity analysis—through a tornado chart or any other form—decision makers might easily fret over the things that won't have much impact on the eventual outcome while ignoring parameters that could have huge and possibly ruinous impacts on NPV. Taking the parameters in the tornado chart for polyarothene, for example, managers might be needlessly concerned with the timing of market entry or working capital, when gross margin and fixed capital investments clearly matter most in this case. Tornado diagrams prioritize the *value drivers* of a business.

Quick-and-Dirty Models

At this point, the scientist or engineer probably suspects what every experienced financial analyst already knows—that there isn't much point to developing highly detailed pro forma or sensitivity analysis in the very early stages of an R&D project. There simply isn't enough data, and what data exists is high uncertain. Even when plenty of good numbers can be had, the press of business rarely allows time for the detailed comparison of many different alternatives. In the early stages of a project, the object should be not financial precision but a quick-and-dirty assessment that identifies the real drivers of value and how changes in them can change the bottom line of our projects. A quick-and-dirty model puts the analyst in the right ballpark for the type of business proposal being analyzed and simplifies issues that are initially unimportant and that can be dealt with later. The power of simplification is that, given comparable methodology and assumptions, a group of alternative proposals can be ranked for financial attractiveness. The quick-and-dirty method I recommend is sensitivity graphs, such as shown in Exhibit 10.2, for NPV. (A similar chart can be constructed making IRR the Y-axis. Economic breakeven occurs where IRR = WACC, rather than where NPV = 0.) These graphs will show the set of circumstances under which large absolute rewards

(NPVs), or very attractive relative returns (IRRs) can be achieved. Use these four steps:

1. Assume a relationship of price to direct manufacturing cost. This is the gross margin percent.
2. Estimate the gross fixed capital required to support a given volume of sales. This represents the fixed asset percentage.
3. Interpolate on the graph to calculate NPV for the nominal sales target used to calculate the generic curves.
4. Scale NPV using the ratio of the actual sales target to the nominal target.

If quite different businesses are represented in the R&D portfolio, you have the option of creating a separate sensitivity graph for each business model. Each may need more or less working capital, have different selling costs and factory overheads, and so on. You may also wish to apply a different discount factor, if the imputed cost of money is different for the varied business lines.

We can now reexamine the key assumptions we find in most value models.

TIMING

The first critical assumption in building the quick-and-dirty model is the *timing* of sales revenues. Normally, sales start small and ramp-up over a period of months or years as a function of either market penetration or growing market size—or both. Any assumptions about timing should be based, if possible, on company and industry experience.

A useful tool for making a simplifying assumption about revenue timing is the concept of the *mature year,* the month or year in which the product substantially achieves full market penetration. Once full market penetration has been made, growth rates can be expected to resemble the long-term growth rate of the market.

Market penetration is generally rapid in industries accustomed to frequent technology upgrades, such as computers, software, and automobiles. Older models are quickly discontinued, and some or much of product sales may come from cannibalization of a company's own products. Having strong channels of distribution already in place facilitates market penetration.

Other new products and technologies require much longer periods before they achieve market penetration; for example, the development of a new transportation concept or a new process for controlling pollutants. The construction industry, constrained by building codes and liability concerns, is often slow to embrace new products. In these cases, the mature year may be 5 to 10 years beyond initial product introduction, with a slow takeoff.

But what is the trajectory of the sales prior to full market penetration? Here the modeler has several options. One option is to go with a linear model based on the mature year. That is, if the mature year is year 5, one might step up sales by 20% of mature year sales in each of the five years: 20% in the first year, 40% in the second, and so on. Once maturity is reached, sales growth would regress to the growth rate of the market.

A second option is to create a generic S-curve with high growth off a small initial base in the early years, and decreasing growth off a larger base in the later years—a scaled product life cycle. This is the type of model used for polyarothene (see column (2) of Exhibit 9.2). Specifically, we chose a growth rate of 40% between years 1 and 2 (usually quite doable), and reduced that growth *rate* by 12% per year until it converged into a long-term growth rate of 5%. (Note that the year 1 sales target of $14.9 was therefore implied by the year 5 sales target of $48 million.) With an S-curve, the mature year approaches gradually. A third approach is rooted in common sense—build whatever curve the business unit agrees is realistic.

In any case, however, it is wise to perform a sensitivity analysis on the mature year. This, after all, may be partly under business control. For example, if the conventional wisdom in the company is that full market penetration of a new herbicide takes 6 years, do a calculation for both four years and eight years. You are likely to discover that the four-year case is much more profitable, prompting a decision in favor of more aggressive promotion and marketing. The eight-year case may or may not prove financially acceptable. If unacceptable, technical management may be alerted to the vulnerability of the project, to, say, delays in the registration process.

In practice, its is helpful to create a family of sensitivity graphs varying the mature year. Shortening the mature year has a strong positive effect on results, while delay reduces NPV and thus restricts the range of gross margin and capital parameters that still give acceptable returns. After gross margin and capital, time is often the most sensitive parameter.

Market Size

If one is reasonably satisfied that the generic model for market penetration is credible, and if other business parameters such as capital intensity are identical, the target market size becomes trivial for purposes of quick-and-dirty calculation. The NPV of a business with a 5-year sales target of $24 million is exactly half of that for a target of $48 million.[1]

Overheads

Many of the overhead rates used in building our pro forma spreadsheet are characteristics of the business unit that will deploy the technology. These include factory overhead, selling and administrative expense, R&D, tax rates, and working capital per sales dollar. These numbers are known to management and can be easily obtained, although terminology will vary from one business to the next. Use the official numbers. If they seem too high or too low for the circumstances of your project, adjust them, but explain why. For example, if the new product or technology will require a dedicated sales force that does not currently exist, an upward adjustment in selling expense is probably warranted. A plant dedicated to a single customer site should have lower working capital requirements than those that serve many customers over a large region.

Gross Margin/Pricing

Pricing is an enormously important issue and is almost always the most sensitive business parameter—and a politically tricky one for technologists. Fortunately, this issue can be finessed at the early project stage if cost-based pricing is assumed and a preliminary estimate of direct manufacturing cost is available. As mentioned earlier, if one can justify moving up to value-based pricing later, so much the better!

Variable costs can often be estimated in a quick-and-dirty model from the raw materials cost (RMC), and the typical relationship between RMC and direct manufacturing cost (DMC) in the business unit, usually another characteristic parameter of the business.

The business unit will be accustomed to working within a fairly narrow range of gross margin. Its competition is likely to be in a similar range. If that gross margin is 50%, pricing should be 2 × DMC. If a project requires more aggressive pricing to be financially attractive, be prepared to justify that higher price in terms of performance characteristics. Sensitivity analysis can be performed at levels above and below the business's accustomed level of gross margin. If the business is comfortable with a 50% gross margin, returns at 45% gross margin may be financially acceptable; at 50% they will viewed as good, and they will be great at 55%.

Capital Intensity

Capital intensity relates to the ratio of capital dollars that must be invested to support a dollar of sales. We have seen that financial results can be very sensitive to capital intensity. In the chemical industry, this number can vary by a factor of up to five, which is why businesses are described as "capital intensive" or "less capital intensive."

As discussed in Chapter 9, the ratio of annual sales to total capital employed is one form of a *turnover ratio*. Low turnover ratios occur in the petrochemical industry, in pulp and paper, metals, and so on. If a $1 billion investment is required to generate $0.5 billion of annual sales, the turnover ratio is 0.5. Typically, such businesses are commodities and enjoy the benefit of low selling costs. On the other hand if only $400 million of capital is required to support $1 billion of sales, the turnover ratio would be 2.5. This range of capital intensity might be typical of businesses with high service content and high-selling expenses, such as water treatment. At the lowest extreme, supermarkets and other retailers compete largely by seeking the highest turnover possible and minimal accounts receivable and can reach ratios far below those characteristic of manufacturing businesses. They are not capital-intensive, but competition causes them to have very thin margins.

The classic quick-and-dirty method for determining the capital required by your project is to obtain an engineering estimate of the fixed capital (manufacturing plant and equipment) and add to it an estimate of working capital. The engineers will tell you they need data that is typically not available until the early development phase of the project: mass and heat balances, reaction kinetics, and rheological parameters. Happily, this issue, too, can be finessed. Capital intensity is a highly characteristic number of a business, and usually changes very slowly over time. It is readily available from the corporate financial staff, as both sales revenues and capital employed by the unit are key business parameters. Use this number for range-finding if an engineering-based calculation would be premature.

In the pro forma example, we used a ratio of gross fixed assets to sales of 0.7. The typical ratio for any target business can be obtained from internal financial data for similar businesses, or from the annual reports of competitors in that business, which usually list both revenues and gross fixed assets in their financial statements. Then, perform a sensitivity analysis varying the ratio by ± 25%–35% (the minimum range of cost uncertainty before serious engineering has begun) to cover an upside and a downside case, and look at the effect on returns.

Using Quick-and-Dirty Models for Decision Support

The parameters required to reach a "just acceptable" hurdle rate define the minimal conditions that must be met if a project is to proceed, based on financial justification. Some projects should be discarded when unrealistic assumptions are needed to meet financial acceptability, freeing resources to create additional value. Projects whose pro formas show exceptionally high rates of return, however, should not be complacently accepted, even when their assumptions are reasonable. Why? Whatever appears highly attractive to your firm is bound to appear the same to competitors. In these cases, project acceptance is still called for, but with accelerated schedules and increased resources. The full value of these projects may only be obtainable if the firm obtains key patents and/or establishes a market beachhead.

Project Justification

Assuming you have made an estimate of NPV using the preceding methods, there are three approaches to deciding whether or not to move forward with the project.

1. Eliminate any project with negative NPV, where NPV comprises the value of (1) the projected cash flow of the business, and (2) the (negative) cash flow of the R&D required to create the business. In other words, no project is *financially* justified if the R&D cost exceeds the NPV of the resulting business *after* both the R&D costs and NPV of the projected commercial cash flows are discounted to a common base year. An immediate benefit of this approach will be to weed out projects whose rewards are too small to justify the R&D resources.

   (Having said this, *we must also recognize that the NPV versus cost criterion is purely financial: it does not, in itself, say the project is not justified.* There may be strategic reasons not captured in the spreadsheet that justify going forward. Such reasons could include opportunities or applications (technology options) not included in the cash flow model, or synergy with other product lines of the company. The distinction is critical, because the unstated motivation of the R&D champions may be less tied to direct cash flow return than the creation of strategic and technological options.)
2. *Add a safety factor.* The simple NPV criterion also assumes a 100% probability of success. In reality, the likelihood of success may be far less. A good rule of thumb is that the project should be started only if the NPV *multiplied by the probability of success* exceeds the projected R&D cost. Using the methodology developed in Chapter 11, that probability might be 10%

or less for projects in the conceptual stage, so a minimum multiplier of 10-fold might be implied for a new idea. A more rigorous approach considers the value of an option to terminate a project when a "fatal flaw" becomes apparent. This valuable option creates additional leeway for early-stage exploratory projects and is discussed in Chapter 12 when we look at decision trees.

3. Prioritize projects based on the best ratio of NPV to R&D cost. The considerations previously discussed presume a classic, financially based "go/no go" decision. In a world of both choices and resource constraints, prioritization among competing, comparable opportunities may be a more realistic approach. After all, at the early stages, the proposed project benefits, its costs, and its time frame are fuzzy numbers and not yet suited to hard-edged decisions. Even with these caveats, some numbers look much better than others, and the use of quantitative tools simplifies the choices that must be addressed by the decision makers.

   Quick-and-dirty valuation is useful for prioritizing a pool of projects or project concepts. Good R&D organizations always have more ideas than resources. They must rank-order them in terms of risk and return and place their money on those with the greatest potential for value creation. This task is aided by the fact that many of the assumptions in the model (depreciation schedules, factory overheads, selling costs, etc.) are common among R&D projects. The differences in return are then created by those factors that differ between projects—potential sales volume, gross margins, capital intensity, and speed to market. A review of projected R&D costs, NPV estimates, and probabilities of success will quickly identify the most (and the least!) promising projects in the R&D portfolio.

   One way to prioritize is to divide the NPV associated with a successful commercial outcome by the estimated total project cost. Those with the highest ratio are deemed the most attractive. This approach is purely financial; R&D directors would also weigh the technology options created.

R&D Management as Risk Reduction

Well-managed R&D incorporates a process of risk reduction. As a project proceeds from one stage to the next, the risk of failure should progressively diminish. From the standpoint of valuation, the R&D manager must keep a constant eye on the ratio of the project value (which may need to be adjusted with new information) and the cost of completing the project. The ratio of risk-adjusted NPV to future cost should be steadily improving in a sound project.

It follows that the R&D manager should be continually identifying and reducing the uncertainties associated with a project, doing so as quickly and cheaply as possible. Ideally this process should leave no large risks (say a 1 in 2 chance of being right) unresolved before a project exits the feasibility stage.

Even so, conditions change and with them perceptions of risk. A periodic check on key assumptions is time well spent.

Finally, in emerging situations, and as time passes, increasing certainty is often accompanied by diminishing opportunity. For example, as this book is being written, there is tremendous uncertainty whether future Internet access will be dominated by (upgraded) traditional telephone services, by cable, or by direct broadcast. A well-placed bet on the direction could still be very rewarding. Parties with vested interests will attempt to hedge. As the direction becomes clearer, these concerned parties will change their own betting patterns. Competition ensures this. Eventually, the winner (or winners) will be obvious, but the opportunities for "unfair advantage" will also be highly restricted. In the end, obsessiveness about risk reduction to the exclusion of opportunity may prove as fatal as disregard of risk.

Treating Sunk Costs

Many managers wonder how to handle past expenditures when they look to the future prospects of a project. "This project will draw on research that has already cost us $2 million over the past few years," they say. "Shouldn't those sunk costs be included in our pro forma analysis?" In considering the larger profitability of the firm, those costs surely matter, but in making decisions about the future, sunk costs are simply history. A financially sound approach to making a "go/no-go" decision is to look at the internal rate of return of the project *from that time forward.* If that rate is attractive, proceed. Ignore sunk costs. Conversely, a project should be dropped if the internal rate of return is unacceptable, even if millions of dollars have been spent. Again, ignore sunk costs. In fact, a well-planned project that is on time and on budget should show an increasing rate of return versus the original project, as it gets some sunk R&D costs behind it, and the rewards move closer in time. Improving return projections are an excellent sign. Conversely, diminishing returns are telltale warnings of an "ever-receding mirage."

Terminating Projects

Inevitably, most early stage projects are terminated by research managers. Indeed, they must be, since there are generally more promising ideas than there are resources to pursue them. Maintaining a diversified portfolio of early stage projects is an important source of value, but requires the termination of weaker projects to maintain portfolio balance. Nevertheless, cutting these projects is rarely painless, as each represents the effort and inspiration of individual contributors, who are bound to be disappointed. And, as every experienced researcher knows, many terminated projects contain hidden values: opportunities for serendipitous discovery and seeds of technological progress that will strengthen the firm in other ways. As noted in Chapter 9, medical technology evolved from a sweetener project. Indeed, the discovery of aspartame is said to have occurred when a research chemist working on quite a different objective licked his fingers (a safety violation!) and found them to taste sweet. There is also growing recognition that, in their initial stages, many of the breakthrough innovations on which the fortunes of great industrial companies currently rest would not have survived the financial tests advocated here. Research by a multidisciplinary team at Rennselaer Polytechnic Institute has, in fact, confirmed a familiar pattern in the progress of industrial technology: truly powerful breakthroughs generally evolve from projects that were repeatedly axed and restored.[2] The value-creating power of these innovations emerges only over time, and only through the dogged persistence of dedicated employees.

And so, the research manager asks him- or herself: "Am I killing a project that might assure the future of the corporation? Could unforeseen future technology, or the discovery of an overlooked market, change this project's economics for the better?"

These are justifiable questions, but questions that should have been answered in the course of project evaluation, through disciplined market investigation and other efforts to reduce project uncertainty. In the end, the manager's questions about killing a project can never be answered with certainty, since the future is inherently uncertain. A more important question would be, "Will more valuable projects be impaired if I allocate resources to this project?"

Finally, a terminated project often has some hidden values: The company has explored a piece of intellectual terrain and now knows there is no gold there. It may be able to use that knowledge to achieve greater focus. Technical skills of researchers have probably been sharpened. Unexpected leads and new ideas may continue to be pursued, even when the original targets have been abandoned.

Economic Profit Calculations

Economic profit provides a second shortcut to project valuation. The economic profit model discussed in Chapter 5 stated that an investment adds no value if it only earns the cost of money (WACC). However, if the investment return earns a premium above the cost of money, it produces economic value added, or economic profit:[3]

$$\text{Economic profit} = \text{Invested capital} \times (\text{ROIC} - \text{WACC}),$$

where ROIC is the return on invested capital and WACC is the weighted average cost of capital.

If the investment is $1 million and the return on invested capital is 21%, the premium above a 12% cost of capital represents an economic profit of 9%. In other words, though the "book profit," which fails to account for the cost of capital, is $210,000, the economic profit is $90,000. This is an annual number. If the 21% return is earned in perpetuity, we can use the simple perpetuity formula to give the long-term added value:

$$\$90,000 \div 0.12 = \$750,000.$$

(If the return changes with time in a predictable way, a simple NPV calculation will give a better answer.)

This shortcut method is best employed when the technology investment is a direct and linear add-on to an existing business. If it has secondary and complex effects involving changes to the marketplace and competitive conditions, a fuller pro forma approach is recommended.

The economic profit model is particularly appropriate in valuing cost savings projects that are not expected to affect revenue growth or pricing, but that will have an impact on the bottom line. For example, an investment of $1 million in new energy saving lightbulbs could reduce utility bills by $350,000 per year and give an after-tax profit of $210,000, as in the preceding example. Engineers would have described the adoption of this technology as a less-than-3-year payback on a risk-free investment. The DCF methodology would give a present value to the revenue stream of

$$\$210,000 \div 0.12 = \$1,750,000,$$

and a net present value of $750,000 (net of the original $1,000,000 investment). Note this is the same value we calculated with the economic profit model above. Any way you cut it, it is a good deal and simple to calculate.

Other cost-savings projects can save on capital. For example, software could be added to a plant's distributed control system (DCS) that adjusts production rates so that in-process inventory is kept at the minimum prudent level. If the one-time cost of the software is

\$100,000 and the reduction in inventory is \$500,000, the annual savings will be the after-tax cost of capital (12%) times \$500,000, or \$60,000. Expressed as a return on capital, this is 60% (\$60,000 ÷ \$100,000). If this savings is immediate and will occur in perpetuity, we divide again by 12% and obtain an NPV of \$500,000 (this should be no surprise, since it is the amount of inventory capital we no longer require). But if it takes a year to write the software, the value is delivered one year later and must be discounted at 12%:

$$NPV = \$500{,}000 \div 1.12 = \$446{,}000.$$

Market-Based Approaches to Value

Thus far, we have dealt with a strictly financial approach to value. This is the approach that a rational investor with no strategic intent[4] would take in making a decision. There are always such people around, so their approach defines a minimal value. The real marketplace, however, includes both financial buyers and strategic buyers. As at an auction, and virtually by definition, the strategic buyers, *if there are any*, will outbid the financial buyers. Their market-based valuations are invariably higher than those provided through strictly financial analysis. Because so many new products depend on the integration of two or more technologies, the invention of one corporation may be the key to unlocking the market potential of another corporation's innovations, and have significant strategic value.

Because the real marketplace includes both financial and strategic buyers, market-based valuations should be equal to or higher than purely financial ones, and as such represent an upside to the technology seller. The remainder of this chapter describes two technology markets in which these valuations occur: the worlds of venture capital and of licensing.

Finally, pro forma DCF analysis can play a role even in market-based valuations. Typically, venture capitalists and other investors will insist the founders provide a *business plan*, which will include a projection of the future revenues on which their valuation will in part be based.

The Marketplace for Technology

A vigorous and healthy marketplace for technology has emerged in the past 40 years wherein the sellers are typically scientists, engineers, and entrepreneurs, and the buyers are venture capitalists, established companies desperate to acquire technologies they have been unable to create, and general investors with a taste for technology speculation. Investment bankers specializing in technology mediate many of these deals.

The marketplace for technology in the United States was first actively developed in 1946, when Harvard professor, General Georges Doriot and a small circle of Boston-area bankers and industrialists founded ADR, the first U.S. venture capital firm. Using start-up funds of less than $5 million, ADR began mining the fertile fields of postwar technological development then flourishing in and around Harvard and MIT. ADR's great moment of fame occurred in 1957, when it invested heavily in a venture spearheaded by a young MIT researcher named Kenneth Olson. Olson called his company Digital Equipment Corporation. By 1971, ADR's investment of $70,000 had grown almost 5,000-fold, and hundreds of other venture capitalists were eagerly attempting to clone its success.[5]

Since the founding of ADR, venture capital has been associated with virtually all the major technology-driven new industries in the United States: semiconductors; super-, mini-, and microcomputers; medical devices; software; biotechnology; and more recently, wireless telecommunications. Venture capital industries have emerged in Great Britain, Europe, Asia, and Australia, but all on a much smaller scale.[6]

A major event in the valuation of technology-based companies was the initial public offering of the stock of Genentech, a company without a single salable product and only the prospect of running at a loss for several years. Nevertheless, its market capitalization on going public exceeded that of American Can Company, a Fortune 100 company with over 100 years of operating history.

For technologists, venture capitalists and others like them represent just the first stage of a potential series of markets for invention. To understand how values are determined in this marketplace, it is worthwhile to review the typical stages of financing for successful technology start-ups.[7]

Stage 1. Seed Capital: Sweat, Angels, and VCs

In the first step, an entrepreneurial group creates an informal strategic intent and identifies the technology assets, ideas, resources, and markets it intends to exploit. Much of the initial investment is represented by so-called "sweat equity" contributed by the founders on speculation that the concept will succeed. The founders may also contribute some cash to fund start-up activities such as incorporation costs, design expenses, the building of prototypes, equipment, and the like.

Other cash is often contributed by friends or associates: so-called angels. A venture capitalist ("VC") may be associated with the founding group as well. While other founders contribute technical knowledge and inventive ideas, the venture capitalist contributes money and, in many cases, contacts in the business and supplier communities.

Venture capitalists do not share their funds with a start-up company out of the goodness of their hearts. Because the risks are high, they demand in return a significant ownership stake in the venture, evidenced by shares of founder stock. In some cases, they will advance new money to the firm in exchange for convertible preferred stock, a bondlike hybrid security that can be converted into common stock at the option of its holder. Both common stock and convertible preferred stock give their holders an interest in the future fortunes of the enterprise. The goal of the venture capitalist is to help the struggling start-up develop its technology and business to the point at which it becomes either (1) an attractive item for purchase by a larger company, or (2) capable of selling its shares through an initial public offering of stock (IPO). In either case, the venture capitalist hopes to "harvest" his or her investment for much more than its initial cost. The VC will almost always focus on an "exit strategy" and cannot be regarded as a long-term investor.

Venture capitalists know from experience that many start-up investments fail to pay off. By investing in a number of start-ups, however, they expect to achieve superior average returns. A few outstanding successes will make up for a number of failures. ADR provides a perfect example. Over the years of its activities and many investments in small companies, ADR earned a compound annual return of 14.7%; almost half of this was due to its big success with Digital Equipment.[8]

Since almost no banks or other institutions will provide capital to a start-up company, VCs are often in a good position to get what they want (hence, the term "vulture capitalist"). The high cost of VC financing induces some company founders to hold out for other sources of capital. In their view, giving large blocks of dirt-cheap founders stock to the venture capitalist is like giving away much of the upside potential of their company and their ideas.

We can take a simple approximation of value at this stage by simply looking at the R&D funds the company (let's call it MiracleCure) has spent and the amount contributed by early-stage investors, including the founders. Assume that MiracleCure has already spent $1 million on R&D in its quest for the ultimate cancer cure. Early investors have invested an additional $3 million, for which they have received 3/7 of the stock. The founders retain control, holding 4/7. Using a "step-up ratio" of 4, which is the ratio of premoney valuation to cumulative R&D, we can make a rough estimate of MiracleCure's value using the following formulas:

$$\text{Cumulative R\&D\\$} \times \text{Step-up ratio} = \text{Premoney valuation,}$$

$$(\text{Cumulative R\&D\\$} \times \text{Step-up ratio}) + \text{Financing} = \text{Postmoney valuation,}$$

or

$$(\$1 \text{ million} \times 4) + \$3 \text{ million} = \$7 \text{ million.}$$

Where did the $4 million premoney valuation come from? In the last analysis, it was determined by negotiation between the seed round investors and the founders; but if the parties were sophisticated, they might have looked at the step-up ratios in previous deals for guidance. Just such a proprietary deal database has been created by Recombinant Capital as a guide for biotech investors,[9] and a step-up ratio of 4 is in the middle of the range for initial financing rounds.

From the point of view of technology valuation, a nominal $1 million of cumulative R&D in sweat equity and founder's cash has been monetized fourfold. While this may seem high, the risk of project failure in the first stages of research was also high, and these risks have now been overcome. To use the vocabulary of Chapter 2, there was initially a 25% chance that $1 million of research would meet the conditions to advance to the feasibility stage of R&D. The goal was met and the probability is now 100%. A value of $4 million was created through a process of risk reduction.

Stage 2. Private Placement, Round I

Now assume that the MiracleCure management group has spent some time developing its ideas into a demonstration of technical feasibility and that these ideas continue to look promising. However, the initial funding is starting to run out, and more capital will be needed. A total of $4 million has now been invested in R&D activities, and the founders have enough data and confidence to write a credible business plan, which will include financial projections and a description of markets, customers, manufacturing plans, patents, competitors, and risks. This business plan will be presented to venture capitalists and other sophisticated investors.

Venture capitalists typically not only invest their own money but manage funds placed with them by sophisticated private and institutional investors. Some of the latter may be large corporations that are cash rich, but idea poor, and looking for a "window on technology." Venture capitalists, in fact, often syndicate their investment favorites with other venture capital funds to diversify their risks. They say, "I'll buy a piece of your company if you'll buy a piece of mine." Many of these investors do not intend to stay for the long term; rather they expect to sell their shares at high markups at the initial public offering or sooner.

The amount raised in the first round of financing is usually only enough to see the company through a few years of its business plan and negative income. In fact, people describe start-ups in terms of "burn rate"—a company with $9 million in cash and a burn rate (projected negative cash flow) of $3 million will be out of cash and back for a new round of financing in three or fewer years.

MiracleCure has "burned" through the $3 million invested by its angels and is eager to begin developmental stage R&D. The strength of its business plan, the prominence of its scientific staff, and general optimism about the anticancer drug market induces one of the behemoths of the pharmaceutical industry to make an investment of $7 million—"chump change" for the big drug companies, but significant money for our friends at MiracleCure.

We can once again estimate the value of MiracleCure, using the same formula used earlier. Now, however, cumulative R&D has grown to $4 million. And while the continued R&D progress in the feasibility stage is very encouraging, it not as startling as the brilliant discovery that quadrupled the investors' original stake. A premoney valuation of the company of $10 million is negotiated, representing a step-up ratio of 2.5:

Postmoney valuation = ($4 million × 2.5) + $7 million = $17 million.

Exhibit 10.5 categorizes MiracleCure's financing through this and subsequent rounds, indicating the postmoney valuation in table and graphic form.

Stage 3. Private Placement, Second or Mezzanine Round

Many start-ups go through several rounds of private placement, and engage in other financing activities, such as spinning out technology to joint ventures and limited partnerships, or performing contract research for industrial partners.

For simplicity, we assume that none of these other financing techniques is needed, and that after a time MiracleCure has spent the $7 million it raised from the first-round investors on more R&D (cumulative R&D is now $11 million) and that its prospects look sufficiently attractive to warrant a second round of investment. In our example, we have assumed that investors now view the company as being worth $30.8 million, the new premoney valuation, representing a step-up ratio of 2.8. Another $17 million is raised and postmoney valuation rises to $47.8 million, as shown in Exhibit 10.5.

Although these are increasingly larger sums, the company is also growing and has a higher burn rate. Presumably, the burn rate of our company has increased from the initial $1–2 million per year to $5

Stage	Name	R&D	(M$) Cumulative R&D	Financing	Stepup Ratio[a]	Premoney Valuation	Postmoney Valuation
1	Angels and Sweat	$ 1.0	$ 1.0	$ 3.0	4.0	$ 4.0	$ 7.0
2	Private Placement 1	3.0	4.0	7.0	2.5	10.0	17.0
3	Private Placement 2	7.0	11.0	17.0	2.8	30.8	47.8
4	Initial Public Offering	17.0	28.0	40.0	2.5	70.0	110.0

[a] Premoney Valuation/Cumulative Expense

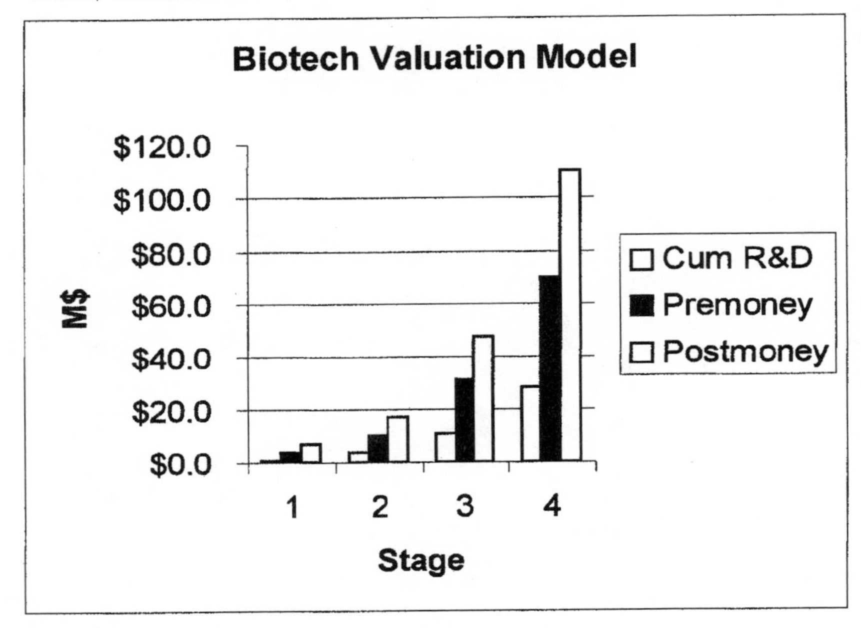

Exhibit 10.5 Start-up Investment Model

million or more as early products move into clinical trials, and new promising second-generation products are added to the R&D pipeline.

At this point, a new and useful piece of nomenclature enters the vocabulary: "mezzanine financing"; this is the financing round immediately preceding the IPO.

STAGE 4. INITIAL PUBLIC OFFERING (IPO)

With its credibility established by successful test results in human subjects, MiracleCure is ready to "go public" with its first public offering of stock through what is called an *initial public offering* (or IPO). Indeed, it must attract more capital if it hopes to complete the clinical trials and build the manufacturing capacity it needs to produce, sell, and deliver finished products. At this stage, $28 million has been spent on cumulative R&D, and the firm is valued by the investment bankers promoting the offering at $70 million. New funds of $40 million are being raised to support growth, but in all likelihood some of the initial investors—including the founders and early employees—will be using the IPO as an opportunity to "harvest" part of the profits on their early investments. This is when people who were recruited for modest salaries plus stock options at 20 to 30 cents per share see a chance to cash out at $20 or $30 per share and become instant millionaires.

Secondary Offerings

If MiracleCure continues to grow, it will need additional capital. If cash flow remains negative, it may be necessary to offer more shares of stock to the public through what are called *secondary* offerings. As tangible assets grow and operating cash flow turns positive, MiracleCure may be able to raise new capital through borrowing.

In the real world, not every successful start-up makes the transition through these many financing steps. Promising start-ups are, instead, often acquired and absorbed into larger companies eager to capture their technology and human talent. MiracleCure, for example, may be the target of a strategic acquisition by a larger firm attracted to its technology and to the markets it is addressing. That larger firm might be the large drug firm that provided much of its early funding. Acquisition may be the best possible exit strategy for the venture capitalists, founders, and early employees. Although some may mourn the loss of MiracleCure's independence, these individuals may celebrate its acquisition as the biggest payday in their lives.

Technology start-ups, often companies with no real operating components, have been highly rewarding for sophisticated investors. Step-ups in value of over 100% are common between first and second financing rounds, another 80% step-up can occur in the third round, and another 35% at the IPO level. IPO returns on some companies have been sensational: Apple Computer rewarded early investors 235 times over.

One of the most important determinants of returns, according to a study by William Bygrave and Jeffry Timmons, is the health of the IPO market at the time a company goes public. That market is notoriously fickle, enjoying, in turn, periods of euphoria and of retrenchment. As they state:

> When the IPO market is buoyant, it's comparatively easy to float new issues of venture capital-based companies at high valuations. This causes venture capital returns to rise, because . . . IPO's, on average, provide the most bountiful harvest of venture capital.[10]

However, they found that during the long periods in which the IPO market has been unenthusiastic, returns were far lower.

Those That Fail

What happens to the majority of start-ups that prove unsuccessful? Failure looms when further financing is impossible, and the company lacks internally generated cash flow to carry on. Financial liquidation makes little sense because most of the property is intellectual property, either patents or know-how in the heads of the scientists and engineers. Shareholders do not want to lose this asset, and competitors do not want to see the scientists and engineers restarting their venture under a new roof. Hence, a competitor is likely to buy the start-up in a merger transaction. Shareholders receive equity in the merged company, duplicative administrative personnel and facilities are eliminated, but the better scientists and projects may well survive in their new homes.

Licensing as Valuation

In addition to the market for reasonably pure "technology-play" companies, a second market exists in which technologies are freely bought and sold. It is called *licensing*.[11]

The licensing market is important in valuation because it provides independent rules of thumb regarding technology values based on decades of experience. It also provides evidence that technology can have enormous value independent of an operating business.

Licensing can generate tremendous revenues. Union Carbide was one of the original inventors of linear-low density polyethylene. It licensed that technology broadly—to Exxon, Mobil, NOVA and others—while using it to produce its own products. At its peak, Union Carbide's annual licensing income from this one technology exceeded $100 million. Texas Instruments won a patent on the large-scale integrated circuit. After a long patent infringement battle with Japanese chipmakers, it prevailed and has earned royalties in the nine-figure range. Stanford University and the University of California shared many millions of dollars in royalties from basic gene-splicing patents, which were broadly licensed to companies using genetic engineering methods on standard terms. These two universities alone earned $100 million from royalties in 1995; all U.S. universities earned $300 million.

What is licensed in these deals is intellectual property, commonly a mixture of *patent rights* and technology; the latter often referred to as *know-how*. In exchange for a fee, the right to use this property is granted for a period of time, typically until patent expiration. The owner of the licensed technology, often but not always the inventor, is called the *licensor*, whereas the party who seeks to employ the technology is called the *licensee*. The terms of license may be *exclusive*, as when a chemistry professor licenses all rights to a molecule he has patented to a single drug company, or *nonexclusive*, as in the examples of polyethylene and gene-splicing previously mentioned. To the degree that rights are exclusive, a monopoly power is granted, which commands a premium. The fee paid for the license may be in the form of a lump sum, an annual minimum payment, and/or a scheduled percentage of annual sales, called a *running royalty*, or a combination of all three.

The characteristics of the license market differ from securities markets in that most transactions occur on a bilateral basis between buyer and seller. However, bidding processes occur when more than one buyer is interested in licensing a property.

Unlike security sales, in which the seller has no further interest in the buyer's fortunes once the deal is struck, many license deals are structured as "win-win" situations between buyer and seller. In these cases, where an inventor is counting on a licensee to commercialize his or her invention, it is not in the inventor's interest to exact onerous terms from the licensee, because value will be maximized for both parties only if the licensee maintains the license and has a strong financial incentive to diligently pursue the commercialization of the licensed technology. As in the earlier discussion of product pricing, the value chain, from inventor, to licensee, to licensee's customer must be analyzed and the rewards apportioned down the chain.

Other licensing scenarios are hostile. A patent holder believes a competitor is infringing the company's patent, and after litigation or the threat thereof, the infringer negotiates or is compelled to pay a royalty.

The case study at the end of Chapter 9 showed how pro forma analysis and net present value proved excellent tools for determining the value and fairness of a licensing arrangement. The pro forma model in effect laid bare an unworkable value chain from licensor (Genentech) to licensee (W.R. Grace), after which the two parties could focus on win–win outcomes.

Finally, in a licensing negotiation there is a clear separation of value added from technology from value added from other business processes; a separation that is more problematic inside an ongoing business. The question of whether the profits from a new product are attributable to the technology created by the R&D department or to the wisdom of the business management gets resolved; since only the technology is licensed, the two contributions are uncoupled.

Licensing Scenarios

Although no two licensing situations are alike, they fall into three broad classes: (1) proven technologies; (2) unproven or partially proven technologies; and (3) patent rights only.

Proven technologies are essentially the standard by which unproven technologies must be measured. Excellent examples have occurred in the plastics business, for example, the polyethylene technologies described earlier. These technologies provided both improved product performance and manufacturing efficiencies over previous generations of technology. They were broadly licensed both domestically and overseas. Royalty rates were determined largely by competitive factors: prospective licensees always had the option of using the next best technology but at a lower price. In addition, both licensors and licensees were forced by the market to recognize the competitive force of alternate technologies, such as those developed by Dow Chemical, which were not made available for licensing. Hence, setting the right royalty rate for commercially proven technology largely came down to a question of balancing technology gains with market forces. For the licensee, there was little risk that the technology could not be applied at the manufacturing level since plant design and operating procedures were part of the licensing package.

The case of unproven or partially proven technology is more interesting. The royalty demanded for it should be less than that for proven technology. The licensee understands that there are risks associated with the unproven/partially proven technology and that he or she must assess those risks and the research and development costs needed to minimize them. If the licensor is commercially inexperienced—a lone inventor, a university, or a national laboratory—he or she may fail to appreciate these costs and mistakenly seek royalties in line with proven technologies.

In addition to development costs, the licensee must make a capital investment and be compensated for the cost of money plus the appropriate premium for risk. Using the economic profit model, the annual profit from the deal must cover the sum of:

$$(\text{Capital investment} + \text{R\&D investment}) \times (\text{ROIC} - \text{WACC}) + \text{License fees.}$$

The licensor's economics are simpler; after taxes the licensing revenues drop to the bottom line and no capital investment is needed. The economics for either party can also be captured by DCF methods as a net present value.

More experienced technology licensors recognize that a great deal of added value can be provided by the licensee, and compare the rewards of licensing to the potential rewards of a go-it-alone strategy. If doing your own thing is not viable owing to high entry barriers, licensing is the only answer. But if the inventor is a corporation with significant resources and going it alone is a real option, the situation is analogous to a "make-or-buy" decision, which depends on the price and terms of the deal.

Strategic considerations also come into play; if the technology owner has better investment opportunities, licensing is the right course of action.

A clear and simple example of value added by a licensee is the case where the licensor is an inventor who synthesizes a new-to-the-world molecule and receives a patent on this property. Hypothetically, this molecule could be a new drug, a pesticide, or a food additive. The inventor has neither experience in screening for any of these properties nor the commercial infrastructure to play in these markets. Licensing the rights to each potential application to a pharmaceutical company, a pesticide company and a food company is a logical alternative, since each can add value in its field of expertise. The inventor, depending on his or her pocketbook and sophistication, may, however, wish to invest or gamble on some screening tests available commercially from consultants to target the licensing effort and to strengthen the technology package. If these test results are promising, the inventor's negotiating position will be much stronger than the chemical patent taken by itself. With a stronger technology package, potential licensors will not only consider what this molecule may do for them, but also what it could do to them in the hands of a strong competitor. Interestingly, this case also highlights the high value of the proprietary testing procedures in the technology portfolios of drug, pesticide, and food additive firms, and of independent consultants. Test methods constitute an important part of their technology assets.

Licensing Revenues

Apportioning the balance of rewards for the inventor, the developer, and financier of new technology is the key to the licensing marketplace. We have looked at the value considerations for two key cases: proven and unproven technology. In the case of proven technology, the licensor is the *inventor and developer,* but in the case of unproven technology, the licensee takes on development as well as capital investment, and will seek a higher rate of return to cover the increased risks. The right balance of rewards must follow this reality.

Assume that shortly after R&D has invented and patented polyarothene, PMI senior management decides that it does not want to be in the plastics business long-term and wishes to focus its resources on electronics materials. They request R&D to evaluate licensing the polyarothene technology. Exhibit 10.6 again shows the projected sales for 20 years of the hypothetical polyarothene business per the pro forma example of Chapter 9. (Recall that we had calculated a net present value of $76.9 million for this business.[12])

We must now ask, "What if we—as inventor—licensed this patented but unproven property instead of developing it ourselves?" And assume for now that our most likely licensees can achieve the sales and income projected by our own business team. (If the licensee could do even better, the licensing case would be proportionally more compelling.)

Also assume that, owing to the length of time required for development, there will be only 10 years of remaining patent rights, although the business would be expected to continue to grow indefinitely. In that case, as shown in Exhibit 10.6, the net present value of a 5% running royalty for 10 years (the remaining life of current patents) is $13.9 million. We can quickly run a sensitivity analysis. For 3% it is $8.3 million and for 7% it is $19.5 million.

Experience indicates that a fair price for unproven but patented technology is 20% to 30% of the NPV to the inventor, and 70% to 80% for the developer and financier, who is assuming most of the risk. A 7% royalty to the inventor would roughly correspond to 25% of project NPV ($19.2 million), and we might instruct our licensing manager to set that as his target. Let us further assume that the licensee feels he cannot afford this rate during the tough initial period during which he is developing the business. Exhibit 10.6 shows that for this growing business we could counter with an offer of a 3% royalty for 20 years and capture the same net present value.

Proper allocation of rewards—a "win–win" agreement—helps avoid disputes and litigation between licensor and licensee. Inventors

Year	Sales	5% Royalty 10 Years	5% Royalty 20 Years
1	$ 14.9	$ 0.7	$ 0.7
2	20.9	1.0	1.0
3	28.3	1.4	1.4
4	37.4	1.9	1.9
5	48.0	2.4	2.4
6	60.2	3.0	3.0
7	73.9	3.7	3.7
8	88.8	4.4	4.4
9	104.9	5.2	5.2
10	121.9	6.1	6.1
11	139.4		7.0
12	157.4		7.9
13	175.5		8.8
14	193.5		9.7
15	211.2		10.6
16	228.5		11.4
17	245.2		12.3
18	261.2		13.1
19	276.5		13.8
20	290.8		14.5
License NPV(12%)	@ 5%	$13.9	$32.3
	@ 3%	$ 8.3	$19.4
	@ 7%	$19.5	$45.2
Project NPV			$76.9

Exhibit 10.6 Licensing Model for Polyarothene

who get too little (say less than 5%) may feel that they have been taken advantage of and seek legal redress. Or university inventors who feel aggrieved may create difficulties for future research collaboration or campus recruiting.

Using Royalty Rates for Valuation

The perceptive reader will understand that the basic logic developed in Exhibit 10.6, using NPV to determine royalty rates, can be used in reverse to calculate the values of technology, that is to estimate NPV based on prevailing royalty rates. There are well-established rules of thumb in various industries as to what constitutes a fair royalty rate.

Such rates may vary from 2% to 3% for some chemicals, to the mid-teens or higher for some proprietary drugs. If the customary royalty rate is applied to the size of the market target to be covered by the technology, a ballpark number for the value of the technology can be calculated by capitalizing the royalty stream.

The Discount Rate for Royalty Revenues

When licensors attempt to find the net present value of an anticipated royalty stream, they must use an appropriate discount rate. But what rate is appropriate when a firm is licensing, and not developing, its own technology? Generally, a discount rate equal to the licensor's own cost of capital is appropriate for valuing a stream of royalty income if the riskiness of that income stream is comparable to the risks the company encounters in its normal business. After all, the licensee's business development team may fail to produce the forecasted stream of revenues just as easily as might an internal business development team. A lower discount rate would be appropriate if the risk was perceived as being lower or risk-free. For example, if an electric utility licensed your company's environmental technology and was committed *by law* to employing it at a generating plant, the discount rate for royalty income might be comparable to the cost of money for a utility, which is materially lower than for general manufacturing. The opposite might be the case if your company's technology were licensed to a high-tech start-up. In that situation, a discount rate equal to the high cost of money experienced by risky start-ups would be more appropriate.

Evaluation Fees

Licensors are strongly advised, except in the case of trusted partners, to ask prospective licensees to pay a refundable evaluation charge. This charge should, at a minimum, pay for the time of those charged with explaining the technology and discussing licensing terms. Evaluation fees eliminate "tire-kickers" and some of the firms that conduct industrial intelligence mission under the pretense of licensing. Psychologically as well, an early payment may break the ice and create a positive bias toward a successful conclusion.

Up-Front Fees

A royalty agreement is only as valuable to the licensor as the licensee's commitment to follow through with development and commercialization. It is often wise to see some monetary evidence of that commitment in the licensing agreement in the form of an up-front fee—especially if an exclusive license has been granted. In the absence of such a fee, the licensee has merely taken an option (at no cost to itself) to develop and commercialize the technology. Requiring the licensee to evidence its commitment with a cash payment forces it to put some "skin in the game."

And, of course, an up-front fee is good for the licensor because the present value of money received sooner is always higher than the same amount of money received later. An extreme case of the up-front payment is the outright sale of the technology. This should not be a problem if the sum captures an appropriate fraction of the net present value.

Minimum Royalties

Many licensing agreements specify a schedule of minimum annual royalty payments. If these are not paid, the license lapses. The intent of these payments is to ensure that the licensor is diligent in commercialization and has not simply bought the technology cheaply to prevent a competitor from using it. The minimum royalty provision can backfire, however, if the licensee is making a serious effort at commercialization but encountering problems. Faced with certain minimum payments and uncertain revenues from a technology that may take more time than anticipated to master, the licensee may simply throw in the towel.

Running Royalties

These are usually structured as a percentage of sales revenues, since the accounting difficulties and potential misunderstandings in attempting to calculate a percentage of profit are formidable. Running royalty rates can be simply a set percentage, but in many cases more complicated schedules are negotiated, where rates increase or decrease above stipulated levels.

Exclusivity

Another key issue in licensing is exclusive versus nonexclusive strategies. Naively, one might conclude that a nonexclusive license would be more profitable than an exclusive license. After all, the nonexclusive license can be sold any number of times. Bill Gates followed this strategy, selling nonexclusive rights to various generations of the Microsoft operating system software. However, when one considers the opportunity to create a monopoly behind a wall of patents and technology, the exclusive strategy may be more advantageous to the licensor. An exclusive license that keeps out competition and gives the technology holder a competitive edge will almost always justify a much higher royalty rate, and may generate more revenues if the licensee's share of the market is high. In fact, a level playing field in which technology is equally available to all players creates economic value for only two parties: the technology licensor and the customers of licensees. Licensees are not advantaged in any way when all competitors have the same technology.

If an exclusive license is granted, the best licensee is likely to be the number one industry player, since this firm will have the largest revenue base on which to charge royalties, and probably the best odds of commercial success. The number two player may be hungrier for an edge and should be considered, especially if that hunger translates into higher royalties or a better arrangement for the licensor. Small companies that are looking for a technology life raft are likely to waste precious time and produce few revenues. Don't license on the basis of first to show interest!

An additional principle for maximizing value is to restrict licensees to fields in which they have demonstrated competence (e.g., to a drug company for medical applications alone; to a chemical company for agricultural applications alone). Similarly, global rights should only be licensed to global players. This art has been described as carving the salami into many pieces and charging nearly what the whole salami might bring for each piece.

Patented technologies with broad applications are excellent candidates for nonexclusive licensing on reasonable terms, since no single licensee can expect to exploit the entire potential of the technology. The gene-splicing patents generated by Stanford University and the

University of California are good examples: they covered techniques that had broad applicability in biotechnology in many end-use markets (drugs, diagnostics, gene therapy, agriculture, specialty chemicals, etc.). It would be impossible to predict which of hundreds of potential licensees might make useful downstream inventions and bring them to the marketplace. Giving monopoly rights to a few licensees could well have retarded the growth of the entire industry.

The impact of patents as a value driver is discussed in Chapter 15.

Technology Appraisal

The preceding discussion may be usefully summarized by analogy with appraisals as they are commonly performed in the real estate industry—anyone who has had a home appraised is familiar with these. Real estate appraisers use three methods: income capitalization, sales comparison, and cost. The technology appraisal analogs are shown in Exhibit 10.7.

The analogy with the *income capitalization* approach is straightforward. Appraisers look at the rental income a house or office can generate and use conventional financial yardsticks in the real estate market to assign a capital value. This is in principle no different from the discounted cash flow approach we developed using pro forma models, where technology enables income-producing businesses. Potential licensing revenues can be capitalized in the same way.

The use of comparable transactions is a powerful method in both the real estate and technology marketplaces. In real estate, appraisers will find recent sales of properties in the general neighborhood and choose several that seem similar to the property being appraised. They

Real Estate	Technology
Rental income	Discounted cash flow
Comparable transactions	Comparable transactions
Replacement cost (less depreciation)	Ratio to R&D costs

Exhibit 10.7 Appraisal Methods

will next adjust for differences—house size, lot size, condition, and a host of unique factors.

The appraisal of a technology business through comparables is similar. If it is already an income-producing business, buyers and sellers will calculate the ratio of the value realized in comparable transactions to income, for example as a multiple of EBITDA. If the last company sold at 8 × EBITDA, this will be the starting point for the next transaction. If there is no income, comparables can still be a starting point for negotiation: for example a biotech boutique of comparable quality with twice the scientific staff and twice the number of patents might be valued at twice the previous comparable transaction.

In the cost-based approach, real estate appraisers look at replacement cost less applicable depreciation. They start with what it would cost to buy a comparable lot and erect a new home on it, and depreciate for the age and condition of the house. The use of step-up ratios does the same with one important difference. The method is similar to the extent that the costs expended on R&D are the basis for valuation. But the step-up ratios must take into account success, failure, and technology depreciation, if applicable. R&D that has failed must be written off. R&D that has succeeded should be "written up" to reflect that it has succeeded against the odds; if the chances of success of hitting a valid target were 1 in 5, the value of having hit it is at least 5 times the R&D cost. All other things being equal, a buyer for this technology would have the option of doing it him- or herself, but would face the same odds. All other things are never equal, so step-up ratios vary broadly, but they should always be considerably greater than the cost for successful technologies.

Finally, technology has value only in context, never in isolation. And in context, technology is never the only value driver. The recruitment of a high-quality management team will be reflected in company value. Strategic alliances with major corporations are also important value drivers. So, too, is the crowd behavior that periodically makes one industry "hot" and another not.

CHAPTER 11

Managing Value and Risk in the R&D Portfolio

The high-risk factors involved in individual projects, and the lengthy period between invention and bottom-line profits, makes the quantitative aspects of R&D portfolios difficult to track and to manage. As a result, many executives rely on rules of thumb when making decisions about portfolio dynamics: the need to maintain a balance of projects, to accept and even embrace prudent risk, and to diversify the R&D effort over a variety of opportunities.

This chapter aims to help the manager and the R&D professional make sense of the R&D portfolio in analytic terms, and to provide a set of tools for managing its value-creating potential and attendant risks.

Corporate Growth Goals and R&D

Managing the value and risks of the R&D portfolio begins at the top, with corporate strategy. R&D strategy must be consistent with this strategy. Very typically corporate strategy will include an earnings growth target, often more than 10% per year. To be truly sustainable, it must mean growth in revenues as well. Another common goal that should ensure continued corporate health is a stipulation that some percentage of revenues be generated by new products. Say this is 30%— a figure achieved by Gillette, 3M, Rubbermaid, and other leading new product companies. And we can define a new product as one introduced to the marketplace during the past 10 years. But why a separate goal for new products? One reason is that the sales revenues of new products grow faster than do those of older products. Another is that innovative new products are required to sustain or improve profit margins. Profit margins are generally ample for new products but thin for mature products.

We have learned that growth by itself does not equate to value creation. An economic profit must be earned. Therefore, we must assume that the CEO of our example company, PMI, has set a hurdle rate for new investment. If that rate is 20% and the company's cost of capital is 12%, the company will earn an economic profit of 8% on each of its investments. In our model, we assume any project that cannot meet its hurdle rate will be terminated.

Our goals, as stated here, are all problematical. Is 10 years the right time frame for defining a "new" product, or is 5 years more appropriate? Is 30% of total corporate revenue from "new" products too high or too low?

And what is new *really* new? Is a PC new if its microprocessor is upgraded from 400Mhz to 500Mhz? Is the current year's Taurus automobile really a new product? Is a dental floss with a new flavor and in a new package new? Many do not think so, but marketers and managers, being human, like to count these incremental improved products as "new" even when they merely cannibalize the company's previous products.

Is the hurdle rate too high or too low?

The answers to these questions vary with the business, and the reader who tries to apply the methods described here is urged to apply time frames and performance standards meaningful in his or her industry. Clearly, however, the first laptops with CD-ROMs, active matrix displays and full multimedia capability were new products: they created new capability and new demand and threatened competitors that could not match their performance. Ten years, moreover, is not unreasonable for truly new products in much of the industrial marketplace where customers need time both to evaluate the innovation and to adjust their products and processes to exploit it. It also is reasonable in view of patent life. Patents run 17 to 20 years, but the first 7 to 10 years may well be spent in developing and testing the product prior to first commercial introduction. It is a truism in industry that more than half of the revenues from a patent will be earned in the last 3 to 5 years of its life!

Having set our goals, we need next to examine their implications. As noted, a survey of industrial products I have worked with over several decades indicates that new products (using the 10-year definition) have growth rates of about 20% to 25% per year. Analysis indicates that the rate of growth is usually fastest in the early years (since the base is small), and that growth starts to slow in the last half of the 10-year period. Older, established products in the portfolio continue to grow at an average of about 7% (see Exhibit 11.1). These observations were the basis for the smooth curves introduced in Chapter 8 (Exhibits 8.8 and 8.9).

What does this mean for overall growth? Our model company, PMI, has sales of $5 billion, of which 30% are new products. Its blended growth rate will be a very respectable 12.4%. This growth goal creates tremendous implications for the organization in general and for the R&D portfolio in particular. If the corporation takes that 12.4% as its goal for future annual growth, Exhibit 11.2 shows the implications for corporate revenues. Here they advance in 10 years from a base of $5 billion to $16.1 billion. If 30% of sales must come from new products, then year 10 sales of new products (those introduced within the past 10 years) must be $4.828 million—almost equal to the company's current sales!

Modeling the R&D Pipeline

Given the quantitative goals projected in Exhibit 11.2, we must ask, "How many projects must PMI have in its R&D pipeline to achieve

	Percent	Growth Rate (%)	Contribution (%)
New products	30.0	25.0	7.5
Other products	70.0	7.0	4.9
Total	100.0		12.4

Exhibit 11.1 Corporate Growth Model

Year	Corporate Revenues	From New Products	From Established Products
0	$ 5,000	$1,500	$ 3,500
1	5,620	1,686	3,934
2	6,317	1,895	4,422
3	7,100	2,130	4,970
4	7,981	2,394	5,586
5	8,970	2,691	6,279
6	10,082	3,025	7,058
7	11,333	3,400	7,933
8	12,738	3,821	8,917
9	14,317	4,295	10,022
10	16,093	4,828	11,265

Exhibit 11.2 Corporate Revenue Growth Model (in $millions)

these results, and what will these cost?" To answer these questions, we must do a calculation based on three elements:

1. A sales projection for each project.
2. The cost and duration of each project.
3. The probability of success for each project.

Each project is different and will have its own revenue stream, rate of market penetration, capital requirements, costs, starting time, time to completion, and so forth. Nevertheless, if we have a good corporate database, with plenty of information about past projects, we can make a reasonably good estimate for each of the three vital elements.

Sales Projection

The first step is to establish a sales projection for each R&D project in the pipeline. Readers performing an analysis of their own portfolio will obtain the necessary information from the corporate database. But since our primary purpose here is to illustrate the method and to understand the implications of the model for R&D strategy, there is no learning benefit to be had from working with a complicated ficti- tious R&D portfolio. We will assume for simplicity that there is only one kind of project—a typical or average project based on the familiar polyarothene pro forma model of Chapter 9. The typical project should have revenue levels and growth characteristics similar to the average of successful projects commercialized in the past by the PMI Corporation. Recall (Exhibit 9.1) that polyarothene is a new product with sales growing at an average rate of about 25% per year for the first 10 years, from $14.9 million in year 1 to $48 million in year 5, and reaching $121.9 million in year 10.

These assumptions enable us to calculate how many successful typical new products PMI needs each year to meet the corporate growth goal. Exhibit 11.3 illustrates the dynamics. The first column shows what would happen if PMI introduced one polyarothene-class project per year (in the exhibit, this is called a project unit[1]). It is clear that in year 10 new product sales will constitute those products introduced in year 1, year 2, up through year 10. When we assume a steady flow of one typical successful new product per year, the largest expected contribution will be from the product introduced in year 1, now in its10th year and enjoying revenues of $121.9 million. (It will no longer be counted as "new" in year 11!) The next cells in this column show contributions from each succeeding year, with the most recent addition

Source Year	Per Project Unit	Per Lab Unit[(b)]	Growth Factor	Per Lab Unit with R&D Growth
1	$121.9	$ 609	1.000	$ 609
2	104.9	525	1.124	590
3	88.8	444	1.263	561
4	73.9	369	1.420	524
5	60.2	301	1.596	480
6	48.0	240	1.794	431
7	37.4	187	2.016	377
8	28.3	142	2.267	321
9	20.9	104	2.548	266
10	14.9	75	2.863	213
Total	**$599.1**	$2,996		$4,373
Units[(a)]	**8.06**	1.61		1.10

[(a)] To achieve sales of $4828M.
[(b)] In this model, a lab unit produces 5 commercial projects/year.

Exhibit 11.3 New Product Sales in Year 10 (in $millions)

being smallest. The total sales of new products in the tenth year will be $599.1 million. Obviously, the annual introduction of only one new product unit per year will fall miles short of meeting the CEO's goal of $4.828 million from new products in year 10. In fact, we need an output of 8.06 such projects a year to reach that goal, (or, in the real world, any combination of nonidentical projects summing to 8.06 project units).

At this point, we are introducing the fictitious concept of a *model laboratory unit*—one defined to be of sufficient size to commercialize five project units per year. This size is selected so that for the sake of transparency we can deal in round numbers of projects, rather than fractions, in analyzing the portfolio dynamics. Because our calculations show that PMI needs 8.06 successful projects per year to reach its revenue goal, the CEO can calculate the need for 1.61 lab units (or a laboratory scaled as 61% larger than a model lab unit) to achieve the growth goal, given the output assumed in the model. This defines the magnitude of the R&D investment that must be made. Some of this investment can be deferred, however, as shown in the last two columns of Exhibit 11.3. If the CEO decides in principle that the R&D department should grow as fast as the company as a whole (at 12.4% per year), and R&D productivity is maintained, R&D output should increase with each passing year. Consequently, R&D should reach its goal of $4.828 million in new product sales with only 1.10 lab units, a smaller initial investment. In addition, the growing R&D organization will be in far better shape to sustain growth in the ensuing decade, years 11–20.

R&D Costs and Project Duration

Assume, based on a review of historical performance, that the average project reaching commercial status takes 8 years from the time the underlying concept is first defined as a formal R&D project (the actual range may be 3 to 13). Laboratory operations are assumed to cost $250,000 per *professional* man-year, an arbitrary unit widely used in industrial R&D organizations, usually consisting of a degreed (usually Ph.D. or M.S.) professional, technician support, and general support. General support would include management, library, maintenance, computer department, secretarial, security, and so forth. This cost may seem high, but must be understood in terms of some common rules of thumb—that in a typical laboratory (1) two-thirds of the costs will be salaries and benefits and one-third operating expenses, and (2) one-third of the staff will be scientific professionals, one-third technicians or equivalent, and one-third support personnel. If the research operation has high capital costs (e.g., a pilot plant) or requires large amounts of raw materials, per-man-year costs will be considerably higher.

We treat each project as having five distinct *stages* (as described in Chapter 2). Stages are vital to understanding portfolio dynamics and the allocation of R&D funds. The following list reviews the five stages and fleshes them out with some assumed costs:

0. *Finding and Screening Ideas.* We will ignore this stage in subsequent analysis and assume that its time is indeterminate and that its costs are absorbed in laboratory overhead and allocated to defined projects.
1. *Conceptual Research State.* Early stage projects are called *conceptual* projects. They explore research ideas and alternative approaches, obtain patents, and define issues that need to be addressed in the future. We assume that projects remain in the conceptual stage for an average of two years, and that, on average, they are staffed by one or two (1.5) professionals.
2. *Feasibility.* Projects in the *feasibility* stage seek to resolve issues defined in the conceptual stage, and provide the data required to undertake development. We assume that this stage is twice as costly as conceptual stage work—involving three professionals at an annual cost of $0.75 million, for an average period of two years.
3. *Developmental Stage.* Entry into the *developmental* stage is heralded by an increasing level of commitment. A clear sign that development is underway is the exposure of the technology, including prototypes, to customers for feedback. Pilot-scale production is likely to occur in this stage, its aim being the resolution of the issues associated with scale-up, quality requirements, and fine-tuning of performance specifications to

customer needs. This work is expensive, and we assume that costs double in this stage to $1.5 million per year, representing a level of commitment of six man-years.

4. *Early Commercialization.* The fourth stage is called *early commercialization*. For an industrial product, product confidence is now high enough that the customer is beginning to purchase the product. Uncertainties in product specifications, quality, and delivery nevertheless exist, and the project is far from a routine operation. For the consumer product, test-marketing is a clear indicator of entry into the early commercialization phase. We have assumed two years for the early commercialization phase at an annual cost of $3 million per year.
5. *Commercialization.* The term "returned to operations" has been used for the entry into this stage. While the transition from early commercialization to full commercial status is gradual, the phase is marked by the assumption of full authority by manufacturing, marketing, and sales (and the absence of further charges to the R&D budget).

In doing DCF calculations, it is important to be consistent about what year one is in. The year of first commercial sales is deemed to be year 1. In this sense, year 0 is the last year of early commercialization, in which sales and earnings are assumed to be zero (on the assumption that R&D is still absorbing net costs).[2] No discount is applied to year 0 costs, while year 1 is discounted at the cost of money.

Warning to computer users! NPV functions contained in spreadsheet programs usually discount the first number in the series. You must adjust for this if it is not your intent.

Success Rates

We use historical data to establish probabilities of success. Getting historical data requires the existence of a corporate database, access to an industry database, or direct benchmarking. We also need a fairly consistent definition of stages to interpret these data. What is a feasibility stage for one company or business unit may be another stage for a different organization. Fortunately, with broad adoption of stage-gate methods, such data are likely to be available.

We find it useful to track each project through a series of stages involving progressively higher probabilities of success but also increase costs. The improvement of probabilities of success in each stage is consistent with sound management of project risks.

There are two alternatives to the historical approach. The first is to use management estimates of project success probability. Management estimates may be subject to bias, and there are certainly indications that intuition is not a good guide to estimating R&D probabilities. Psychologically, it is difficult for a project champion to admit the chances for success for his project are less than 20%, though historical experience may show even worse odds. But he is rightly concerned that the project will be killed on that basis alone. The champion has a positive bias. Other managers have various negative biases; for example, there is a natural tendency to avoid areas of past failure, even though the nature of the opportunity has since changed.

Some companies and consultants favor scoring projects on the basis of "attractiveness," applying a 1-to-10 scale to qualitative parameters such as inventive merit, durability of competitive advantage, odds of technical success, time to completion, and odds of commercial success. They believe that the issues are too complex to be reduced to probabilities.[3] This argument may be circular, however, since initial ratings of attractiveness can, in principle, be later validated against historical experience. Without validation, they are a more sophisticated form of intuitive rankings.

In any event, probability of success at each stage is industry dependent and dependent on the definition of terms. In the automobile industry, a complete failure at the early commercialization stage, as occurred with the Edsel, is very rare, while in the pharmaceutical industry where 10,000 or more molecules are synthesized to find one successful new drug, failure at the conceptual stage is the norm. Probabilities of success at each stage will also be determined by the willingness of R&D managers to prune marginal projects.

The probabilities of success used in our model laboratory unit are based on metrics gathered in my own experience at a major industrial laboratory with a database of about 70 projects. Rounded off for convenience, they are:

Concept to feasibility	33.3%
Feasibility to development	50%
Development to early commercialization	75%
Early commercialization to full commercialization	83.3%
Full chain (cumulative): concept to commercial	10.4%

An effort to develop a universal algorithm for probabilities of success in R&D has also been published.[4] Its definitions differ considerably from the ones used here (with more stages and lower success rates), but the database is much larger and is based on a broader range of industries. The reader may want to study it for further insight on this key topic.

The entry point to the concept stage, where a project is formally defined and tracked, can be loose or tight. If it is loose, that is if relatively raw ideas are included within the tracking system, the odds of success in moving to the feasibility phase will be accordingly lower. The other alternative, where researchers have to present a strong case to authorize a defined project (but must correspondingly be given considerable latitude to explore projects before being required to track them) will give higher overall odds of success for defined projects. In addition, the data shows a progression of improving odds of success in each successive stage. Since the stages are also progressively costlier, this is again consistent with sound risk management.

Robert Cooper[5] has effectively summarized five rules for risk management in the development of new products. The first rule is that when the uncertainties are high, keep the amounts at risk low. This situation especially applies to the conceptual and feasibility stages of research. The converse (second rule) is that when uncertainties decrease, the stakes can be increased. This rule applies especially to development and early commercialization. The third rule is to incrementalize decision making—break the process into a series of stages and decision points. This allows needed flexibility as perceptions of opportunity and risk change over the course of the project. These stages correspond to the five stages of R&D we have just discussed. The fourth rule is to be prepared to pay for information that can reduce risk, for example, market research. The fifth rule is to provide ample bailout points. Our simplified portfolio model provides for a bailout only at the end of each of four R&D stages. More would be better. And as we shall see in our discussion of options below, the skillful use of the option to terminate adds substantial value to the overall R&D process.

There is a sixth rule for managing risk that applies to the portfolio as a whole—*diversify*. It is important and we shall discuss it in Chapter 13.

Project Flow in a Model Laboratory Unit

Exhibit 11.4 summarizes the project flow, productivity, and costs in our model laboratory unit. This exhibit is based on the concept of *cumulative probability of success*. Simply defined, it indicates the probability of a project successfully reaching successful commercialization. For projects in early commercialization, this is 83.3%; but for projects in the conceptual phase it is only 10.4%. It therefore follows that there is a significant need to fill the pipeline each year with enough

Stage	Probability to Advance	Cumulative Probability of Success	No. Projects	% Projects	Cost per Project ($M)	Annual Cost ($M)	% Cost	R&D Yield ($M)
Conceptual	33.3%	10.4%	96	57.8	0.375	36	30	3.7
Feasibility	50.0	31.3	32	19.3	0.750	24	20	7.5
Development	75.0	62.5	16	9.6	1.500	24	20	15.0
Early commercialization	83.3	83.3	12	7.2	3.000	36	30	30.0
Commercialization	NA	100.0	10	6.0		NA	NA	NA
Total			166	100%		$120	100%	$56.2
					Sales (5)	$240		46.9%

Exhibit 11.4 Projects and Cost in Model Laboratory Unit

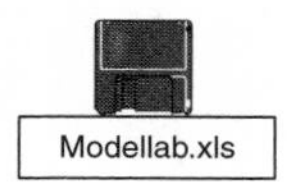

new projects (48 in this example) to maintain an output of 5 projects per year. There is also a significant obligation to kill losers early. If too many projects advance into the expensive development stage, available resources will be overwhelmed, vitality and productivity will be sapped, and the number of new conceptual projects allowed into the front end of the pipeline—including tomorrow's potential blockbusters—will have be reduced. "Culling the herd" isn't always easy, since management is tempted to think that the more projects in the development stage the better, since the rewards are closer. And some are pet projects of one or another researcher or manager. Careers and reputations are also on the line.

The dynamics of steady-state project flow through our model lab are illustrated in Exhibit 11.5. It shows the following for a given year:

- 48 new conceptual projects are begun.
- 48 conceptual projects continue (the conceptual stage lasts two years on average).
- 16 (33% of the annual turnover) advance successfully to the feasibility stage.
- 32 projects (67% of turnover) are dropped.

The first number in the feasibility row is equivalent to the 16 projects successfully exiting the conceptual stage. The process continues until we find 5 early commercialization projects advancing per year to successful commercial status. This process of advancement is often referred to as the R&D pipeline.

Stage[a]	New	Continue	Advance	Kill	Total	Time	Percent	Cumulative Percent
C	48	48	16	32	96	2	33.3	10.4
F	16	16	8	8	32	2	50.0	31.3
D	8	8	6	2	16	2	75.0	62.5
EC	6	6	5	1	12	2	83.3	83.3

[a] C–Conceptual Research, F–Feasibility, D–Development, EC–Early Commercialization

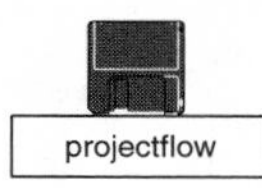

Exhibit 11.5 Project Flow in Steady State

The steady state is only a model; in a given year actual results will vary, perhaps substantially. The reason has to do with both statistical fluctuations among small numbers and the level of receptivity new products find in the marketplace. Tracking project advancement is an important metric, to be discussed in Chapter 14.

One lesson of this project flow exercise is clear: when cumulative probabilities of success are low, a large number of projects must be started annually to maintain R&D momentum. In our example, 57.8% of the projects are in the conceptual stage. This fact is appreciated by the pharmaceutical industry, which annually synthesizes and screens very large numbers of new molecules for biological activity; each molecule individually has a very low probability of success.

Costs, Productivity, and Value in the Portfolio

One look at cost distribution in the pipeline (Exhibit 11.4) indicates how costs fail to correlate with project numbers. Although conceptual projects represent 57.8% of project numbers, they only account for 30% of overall R&D costs. Conversely, early commercialization, which accounts for less than 8% of projects generates the same level (30%) of overall costs. The balance of overall R&D costs is found in feasibility (20%) and development work (20%).

The example shows, however, that resources must be balanced to have strong capabilities in each stage: a laboratory program cannot succeed if it has too few inventors and creators, nor will it succeed if it cannot support early commercialization. Certain resources can be outsourced (e.g., conceptual research to universities), and the manufacturing portion of early commercialization can be handled by contract, or "toll," manufacturing.

In total, the pretax annual cost of this R&D program is $120 million ($72 million after-tax if a 40% tax rate is assumed).[6] We can tie these results to the "productivity factor" we defined in the corporate growth model in Chapter 6 (Exhibit 6.6). This laboratory generates five typical commercially successful products per year, which should each have sales of $48 million in the fifth commercial year, for a total of $240 million. This is a productivity factor of 3.33 ($240 million/$72 million), and represents excellent performance.

The last column in Exhibit 11.4 is labeled R&D yield, and is calculated for each stage by multiplying the annual cost for projects in that stage by the cumulative probability of success. We see that on average only 46.9% of the total R&D effort will be associated with successful products, given the distribution of success rates used in the analysis. Most of the effort that does not pay off will occur in the conceptual and feasibility stages, where risk is high.

Another important observation is that, when R&D spending is adjusted for the time value of money, the conceptual projects at the beginning of the pipeline rise in financial importance even above the 30% of costs allocated to them in the annual R&D budget. The time-adjusted number is about 40.6%. These investments are 7 to 8 years away from being realized in commercial products, whereas money invested in on projects in early commercialization will be returned in 1 to 2 years. Hence, managing and prioritizing the long-term part of the portfolio is of both financial and strategic importance and must go far beyond benign neglect or the slogan "Let a thousand flowers grow."

R&D Contribution by Net Present Value Analysis

We can next determine the contribution of our pipeline of products to the company through net present value analysis. This is shown in Exhibit 11.6, both for the laboratory unit, and for PMI as a whole. Recall

	Model Lab Unit		PMI Case	
IRR	26.0%	20.0%	26.0%	20.0%
NPV per project ($M)	$76.9	$41.0	$76.9	$41.0
Number of project units	5.00	5.00	8.06	8.06
NPV	$384.5	$205.0	$619.8	$330.5
After-tax R&D cost	$72.0	$72.0	$116.1	$116.1
Annual R&D contribution	$312.5	$133.0	$503.8	$214.4
Value as perpetuity	$2,604.2	$1,108.3	$4,197.9	$1,786.6

Exhibit 11.6 Net Present Value Analysis of R&D Portfolio

that for the polyarothene project the pro forma NPV was $76.9 million and the IRR was 26.0%. There are five such projects in the lab unit and 8.06 in PMI, contributing NPVs of $384.5 million and $619.8 million respectively. But we must net out the after-tax cost of running the research program, $72.0 million and $116.1 million respectively. The annual R&D contribution is therefore $312.5 million for the lab unit and $503.8 million to PMI respectively.

What does this mean to PMI? Not only are we supporting PMI's aggressive goal of growing its sales at 12.4% per year, with a minimum 20% return on capital, we are adding shareholder value at the project level. To put the number in perspective, if PMI is earning 20% or $700 million, on invested capital of $3.500 billion, it might be fairly valued at 20 times earnings or $14 billion. If the company is growing at 12.4%, its annual expected increase in shareholder value will be $1.736 million and in invested capital $434 million. PMI has a powerful growth engine driven by opportunities to invest profitably in its mix of existing products, some (70%) old and some new (30%). The "brand-new" product stream from R&D is adding value of $503.8 million per year on an investment of $116.1 million. This value is largely hidden from shareholders: its current cash flow contribution is negative, and the first-year sales are too small to be apparent. But the $503.8 million is an impressive annual contribution. If R&D can perform at this level each year, we can capitalize this contribution, for example by treating it as a perpetuity[7] (see Chapter 5). Dividing by the cost of capital, 0.12, the value of R&D's sustained contribution to PMI's total value would be $4.198 million.

It might be fairly objected that this result is unrealistically optimistic, because polyarothene, with an IRR of 26.0% is an excellent project, and not typical. In the real world, great projects are scarce, and the R&D department might have to stretch to make PMI's 20% rate-of-return hurdle while maintaining a productivity factor of 3.33. The fourth column of Exhibit 11.6 illustrates this case. We have adjusted the typical project to an IRR of 20% and an NPV of $41.0 million.[8] Under these circumstances, the annual contribution to value drops to $214.4 million, still a very respectable number.

This method for estimating annual R&D contributions can be applied to any R&D portfolio—what are needed for each project are estimates of time to completion, NPV (which may be estimated using the quick-and-dirty methods discussed in Chapter 10), and probability of success. The results will be subject to a high degree of uncertainty, but will still be useful in identifying which projects will contribute value when, and in analyzing the difference in the value position compared with previous estimates.

Summary

A general approach has been outlined to project the output of an R&D portfolio in terms of both growth targets and value creation. Key inputs are estimates of success rates (best based on historical data from a corporate information base) for projects that meet the corporate hurdle rate and projected sales for these new products. In Chapter 12, we apply some of the probability–based methods introduced here to individual projects, including an analysis of decision trees and technology options.

CHAPTER 12

Decision Trees and Options

An understanding of decision trees and options provides critical insights into how the industrial R&D process is managed to minimize risk, and more importantly, how it can be used to maximize opportunity. The pro forma DCF tools introduced in Chapter 9, when coupled with probabilities of success, can be extremely useful in arriving at a first estimate of the value of a project, or the value being created in the R&D pipeline. However, it overlooks some important options: the option to terminate; the option to create second-generation products; and the option to accelerate a project when skillful R&D managers identify the opportunity to extract much more value than apparent from the most likely case. These options are analogous to those available to the skillful card player—to fold or lie low if the best cards are gone, or to increase the bet when the odds turn in the player's favor.

Pro forma models tell us little about some of the more spectacular examples of how real wealth and enormous shareholder value have been created through technology. Microsoft, the most visible example, leveraged a first-generation operating system into dominant positions not only in operating systems, but also in office productivity software and potentially in Internet communications. Control of a radical new technology to produce propylene oxide made Arco Chemical a major player in the global urethane polymers business. Genentech leveraged a portfolio of conceptual-stage projects and a stable of biotechnology stars into an instant market capitalization of nearly a billion dollars.

Netscape went public with a comparable valuation with no secure source of cash flow in sight. None of these phenomena can be explained simply by the reinvestment of free cash flow into a business earning solid returns.

Decision Trees

Even though most projects fail, the overall results of R&D can be profitable when the combination of R&D productivity and average project return is sufficiently high. The same logic can be applied to individual projects, taking the probabilities of failure into full account. The *decision tree* is a valuable tool for estimating probabilities for success and failure in these individual cases.[1] Decision trees capture a set of possible outcomes for a project where probabilities can be defined at each node or decision point. The methodology described in this section is often referred to as "Decision and Risk Analysis" and is routinely employed by many corporations. In our R&D case, the decision points are the four stage gates representing the transitions to feasibility, development, early commercialization, and full commercialization. A decision to continue or terminate a project must be made at each of these gates.

The term *event tree* is sometimes used when there is no decision to be made, only a logical reaction to events one cannot control. Strictly speaking, the model in this chapter is a combination of the two. Here we face decisions at each stage gate, and the outcome is subject to events we cannot initially predict, but to which we can assign probabilities, such as competitive reaction.

The decision tree exercise contains two critical lessons. First, it highlights the value of being able to *terminate a project* at each point in its development. This substantially reduces cost and risk. Second, it highlights the distribution of value in various possible project outcomes, in particular, the concentration of value in upside cases. Put another way, many R&D projects have an *asymmetric distribution of returns*—a huge potential upside with very limited downside exposure.

The project termination option is an absolutely critical tool to value creation in R&D management. We calculated that our model polyarothene project had an NPV of $76.9 million, and a 10.4% cumulative probability of success in the conceptual stage. We also estimated in Chapter 11 that it would cost $120 million pretax or $72 million after-tax to commercialize five projects of this size per year, that is $14.4 million per project if the cost of all failed projects is allocated against those that succeed. This creates a paradox: No one would invest $14.4 million for a 10.4% chance of making $76.9 million down the road. Yet we have demonstrated that both the individual project, and the R&D portfolio as a whole, are financially attractive. The option to terminate explains the paradox, making it possible to turn negative value into a positive outcome.

Also, in all our NPV calculations regarding the polyarothene project, we have not charged the enabling R&D investment to the project. We will do so now, while considering the full cost of money and the cost of failed projects in the analysis.

We also have not yet used the upside case and downside cases explicitly in our financials, although our pro forma analysis in Chapters 9 and 10 provide the sensitivity analysis required to do it. This decision tree exercise will incorporate them. First, we will consider project dynamics in terms of the eight-year development program for the typical project discussed earlier, with conceptual, feasibility, development, and early commercialization stages. For the project outcome, however, we will consider not only the familiar base case with a net present value of $76.9 million, but two other cases as well: upside and downside (labeled "high" and "low").

The Upside and Downside Cases

The upside case assumes that our proprietary position turns out to be stronger than anticipated; a powerful technical advantage protected by patents and trade secrets has significantly deterred competition. This improved competitive position has many good financial outcomes. We will assume our gross margin will be 55% instead of 50% and that sales will be double those in the base case. (We construct the case by changing the gross margin parameter in the spreadsheet exercise, then double the resulting NPV to reflect that project revenues and profits have doubled as well.) The downside case is the mirror opposite; we assume that gross margin drops to 45% and that sales are half of the base case.

We have already calculated the NPVs for the 45% and 55% gross margins cases in the sensitivity analysis exercise done in Chapter 10 (see Exhibit 10.1). The results were $44.8 million and $109 million respectively. Because we now assume that sales have either halved or doubled in our two scenarios, these numbers are halved and doubled respectively to $22.4 million for the downside case and $218 million for the upside case. We also assume that the base case has a 50% chance of occurring, while the probability of the upside and downside cases are 25% for each.

In the real world, each of these cases would be more complicated and present a number of business options. The upside and downside cases could be created by various combinations of price, volume, and cost assumptions. These would be affected nonlinearly by price elasticity relationships and economies of scale. In our simplified case, we have used gross margin as a surrogate for the price/cost relationship, anticipating that higher volumes would be associated with lower unit costs and vice-versa, rather than adjusting price or volume explicitly.

Results

We must first set up the calculations we need to construct a decision tree (Exhibit 12.1). To do so refer back to the model laboratory unit (Exhibit 11.4), where we list the annual cost of an R&D project in each of four stages (i.e., $375,000 per year for a project in the conceptual stage). These costs are shown as losses in the gain/loss column. They are discounted back to present value (in year 0) by the 12% discount factor in the second column, and tabulated again as "discounted gain/(loss)." Future R&D costs will be discounted, just as will be the rewards.

Let's now look at the rewards. Time frames must be kept in reference. This project is supported by R&D through year 8 and begins to earn commercial revenues in year 9. Therefore, the NPV calculated in Chapter 9 for the polyarothene pro forma must be discounted by 12% for 8 years by multiplying it by 0.4039. For the base case, the reward has a present value of $31.0 million. The high and low cases are discounted by the same factor.

The bottom tier of Exhibit 12.1 summarizes the results for each of seven possible outcomes. These include stopping the project after two years in the conceptual stage, and at the end of each stage thereafter. For example, if the project is terminated after year 2, the loss is $335,000 plus $299,000, or $634,000. Each of the first four outcomes represents a progressively larger loss, which is what happens when R&D projects are continued but ultimately fail. If the project fails at the very end, the present value of the loss will be $5.824 million.

The last three numbers assume the project moves forward to commercialization, and they are calculated by subtracting the full cost of the R&D, again $5.824 million, from the net present values of the three commercial outcomes shown in the first tier. There are some

Stage	Year	Case	Gain/(Loss)	Discount Factor	Discounted Gain/(Loss)
Conceptual	1		$(0.375)	0.8929	$(0.335)
Conceptual	2		(0.375)	0.7972	(0.299)
Feasibility	3		(0.750)	0.7118	(0.534)
Feasibility	4		(0.750)	0.6355	(0.477)
Development	5		(1.500)	0.5674	(0.851)
Development	6		(1.500)	0.5066	(0.760)
Early commercial	7		(3.000)	0.4523	(1.357)
Early commercial	8		(3.000)	0.4039	(1.212)
Commercial outcome (NPV)	8	Low	22.387	0.4039	9.042
		Base	76.876	0.4039	31.049
		High	217.955	0.4039	88.028

Total Outcome (NPV)	NPV
Conceptual/Stop	(0.634)
Feasibility/Stop	(1.644)
Development/Stop	(3.255)
Early Comm/Stop	(5.824)
Commercial—Low	3.218
Commercial—Base	25.225
Commercial—High	82.204

Exhibit 12.1 Gain/(Loss) in Polyarothene Project Outcomes

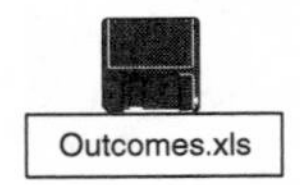

obvious observations. First, the upside case is enormously attractive—with a reward more than seven times the R&D cost. Second, the downside case leaves little value after the R&D costs are subtracted.

The flow can be summarized as a decision/event tree (Exhibit 12.2) showing the probability of each occurrence.

Probability-Weighted Project Outcomes

Armed with our probabilities for the outcome of the base case and its high and low alternatives, we can now look at the full spectrum of probability-weighted outcomes, as shown in Exhibit 12.3.

The first column shows the cost of termination in the conceptual stage. We had spent 0.75 million ($0.375 million in each of two years) in this stage, representing a total of three professional man-years. The NPV for this effort is slightly less, $634,000 because of discounting in the first and second years. But there is a probability of 66.7% of this project not making it even to the next stage. Hence the weighted NPV is a negative $0.423 million, a fairly large number in terms of total

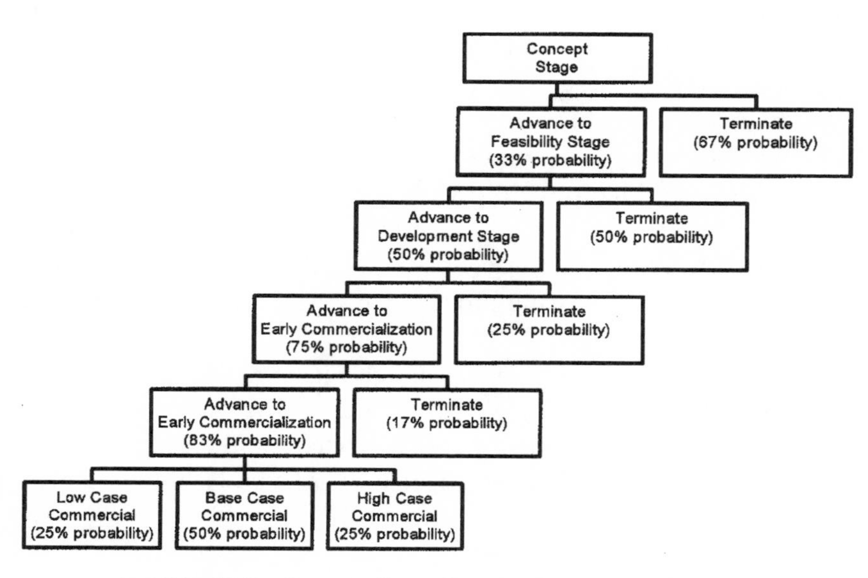

Exhibit 12.2 Decision/Event Tree for Polyarothene Project

Probability Tree ($M)

	Kill after Conceptual Stage (%)	Kill after Feasibility Stage (%)	Kill after Development Stage (%)	Kill after Early Comm. Stage (%)	Low Comm. Case	Base Comm. Case	High Comm. Case	Total (%)
Probability								
Stage C	66.67	33.33						100
Stage F	66.67	16.67	16.67					100
Stage D	66.67	16.67	4.17	12.50				100
Stage EC	66.67	16.67	4.17	2.08	10.42			100
Commercial	66.67	16.67	4.17	2.08	2.60	5.21	2.60	100
NPV	$(0.634)	$(1.644)	$(3.255)	$(5.824)	$3.218	$25.225	$82.204	
Weighted NPV	$(0.423)	$(0.274)	$(0.136)	$(0.121)	$0.084	$1.314	$2.141	**$2.585**
				Memo: Base Case alone				$1.674

Exhibit 12.3 Probability-Weighted Outcomes for Polyarothene Project

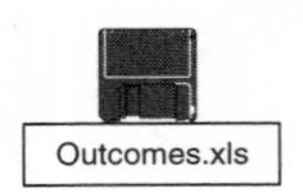

project NPV. The lesson is that the costs of early-stage research are high, and that unpromising projects should be killed as soon as their lack of promise becomes apparent.

The second column of the chart reflects the 33.3% probability of advancement to the feasibility stage, and the 50% probability of project termination. The cumulative probability of this outcome, which has a present value of –$1.644 million, is 16.7% (50% × 33.3%), giving a *weighted* NPV of –$0.274 million.

The exhibit continues in this fashion. Although the absolute cost of terminating the project in the development or early commercialization stages is very high, the probability of that not happening is also quite low, about 4% and 2% respectively, so the weighted costs are fairly modest. This is the result of sound R&D management and a commitment to terminating projects early, when the outcome is in doubt.

The last three columns of the exhibit represent the 10.4% probability that the initial idea makes it to one of three commercial outcomes. As noted, the downside case barely covers the cost of capital after the sunk R&D costs are charged to it. The probability-weighted net present value for all three outcomes is $2.585 million. If the analysis had been done on an unweighted basis (see Memo line), for the base case alone, the project NPV would have been significantly lower, $1.674 million. In other words, weighing a spectrum of outcomes with an asymmetric distribution of rewards increased value by about 50% versus valuing only the most probable outcome.

The lesson here is straightforward: *Never ignore the upside case; value is concentrated there*. In general, upside cases more than outweigh the downside cases, especially if the option for early termination of the project (when downside outcomes seem likely) becomes part of the decision process.

Monte Carlo Analysis

Instead of assigning weights to a small number of discrete scenarios, as in the decision tree above, we can, in principle, simultaneously compute returns using a number of variables, each of which can be assigned a probability distribution. In doing so, we enter the world of *Monte Carlo analysis*, where a probability-weighted calculation is done for a multidimensional grid of possible cases (requiring heavy-duty computing). To do so in a meaningful way, however, would presume knowledge of the probability distributions of outcomes for a significant number of project parameters, which in turn hinges on a sophisticated corporate database.

Monte Carlo analysis is not science fiction. Merck,[2] which has an extensive research database, has found it useful to build a Research Planning Model based on Monte Carlo methods and to employ it in its strategic decision-making process despite its daunting mathematical complexity.

Exhibit 12.4 works a simple Monte Carlo example based on the now familiar polyarothene case. Let us assume that the main unknowns are gross margin and capital intensity, per the tornado diagram in Exhibit 10.4. The base case values for these parameters were 50% and 70% of sales respectively. Further assume that our business research has determined that the standard deviation for gross margin is ±5% and that our engineers believe that the standard deviation for fixed capital is ±15%. We also assume a normal distribution for each parameter. Exhibit 12.4 shows the statistical distribution for Internal

Forecast: IRR

Statistic	Value
Mean	26.7%
Median	26.0%
Standard deviation	7.7%
Variance	0.6%
Skewness	1.20
Trials	29,982

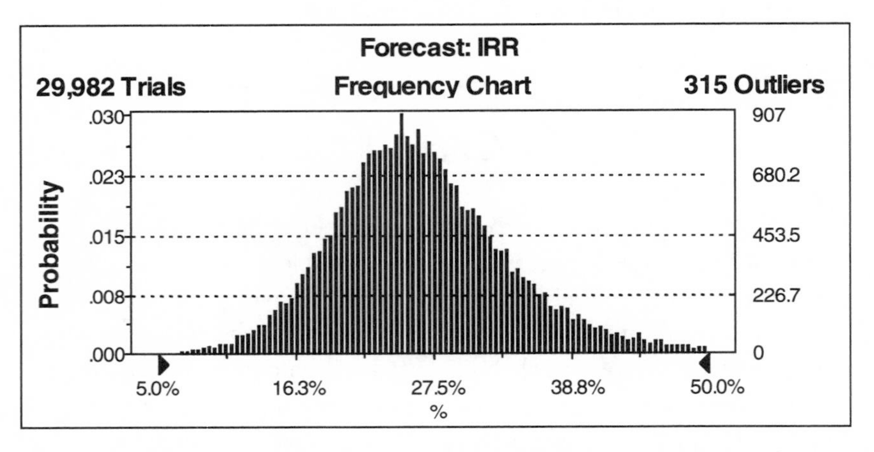

Exhibit 12.4 Monte Carlo Analysis of Polyarothene Business

Rate of Return when we run multiple (30,000 in this case) iterations of the spreadsheet with normally distributed variables. We also chose to treat these variables as independent, but might have assumed a degree of correlation between them.

The most interesting result from the calculation is the standard deviation for IRR: ±7.7%. This tells us there is still quite a wide range of possible outcomes in either direction, given the uncertainties in two sensitive parameters. We also see an upward skew to the curve: the mean is higher than the median, which is identical to the base case return of 26.0%.

This statistically based calculation is no longer onerous and can be made using readily available computer hardware and software. The analysis presented here, in fact, employed a user-friendly add-on to a standard software package,[3] and ran the computations in about 10 minutes on a 300MHz personal computer.

The Option to Accelerate

R&D planning normally involves the pulling and pushing of competing projects, rival champions, and competing constituencies. But every so often, a project idea is so compelling that the need to exploit it to the fullest is obvious to everyone. Indeed, there is a shared understanding that the company's future would be materially diminished if competitors sensed the same opportunity. In these situations, it is time to think about reducing the development cycle by accelerating the project.

Imagine that our company, PMI, has discovered that the polyarothene project, now nearing the end of two years of research in the conceptual stage, is very likely to achieve the upside case outcome. The decision is made to double available resources with the goal of reducing the remaining cycle time from six years to three. (There is an adage that nine mothers can't make a baby in one month, but let's ignore that for a moment.) The question is, how will this acceleration impact the NPV of the project? Exhibit 12.5 shows the results.

The analysis indicates that the acceleration option produces $43.2 million in additional value for the company. This does not even consider a number of other factors that favor shorter cycle time and being first to the market with a new technology or product concept. The effect on project risk may also be positive: changes in the marketplace and the competitive situation are a leading cause of commercial failure.

R&D	Year	Discount Factor	Base Case Current $	Base Case Discounted $	Year	Accel. Case Current $	Accel. Case Discounted $
Stage F	1	0.893	(0.75)	(0.67)	1	(1.50)	(1.34)
F	2	0.797	(0.75)	(0.60)	2	(3.00)	(2.39)
D	3	0.712	(1.50)	(1.07)	3	(6.00)	(4.27)
D	4	0.636	(1.50)	(0.95)			
EC	5	0.567	(3.00)	(1.70)			
EC	6	0.507	(3.00)	(1.52)			
Commercial Stage	6	0.507	$217.96	$110.42	3	$217.96	$155.14
NPV	Total			$103.91			$147.13

Exhibit 12.5 Value of the Acceleration Option

By reducing cycle time, we also reduce our exposure to these uncontrollable factors.

How Realistic?

The thoughtful reader may wonder at the realism of placing a value on any project proposal. After all, a project proposal at this point is merely an idea and has not even benefited from conceptual stage research. There are a number of ways of looking at this concern. First, the $2.585 million at which we have valued the polyarothene project at this very preliminary stage is a small number compared with the several hundred million in sales that a commercially successful project could generate in the future. It became that small because we have discounted its potential rewards by the time value of money and by the low probability that the undeveloped idea will ever become a commercial success. Given these realities, no single project at the conceptual stage represents a case of "betting the company," even if large capital expenditures will be required to turn the dream into reality. At the same time, in this perspective, a brilliant discovery, which will take time to develop, may not justify extraordinary valuations. As money is spent reducing the risks, however, and as the time horizons of the reward move closer in time, the net present value will inexorably rise—say to the $76.9 million represented in the base case for polyarothene, or higher for the upside case. The successful project will then have a truly material effect on the company and its prospects. If it is true that this project only has a present value of $2.585 million, it also follows that 96 conceptual projects of comparable quality in the model laboratory's portfolio would be a material asset, and that the value of the total intellectual property in the R&D pipeline is virtually priceless in the context of the company's growth and its future.

Options Analysis

In considering decision trees, we have already begun to deal with options analysis. The decision tree exercise demonstrates that there is far more to sophisticated analysis of R&D rewards and opportunities than any simple pro forma enterprise model can reveal. The proper utilization of the *termination option* is absolutely critical to maximizing portfolio value. The *acceleration option* is also an extremely powerful tool in both a financial and competitive sense.

But the *option of making follow-on investments* can also be the key to major value creation in business enterprises. Brealey and Myers have demonstrated this in an interesting case involving a first- and second-generation computer; in this case, a financially unattractive first-generation project became a financial winner in its second incarnation.[4]

These types of options, all very important to the value creation process, do not go far enough in explaining the keen interest in technology options. That interest is driven by the desire for a new paradigm for evaluating the opportunity side of technology that does not understate the benefits or put too much emphasis on the risks. Standard enterprise models, such as pro forma models do little to explain the more spectacular examples of wealth creation in the history of technology. They miss the upside and the follow-on investments. They miss the links with other technologies, including those yet to be revealed. Enterprise models are helpful, but only as entry points to more sophisticated forms of decision making.

Certainly, some research executives are frustrated in their attempts to explain the value of their technology projects to financial officers who feel that the risk-weighted returns of R&D simply are not worth the investment. These failures of communication reflect real differences in thinking and language between professionals with distinctly different backgrounds and responsibilities.[5] One of the promises of options analysis is that the characteristics of financial options are well understood by financial professionals and bear important relationships to options as technologists view them. Perhaps the most important of these characteristics is that, *whereas increased volatility reduces the value of an enterprise, it increases the value of an option!*

In this section, we will:

- Define an option in commonsense financial terms.
- Differentiate a technology option from a financial option.
- Introduce the concept of opportunity-weighting in portfolio analysis.

FINANCIAL OPTIONS

To understand the option concept, we will begin with a stock option. Consider an R&D scientist named Julia, whose firm, PMI, granted her an option one year ago to buy 1,000 shares of its stock at $50 per share, no matter what its current market price might be. By agreement, Julia can exercise this option at any time during the 10-year period ending January 15, 2008.

In the parlance of options trading, 50 is the strike price, January 15, 2008 is the expiration date, and Julia has what is described as a "call" option—the right (but not the obligation) to call up to 1,000 shares of PMI stock away from its current owner at the strike price anytime before the expiration date. Stock call options are bought and sold daily, principally on the Chicago Board Options Exchange. So too are their reverse—"puts." A put option allows its holder to put his or her shares to someone else at the strike price any time prior to the option's expiration.

In options trading markets, the value of an option is a function of three things: (1) the strike price relative to the current market price of the stock, (2) the amount of time left before the option expires, and (3) the volatility of the stock price. Thus, if PMI were currently trading at $55, and if the option still had 9 years left before expiration, Julia's option would have an "intrinsic value" of $5,000 (($55 – $50) × 1,000). But even if the market value of PMI shares dropped to $45, a skilled investor would attribute value to the options based on the stock's volatility and the length of time the options had before expiring. (We will evaluate this second case, where Julia's options are "out of the money" by $5 below the strike price.)

USING THE BLACK-SCHOLES FORMULA

Sophisticated investors who trade in options typically use the Black-Scholes formula to calculate option value. Brealey and Myers describe this mathematical formula as "unpleasant-looking" but one that on "closer acquaintance [the user] will find . . . exceptionally elegant and useful."[6] Explaining the formula itself here serves little purpose, but this section examines the underlying concepts and describes the steps required to use it.

Two basic concepts are embodied in the Black-Scholes formula. The first is the assumption that prices of securities perform a random walk about mean values, and that the distribution of values around the mean can be accurately represented by what is disarmingly called a "normal distribution," or bell curve. Many statistically random phenomena, such as a series of coin flips, are normally distributed. This assumption is related to the "random walk theory" of stock price movements. The term random walk derives from the path followed by an inebriated person. There is no pattern to this path, and nothing in past movements is predictive of future movements.

Everyone knows that on average, 100 flips will yield 50 heads. But you must use a normal distribution to calculate the chances of getting 55 heads or more in any experiment involving 100 flips. Normal distributions are quite common in natural phenomena, and scientists, statisticians, and some expert gamblers are familiar with them.[7]

The curve has only two parameters: (1) the mean value about which prices vary, and (2) the standard deviation, which defines the width of the curve, and is a measure of the volatility of the security. Exhibit 12.6 shows such a curve with a mean value of 1 and a standard deviation of 0.20. A statistical characteristic of normal distributions is

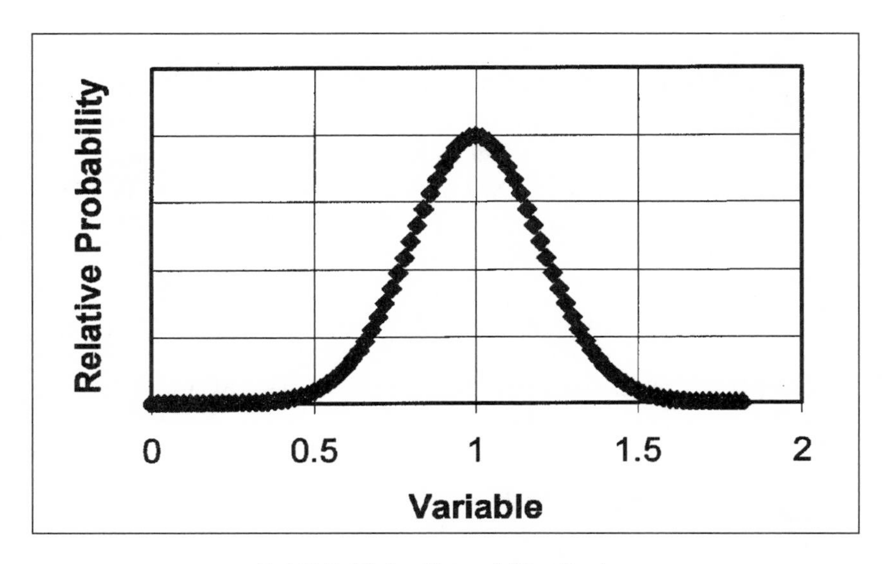

Exhibit 12.6 Normal Distribution

that 68% of the time the value will fall within one standard deviation of the mean, and 95% of the time within two standard deviations. For example, if stocks on the S&P 500 have an average standard deviation of 0.15, they will on average have prices within 15% of their mean value 68% of the time, and within 30% of their mean value 95% of the time.

The second assumption is that the shape (width) of that distribution will be reasonably constant with time. This is absolutely true of coin flips, and true often enough with securities that the theory can be profitably used to price stocks, bonds, and options in real financial markets. Obviously, the standard deviations of *different classes* of securities are not the same. Utility stocks have much less volatility than biotechnology stocks. Moreover, it is mechanically easy to calculate standard deviations for each publicly traded stock (e.g., Exxon), and average standard deviations for all stocks in a class of securities (i.e., petroleum companies), and such calculations are widely published. The published standard deviations will depend on the selection of the data: the time period sampled, the frequency of sampling, and whether outlying data are discarded or not. Standard deviations are not universal constants.

The following steps represent a sample Black-Scholes calculation for Julia's call option (see Exhibit 12.7):

Step 1. We need to first calculate the parameter P/PV(X), where P is the stock price, and PV(X) is the present value of the exercise price *on a risk-free basis.* The stock price P is $45, out of the money by $5. The exercise price X is $50. The risk-free rate is 0.052, and

$P = 45$	$Z = 0.6$
$X = 50$	$D_1 = 0.8848$
$r = 0.052 \quad 1.052^9 = 1.578$	$D_2 = 0.2848$
$t = 9 \quad \sqrt{t} = 3$	$N(D_1) = 0.8119$
$PV(X) = 31.68$	$N(D_2) = 0.6121$
$Y = 1.4203 \quad Ln(Y) = 0.3509$	$W = 0.381 = 38.1\%$
$\sigma = 0.2$	Value $= W \times P = \$17.14$

Exhibit 12.7 Julia's Financial Option

the option has 9 years to expiration. PV(X) is $50 ÷ (1.052)⁹ = $50 ÷ 1.578 = $31.68. The ratio P/PV(X) = $45/$31.68 = 1.42. Call this parameter Y.

Step 2. We next need to calculate the parameter $\sigma\, t^{1/2}$ (call it Z) where σ is the relative standard deviation of the stock, and $t^{1/2}$ is the square root of the time. Julia's stock has a standard deviation of 0.20. The square root of 9 years is 3. Hence $\sigma\, t^{1/2}$ or Z is 0.60.

Step 3. There is a convenient table of coefficients for Call Price as a Percentage of Stock Price (call them W) in Brealey and Myers (Appendix Table 6),[8] as a function of the two parameters, Y and Z, calculated in Steps 1 and 2. Interpolating from this table, we obtain W = 37.8%.

Alternate Step 3. To those of us who prefer working with computers, or detest having to interpolate, the cumulative normal distribution is a standard mathematical function in Microsoft Excel (NORMSDIST) and other spreadsheet software. Call this function N.

First calculate $N(D_1)$ where $D_1 = \ln(Y)/Z + 1/2Z$. {ln is the natural logarithm.}

Calculate $N(D_2)$ where $D_2 = D_1 - Z$. The percent value of the option, W, is given by the relationship $W = N(D_1) - N(D_2)/y$.

Step 4. Per Black-Scholes, Julia's option is worth $17—38% of the stock price of $45. Julia should not worry about having her options "out of the money" when they still have nine years to run.

What would the option be worth if Julia's stock were more volatile? Assume Julia works for a biotechnology company whose shares have a standard deviation of 0.40. In this case, $\sigma\, t^{1/2}$, or Z, is 1.2 and the option is worth 54.5% of the share price or $24.50.

Volatility favors the option holder, but not the shareholder, for whom it represents a higher cost of money, per Chapter 5. On the other hand, the option holder is penalized by dividends, which we assumed were zero in the preceding calculations. If Julia's company paid a $1.50/share dividend each year for 9 years ($13.50), it would severely penalize the value of her options. Figuring in dividends in the Black-Scholes scheme is beyond the scope of this book, but if the issue is important to you, the formulas are known and you should check into it.

The Technology Option

The technological analogy of a call option is an investment in technology that enables entry into a promising field: one that is exercised only *if* that promise begins to materialize.

Today, a company with a major stake in telecommunications may want to create technology options with respect to the critical question of whether data will reach homes predominately via telephone lines, cable, or direct broadcasting. Betting wrong could be disastrous, and the evolution of technology makes the outcome uncertain. Volatility in a technology sense is high—new product announcements and new strategic alliances among potential competitors are announced weekly. Having a position that will work in any of the possible scenarios is valuable. CFO Judy Lewent of Merck put it well:

> When you make an initial investment in a research project, you are paying an entry fee for a right, but you are not obligated to continue that research at a later stage. Merck's experience with R&D has given us a database of information that allows us to value the risk or volatility of our research projects, a key piece of information in option analysis. Therefore, if I use option theory to analyze that investment, I have a tool to examine uncertainty and to value it. . . . To me, all kinds of business decisions are options.

The recognition that volatility is a *positive* factor in research is explicit in Merck's analysis. As Lewent put it,

> . . . a traditional analysis that factors in the time value of money may not fully capture the strategic value of an investment in research, because the positive cash flows are severely discounted when they are analyzed over a long time frame. As a result, the volatility or risk isn't properly valued.[9]

A financial option has two characteristics: (1) a right to exercise, and (2) a possible requirement that capital be available for an ensuing investment. A technology option has these same characteristics. The right to exercise is typically embodied in intellectual property rights such as patents, trade secrets, and the absence of encumbering patents or rights held by others. A capital investment must be made to exploit the option, unless the decision is made to immediately sell the option to a third party. (Stock options, by the way, are rarely exercised by their initial owners. Even when they are "in the money" they are sold to others who may or may not exercise them. And many expire worthless.)

In direct parallel to the Black-Scholes model, Merck uses the capital investment to be made at the time of option expiration as a proxy for the strike price. The present value of project cash flows is a proxy for the stock price, and the time to expiration is the period in which the technology must be judged to be competitive. Volatility is derived from the standard deviation (σ) of biotechnology stocks, and, as expected, the risk-free rate is based on Treasury bills. Merck's logic regarding volatility is strong. We have seen that investors in biotechnology stocks will not pay a premium for unique risk since they can use diversification to achieve an efficient portfolio. Merck, with its large R&D budget and its diversification into a variety of pharmaceutical markets, can also achieve a reasonably efficient (i.e., uncorrelated) R&D portfolio. The most critical assumption appears to be that of timing: when competitive technology positions will be strong enough to preclude market entry. Merck handles this issue via sensitivity analysis—a curve relating value to "time to expiration."[10]

Classic Black-Scholes theory can be used to study the polyarothene business, using the Merck approach.

Polyarothene Technology Option

First calculate X, the exercise price. This is the investment required to enter the business. Assume it is the initial investment in year 9, \$10.4 million in fixed capital and \$3.1 million in working capital (per Exhibit 9.4), or \$13.5 million. From here on, the business will be financed through its free cash flow, which will be used to impute the value of the business, P. To the initial capital investment we must add the R&D investment in the project which totals $\$(2 \times 0.3750 + 2 \times 0.7500 + 2 \times 1.500 + 2 \times 3.000) = \$11.25$ million. With this method, however, we do not impute the costs of other (failed) R&D projects to the option value.

To calculate NPV(X) we must bring both numbers back to a present value, but do so at the risk-free rate of 5.2%. The present value of the R&D investment, which occurs according to the same schedule found in Exhibit 12.1 (but is summed at a lower discount rate) is \$8.326 million. The present value of the initial capital is \$8.555 million for a total NPV(X) of \$16.881 million.

Next calculate P, the value of the project. The NPV of this business at PMI's cost of capital imputed a value of \$76.9 million (Chapter 9). However, following Merck's logic, to this we should add back the initial investment of \$13.5 million, to give a total of \$90.4 million. However, we have yet to consider the odds of success, which are 10.4% in our model. P = 0 .104 × \$90.4M = \$9.402 million.

Calculating Y = P/PV(X) we obtain Y = 0.5570.

R&D projects are highly volatile. Merck assigns σs between 0.4 and 0.6,[11] based on the volatility of biotechnology stocks. Let's use the midpoint for now, $\sigma = 0.5$. We must also assign a time period over which the option can be exercised. In our case, that is 8 years, the time it will take to complete the R&D program; it assumes that there is no additional time delay imposed by the need to build a plant.

The rest of the calculation is mechanical—no different from Julia's financial option (see Exhibit 12.8). The result is an option value of \$3,571 million. This is somewhat higher than the two values we calculated earlier in this chapter using decision trees (Exhibit 12.3), which could be considered as a risk-weighted enterprise model. The decision tree values were \$1.674 million for the base case alone, and \$2.585 million for the blend of upside, base, and downside cases. One could conclude that the option model gives even more weight to

P = \$9,402M	Z = 1.414
X = \$13.5M + 11.25M	D_1 = 0.3095
r = 0.052	D_2 = −1.1048
t = 8 √t = 2.828	$N(D_1)$ = 0.6215
PV(X) = 16.499	$N(D_2)$ = 0.1346
Y = 0.5699 Ln(Y) = −0.5624	W = 0.385 = 38.5%
σ = 0.5	Value = W × P = \$3,622M

Exhibit 12.8 Polyarothene Project Option

the upside case and the value of the asymmetric distribution of R&D rewards.

The two forms of analysis carry some common elements: the reward of the project, based on free cash flow is the same, as are the time frame and the overall probability of success. Nevertheless, caution is warranted because the two analyses also contain some independent assumptions. One is the value and probability of the upside and downside cases in the decision tree model. The second is the risk factor in the option model which is embedded in the parameter $\sigma\ t^{1/2}$. Just as with Julia's financial option, as volatility (standard deviation) goes up, the value of the option increases. As time to exercise decreases, value decreases. In evaluating R&D options, Merck uses a broad range of cases for both σ and t, representing the legitimate uncertainty that research managers have about volatility and risk.

We can look at this more carefully by plotting the value of the polyarothene option as a function of $\sigma\ t^{1/2}$, which is a good surrogate for risk. And for an option, *risk is desirable* (see Exhibit 12.9). The curve is quite interesting and potentially useful since identifying a range of plausible values along the $\sigma\ t^{1/2}$ axis would allow one to calculate a corresponding range of option values. Also, the curve is interesting in showing in the starkest terms that favorable assumptions in the enterprise model—shorter time frames and lower volatility—work *against* value in the option model.

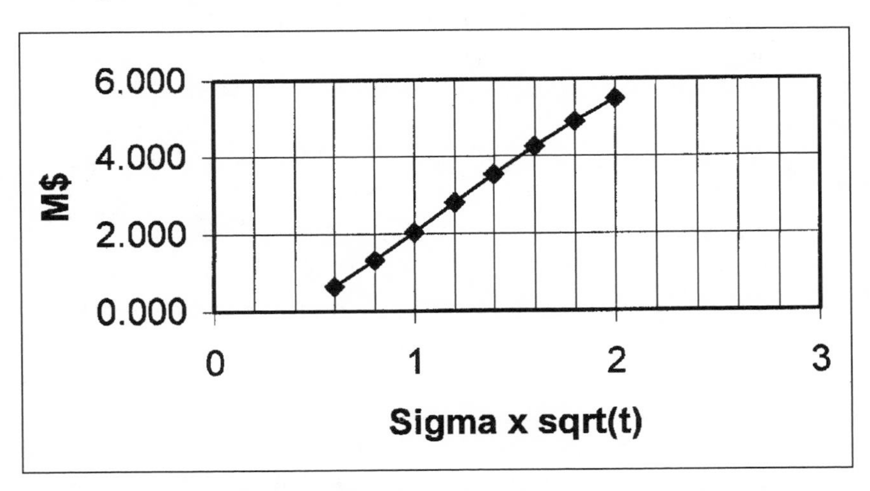

Exhibit 12.9 Option Value and Risk (Polyarothene Case)

Other Characteristics of Technology Options

Technical knowledge and skilled people are two additional characteristics of a technology option. The degree to which one has these elements defines the odds of successful technology transfer and hence the probability of ultimate commercial success. The technology option cannot be effectively exercised without them.

In principle, a new technology can be combined with a set of existing technologies to create a corresponding set of technology options. In practice, that set might be very narrow (perhaps zero) or quite broad. Each of these options, if exercised, can lead to even broader second-generation possibilities. An example of a technology that creates very narrow options would be a new paint formulation. This could be of considerable commercial importance and rely on a substantial body of technical skills, but it would likely be aimed at a particular market segment and create few possibilities outside that segment. The Windows PC operating system, on the other hand, created a remarkably broad set of technology options. The successful development of Windows led to the possibility of combining that technology with existing technology for spreadsheets, word processors, presentation and graphical software, databases, and networks. It also provided downstream options to create major new opportunities in business software, engineering software, and communications that could not have been envisaged from an operating system designed for small computers alone. This is fortunate for Microsoft because its enormous cash flow and investor expectations for continued growth demands a flow of major new investment opportunities.

The two preceding examples illustrate potential differences in the value of the two projects. A simple pro forma analysis of either the paint formulation project or the PC operating system would successfully capture the development costs and subsequent operating revenues. It would not, however, capture the potential downstream opportunities for the PC operating system.

Earlier, we suggested using the ratio of net present value to R&D cost as a tool for project prioritization. Options analysis can give quite a different set of results—for example, the development of a technology with a negative NPV as measured by direct enterprise profits, but creating a wealth of future options, could be an excellent investment.

Sources of Value in Technology Options

Consider the following four factors when evaluating the business options created by a new technology:

1. *Technology Pairing.* A new technology can be paired with an existing technology to create value. It is useful to look at a set of such pairs and begin the analysis of where the greatest value may be created. Platforms (see following section) create enhanced potential for technology pairing.

Example
The R&D laboratory proposes a conceptual project to invent a rechargeable battery with a ratio of energy to weight two to three times better than current products. Technology pairs can be identified to electric vehicles, laptop computers, portable entertainment devices, cellular phones, medical devices, power tools, and consumer batteries.

2. *Strength of Linkage.* The value of the innovative technology may be quite different in different markets—the linkage between the technologies can be strong or weak.

Example
A lightweight battery system may be the key to gaining consumer acceptance of electric vehicles. This linkage is strong. Some new portable medical devices could be commercialized if lighter weight batteries are available. This linkage is also strong. Laptop computers could immediately add functionality if weight and space constraints were eased. This linkage is fairly strong. All other applications would benefit from increased battery life, and adoption is likely.

However, existing products are already satisfactory. These linkages are weaker.

3. *Size of Current and Potential Markets.* The value of the technology option will relate not only to the value added per unit, but the number of units in the target market.

Examples
The potential market in electric vehicles is enormous, but it will take many years to develop that potential. The market in portable medical devices is small because of the limited portion of the population affected, but the value added per unit may be very high. The consumer market for rechargeable batteries is small because primary (non-rechargeable) batteries are considered more convenient.

4. *Polarization of the Linkages.* When two technologies are coupled, the owner of one technology or the other is likely to dominate the relationship, and hence gain the major piece of the value-added. Here, the term polarization is used to define the direction in which dominance occurs. Both technology and commercial strengths will be important in determining how the rewards will be shared.

Example
The inventor of a highly superior new battery may be able to dominate the consumer electric vehicle market because the technology is virtually enabling, the competition cannot counter effectively using price, and because many other firms are willing to sell the automotive components and assembly services needed to deploy the technology. Polarization may favor the inventor. In the case of the laptop, it is highly unlikely that any single superior feature will suffice to succeed in this highly competitive market, even though most of the technology is available commercially. Polarization would seem to favor the current vendors.[12]

Technology Platforms

Product developers are fast becoming familiar with the advantages of developing products from common "platforms." Meyer and Lehnerd have defined a platform as "a set of subsystems and interfaces that form a common structure from which a stream of derivative products can be effectively developed and produced.[13] A technology platform is similar, and equally effective as a launching pad for new products. Basing a line of products on a common technology platform makes it possible to leverage R&D costs and manufacturing economies, and vastly reduce parts inventories. The platform creates a large number of technology options that are less expensive and less time consuming to exercise than technology that is designed and optimized to a single end use. Therefore, a technology option that includes an embedded platform is likely to be more valuable.

The platform idea is not new. It was popularized by the K-Car platform at Chrysler, which had low market share and could hardly afford to compete with GM and Ford in designing each new car model from scratch. The K platform was used on low-, medium- and high-end passenger cars, and very importantly on minivans, which became a Chrysler core competency. Software developers have adopted the concept in their design of "Office Suites"—word processing, spreadsheet, graphics, presentation, database, and scheduling programs. The customer has the benefit of mostly common features in the user interface (hastening the learning process) and the ability to incorporate information created in one part of the suite into another. This capability was used in this book to incorporate modified versions of charts, graphs, and text, originally developed as presentation graphics, into the word-processing format required for publishing. These technology platforms, designed for individual users—corporate, small business, and home—when coupled with client-server technology, lead to whole new sets of options to create platform-based software for large-scale business computing. This is still somewhat distant, but far closer than before the platforms came into existence and gained widespread acceptance.

CHAPTER 13

Creating Value through Diversification

As explained in Chapter 5, there is no economic profit unless one earns more than the cost of capital. It is also true that *investors use diversification to eliminate unique risk*. Further, systematic risk cannot be diversified away and will be directly proportional to the correlation of individual investments to the risk in the market as a whole.

These conclusions, which are well understood by investment managers, are far less well understood with respect to technology investments (i.e., R&D projects and R&D portfolios).

In the first part of this chapter, we will discuss the implications of diversification to an R&D portfolio. We will then look at why a broad portfolio of quality ideas, each of them individually risky, may have more value than a narrow portfolio of low-risk concepts. This leads directly to why R&D creativity should be more highly valued and encouraged than the reduction of R&D risk at early project stages.

Finally, we will look at the use of external technology sources, which can broaden technology options and reduce time to commercialization as value-creating mechanisms.

Efficient Portfolios

It is generally assumed that all financial investors want to maximize return while minimizing risk, although they may have different profiles regarding how much risk they are willing to tolerate. This has led to a financial theory of *efficient portfolios*—portfolios that provide the highest expected return for a given amount of risk.[1] The theory of efficient portfolios is beyond the scope of this book, but the important conclusion is that such portfolios have standard deviations (or, variability of return; a proxy for risk) that are substantially less than that of the average security in the portfolio, and usually less than *any* individual security in the portfolio. Standard deviation, or risk, is minimized by:

- Including many stocks in the portfolio.
- Selecting stocks whose correlations with each other are low.

Correlation factors can be determined statistically from historical stock price data, but also correspond to informed intuition. The price history of General Motors and Ford will be highly correlated, whereas that of Duke Energy, Campbell Soup, and Microsoft would be less correlated. An efficient portfolio of 20 stocks would not likely include *both* General Motors and Ford.

Investors naturally seek what financial scholars call "efficient portfolios." Efficient portfolios minimize risk while maximizing expected return, and lie along a curve such as that shown in Exhibit 13.1. In Chapter 5, we saw this type of risk-reward relationship in correlating

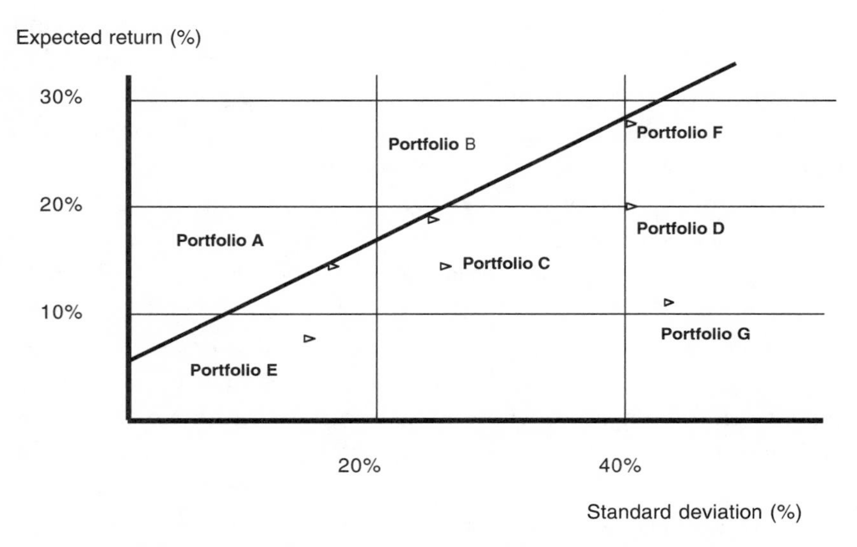

Exhibit 13.1 Efficient Portfolios—Risk versus Reward

the expected return on an equity investment and the sensitivity to market risk (or beta).

In the exhibit, portfolio A has a low expected return, but low risk. Portfolio F has a higher expected return, but also a higher level of risk. But notice the positions of portfolios B and C in the figure. Portfolio C has the same return as A, but takes more risks in getting it. It is as risky as B, but has a lower expected return than B with the result that the portfolio produces less return for the risks taken. Now, if we look at portfolios D and F, we see that they are equally risky, but F has a high expected return for the same risk.

Which of these portfolios would you rather own? The answer would surely depend on your aversion to risk or need for returns. These factors would naturally lead you to either A, B, or F, which give you the greatest expected return for a given level of risk. These portfolios lie along the *efficient frontier.* Only an irrational person would prefer portfolios C, D, E, and G, which have nothing to recommend them. They are "inefficient."

The Problem of Liquidity

The most obvious distinction between financial and technology portfolios is *liquidity*. Financial assets can be bought and sold quickly and with very little transactional friction because they are traded in highly efficient markets. Therefore, one can construct a portfolio virtually overnight, incorporate within it securities with whatever characteristics one seeks. Diversification can be achieved over different classes of assets, within industry groups, between industry groups, over geographic regions, or other parameters.

Because of this fact, individual companies that attempt to diversify for the benefit of their shareholders are seldom rewarded. Conglomerates—companies with major holdings in several different industries (e.g., airlines, hotels, restaurants)—may arguably have more earnings stability than companies whose fortunes are tied to a single market. However, investors can achieve a similar degree of diversification by buying separate holdings in an airline, a hotel firm, and a restaurant chain. It is quicker and cheaper for investors to create their own portfolios than to leave that job to conglomerate executives. They don't need someone like ITT or Litton Industries to play portfolio manager on their behalf. The problem of liquidity is significant for the conglomerate. If an airline company decides to shed its restaurant and hotel businesses, it will have to pay millions to investment bankers, and the time and attention of top management will be largely taken away from the main business. These problems explain why investors often penalize diversified companies with a "holding company discount"—a *negative* premium for diversification.

Diversified portfolios of technology assets are even harder to assemble and liquidate than portfolios of either securities or businesses. Technologies exist for the most part in an organizational context. To obtain them, you must buy *all* the assets of the business in which they are embedded. Thus, if you want to create a technology portfolio consisting of a colon cancer research program at Bristol-Myers, an advanced PC from Apple, and an MRI scanner from GE, there is no practical way to do it. (The two cases in which technology assets are bought and sold in the open market—high-tech start-ups and licensable technologies—represent a small fraction of the technology universe.)

ACHIEVING TECHNOLOGY DIVERSIFICATION

Though it may be hard to obtain through buying and selling, diversification in the R&D portfolio nevertheless helps us maximize value for a given level or risk. In fact, we'll get *more* value per increment of risk than if we remain undiversified. People in the investment business call this the "diversification effect," or "the investor's free lunch."

Like a stock investor, a technology firm faces two types of risk: unique risk and systematic risk. Together, these represent the total risk of the company. Unique risk is associated with the activities of the business and is reducible through diversification. For example, a computer company with one or two new product projects in its R&D portfolio has a high level of unique risk. If either or both projects bomb, it will be in a very bad position. It can reduce this risk by adding more projects to the portfolio as long as those project are not highly correlated. Exhibit 13.2 demonstrates how total risk is reduced by reducing unique risk through diversification.

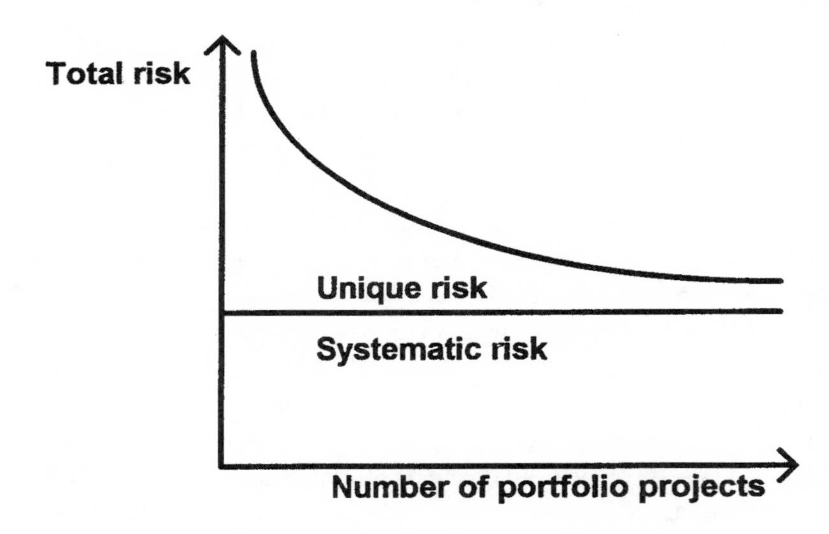

Exhibit 13.2 Unique and Systematic Risk

Systematic risk is the risk associated with the "system" or industry or technology in which the company is immersed. Systematic risk cannot be reduced through diversification as long as one stays in the same "system." For example, no amount of diversification in its computer product portfolio will save our computer company if computers go out of favor or are displaced by a new and heretofore unknown technology.

Let's look at how one can diversify unique risk in a situation where unique risk is very high. Assume drug companies all have similar productivity in drug discovery (i.e., their hit rates in the discovery of new drugs are 9 ± 3 hits per 100,000 molecules synthesized, where "hit" means a drug sufficiently promising to enter clinical trials). Statistically, the standard deviations are known to vary as the inverse square root of the number of trials. Exhibit 13.3 shows this effect.

For a drug company synthesizing 100,000 molecules per year, the year-to-year variation in hits is 33.3%, already quite a high level of risk. A small drug company, say one that synthesizes only 20,000 molecules per year, can expect only 1.8 hits per year, and it will have a standard deviation of ±1.34 or 74.5%. At this rate, it is bound to experience quite a few "hitless" years, which may be an intolerable level of risk for its owners. Certainly, investors should think twice before

Trials	Hits	Std Dev	Std Dev as % of Hits
11,111	1.0	1.00	100.0%
20,000	1.8	1.34	74.5
50,000	4.5	2.12	47.1
100,000	9	3.00	33.3
200,000	18	4.24	23.6
400,000	36	6.00	16.7
1,000,000	90	9.49	10.5

Exhibit 13.3 Diversifying Risk Over a Large Portfolio

investing in a tiny company capable of synthesizing only 1,000 compounds a year (at this level of productivity). On the other hand a large drug company synthesizing 400,000 molecules per year will experience a standard deviation of only 16.7%, in line with the standard deviation of a typical S&P 500 stock. Its research output will be reasonably predictable.

In an industry where value is largely driven by drug discovery, diversification in part explains the forces driving industry mergers. If two companies, each synthesizing 100,000 molecules per year, merge their R&D portfolios, their standard deviations of 33.3% would drop to 23.6%—an enormous risk reduction. The economic power of numbers also explains the rapid rise of combinatorial chemistry, where robots synthesize enormous numbers of new molecules (each in tiny quantities) and other robots screen them for desirable biological properties.

R&D driven companies also face systematic risk. For example, a drug company may have a significant percentage of its R&D portfolio committed to blood pressure reducing drugs. If a competitor discovers, patents, and markets an outstanding drug in this field, it will have a negative effect on the entire portion of the portfolio related to drug pressure reduction. All blood pressure projects have a degree of correlation, and the entire risk cannot be diversified away. But this risk can be substantially diversified by a portfolio that addresses a variety of diseases and is not overly concentrated in the cardiovascular area.

But other risks remain undiversifiable within the context of being a pharmaceutical company. The entire drug discovery portfolio may be subject to systematic risks relating to potential changes in health care reimbursement (as occurred after the 1992 election), regulatory trends at the FDA, changing demographics, global competition, and a host of factors large and small. These too are undiversifiable within the context of being a pharmaceutical company.

Diversification versus Productivity

Research directors have strong reasons for *linking* R&D programs because of synergies between them, and because of the perceived value of focus on core competencies. A company that is good at synthesizing gene therapy drugs carried by viral vectors may focus on that specialty. It can produce more drug candidates at lower cost, and will have better methods of evaluating and improving them. Its research productivity will be higher.

In so doing, however, it will be creating a portfolio of projects that are highly correlated, perhaps dangerously so. If the FDA raised objections to viral vectors, the company would be in serious trouble. The focus strategy creates efficiency but increases company exposure to unique risk. It is a matter of judgment whether focus or diversification is the better value-creating strategy. The financial tools developed in this book can help managers make that judgment.

The behavior of venture capitalists is interesting because many of the companies they support are nearly pure technology plays. Not only do VCs seek diversified portfolios, but they strongly prefer a diverse R&D portfolio *within a single company*—a one-product company may be shunned. However, venture capitalists still are constrained to building relatively inefficient portfolios, since start-up capital is concentrated in relatively few fields, such as biotechnology and software, and the performance of these stocks is highly correlated.

A Value Proposition for Early-Stage R&D

Having established the power of diversification in reducing portfolio risk, we now look at its power to create opportunity. We do so in the context of the highly productive R&D organization of the PMI corporation. This organization is capable of commercializing 10.4% of its conceptual stage projects (in line with empirical evidence), PMI's cost of capital is 12%, and PMI aims to achieve returns on invested capital of 20%. Therefore, PMI is targeting an economic profit of 8% on each new investment.

We need to make these assumptions:

- We have perfect knowledge of economic outcomes (we don't).
- The returns on capital of the portfolio's conceptual projects are normally distributed (this is reasonable).
- The mean return of projects that enter the conceptual stage is 12% (this is arbitrary but conservative because we would not authorize risky projects that would not return a premium above the cost of money).

Somewhat magically, we have now determined the standard deviation of the normal distribution for project returns. It turns out that, assuming a normal distribution, a return of 12% ± 6.4% will meet the criterion that 10.4% of projects have a return of 20% or better.

Let us further assume that we will terminate all projects that do not meet the 20% hurdle rate. Here we are vulnerable to the presumption of perfect knowledge. A Monte Carlo simulation (Exhibit 13.4) shows that the upper tail of the normal distribution (projects with an economic profit of 8% or better) provides a mean return of 11.04% above the cost of capital, or 23.04%.

Voilà! Having enough conceptual projects in the portfolio to generate high-return opportunities is a key to value creation. Keeping them relatively uncorrelated minimizes risk. Options analysis leads to the same conclusion, that a diversified portfolio of high-risk projects should outperform a focused portfolio of correlated, low-risk projects.

In the real world, we do not have perfect knowledge of outcomes: few experienced research managers are confident that they can pick the ultimate winners in a portfolio of early-stage projects. Winners emerge slowly as projects progress. Furthermore, there is no assurance that an R&D organization will come up with *any* winners. That will depend on its quality, creativity, and the arena in which it attempts to innovate.

Forecast Economic Profit	
Trials	18,000
Mean	11.04
Median	10.36
Standard deviation	2.63
Variance	6.90

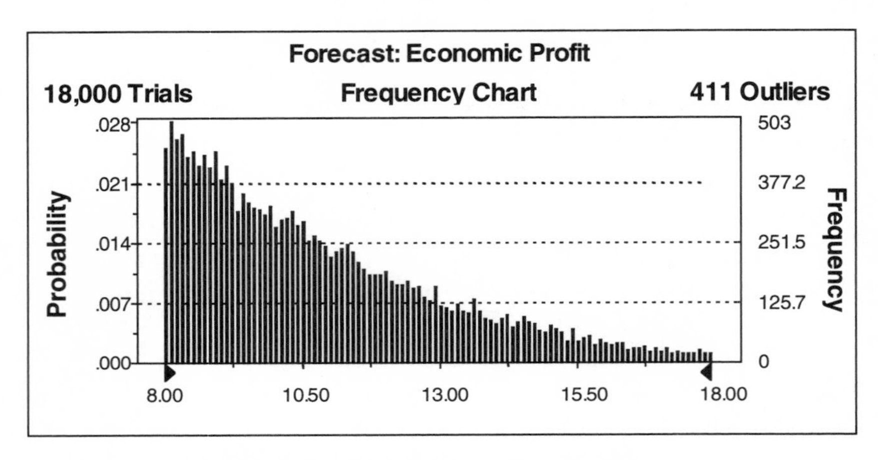

Exhibit 13.4 Distribution of High-Value Projects

Sustaining Innovation

Sustained innovation, diversification, and corporate value are closely linked. This linkage is apparent when we observe the fates of mature companies and industries that have failed to create opportunities to invest above their costs of money. The result is little or no growth in corporate value. R&D productivity in these companies and industries has been slipping, often to the point at which the output of their research laboratories no longer justifies the costs. This is the typical consequence of riding the S-curve for too long, of not knowing when to get off, and of investing at the point of diminishing returns. This situation is bad for scientists who find it increasingly difficult to justify their work. It is even worse for shareholders, since lower growth generally translates into low P/E ratios and the destruction of shareholder value.

The classic management response to this situation has been to seek improved R&D productivity through strategic and operating concepts such as so-called 2nd- and 3rd-generation R&D, stage gate methods,[2] total quality management and tighter screening of new project ideas.[3] These management approaches have in many cases led to significant improvements, and have helped companies pluck low-hanging fruit they would otherwise have missed. They have neutralized some of the effects of the law of diminishing returns, often by reducing uncertainty in the early stages of the R&D pipeline. To the extent that management methods attempt to control risk and raise certainty, they create another risk—the risk of diminishing opportunity. At its extreme, risk reduction can reduce returns to the cost of money and eliminate economic profit. And there is little doubt that most of the changes that have occurred in R&D management in large companies over the past few decades have favored risk reduction over increased opportunity.

Our analysis suggests that this emphasis on control may be self-defeating. Our examination of the research portfolio (Chapter 11) showed why it is necessary to have a large portion of the R&D budget committed to conceptual-stage projects, where ideas are incompletely developed and their implications poorly understood. At this stage, a project and its champion can be extremely vulnerable; almost by definition, few answers exist to penetrating questions. There is a strong tendency to screen against such projects when operating in the risk reduction paradigm motivated by a desire to avoid the costs and embarrassment of going down blind alleys. However, when exploratory activity is limited, the pipeline is dominated by later stage projects with lower potential for economic profit, and too few of these are likely to be terminated. In time, the internal research portfolio will become opportunity-poor and returns will suffer. Or, to put it in the context of our discussion of diversification, to exploit the high-profit end of the normal distribution, we need enough projects to create a distribution in the first place.

Economists Merton Miller and Franco Modigliani recognized the importance of opportunity in their value model of the corporation.[4] They calculated corporate value as current free cash flow divided by the cost of money, *plus* the present value of growth opportunities. Their definition of opportunities is broader than that of the technologist. In effect, they viewed opportunity as the ability to earn more than the cost of money for whatever reason: an unassailable market position, a brand name, a patent, or a magnificent technology. The economics of this view are interesting but beyond the scope of this book. However, the message is profound: *value creation is driven by the creation and exploitation of opportunities.*

Finally, there is abundant empirical evidence for the link between radical (or discontinuous) innovation and wealth creation.[5] Examples include nylon (Du Pont); the techniques for making synthetic diamonds (General Electric); the transistor (Bell Laboratories); aspartame (Searle); personal computing concepts like the mouse and the user-friendly computer interface (Xerox PARC); and hundreds more. Much of the thinking that led to these breakthroughs was nonstrategic: General Electric was not in the diamond business; Du Pont was not in the synthetic fiber business; Searle had no strategic intent to dominate synthetic sweeteners; AT&T/Western Electric (Bell Labs) saw itself as an equipment supplier for captive markets; and so on. These breakthrough discoveries created high-value opportunity for the parent to exploit. Although some of the opportunities were lost by their discoverers and captured by others, there is little doubt that none of this work could have occurred in a corporate environment that emphasized incremental extensions of existing successful products. The common observation that the cutting edge of innovation and growth in American industry has shifted to smaller firms may have its origins in the unintended consequences of a focus on risk reduction in the larger companies.

Accessing External Technology

Not every technology in the portfolio has to be homegrown. In many cases, it makes more sense to acquire them externally. External technology drives value in three ways:

1. It can add to, and diversify, the portfolio of high-value new opportunities, often quite inexpensively.

2. It extends the technology options available to the firm.
3. The company can bring projects to the marketplace faster and with lower risk, implicitly expanding the number of attractive opportunities.

Most businesses have a make-or-buy option for various parts of their operations. General Motors has the option of making or buying several million tires and batteries each year. It has chosen to buy tires from external suppliers but has elected to make some of its own batteries (Delco and Delphi). The choice itself is generally driven by issues of cost, availability, and competitive strategy.

A similar make-or-buy option exists for technology.[6] Here "make" generally translates into internal R&D, whereas "buy" means the purchase of services or licensing. Many corporations today have created external research directors to assist in these decisions, and have even developed databases on successful and unsuccessful transactions to guide future choices. Cost, time-to-market, and strategy are generally among the factors that frame these decisions. The issues, however, are somewhat different when a company is dealing with early-stage and late-stage providers of technology.

Early-Stage Providers

We have seen earlier that conceptual stage research is an expensive portion of the new product development process—expensive because of the low odds of a successful outcome and the time value of money. For these reasons, many companies outsource conceptual stage research to institutions that can do it more cheaply: universities, national laboratories, R&D start-ups, and others. Outsourcing to these technology suppliers is one approach to resolving many conceptual stage research issues and advancing projects rapidly to the feasibility stage. A company committed to this approach may be able to significantly reduce conceptual stage R&D costs, focus its efforts on feasibility studies and development, and still retain a large range of technology options. But there is no free lunch. The benefits of conceptual stage outsourcing are likely to be offset by profit sharing in the form of royalties. In a "win-win" deal, value is captured by both buyer and the seller.

There are several categories of early-stage technology providers: small companies; large companies; universities, national laboratories, and consortia.

Small Companies

These are a preferred source, since they are motivated to establish the concept and solidify intellectual property, but often have no serious intention of commercializing their work. Many of these entrepreneurial companies simply hope to establish a credible commercial presence, then sell out to the highest bidder. Their strategy is rational: the technology seller's core competence is in the R&D phase and is outmatched in the commercial phase. As a result, embryonic or partially developed technologies that emerge from start-up companies are rarely commercialized successfully by them. Instead, larger companies with established manufacturing and distribution act as midwives for these new technologies.

A license represents a typical agreement between the small company and the technology buyer. However, several other options exist. A large company can acquire the technology by simply buying the small company. It may also set up a joint venture that contains only the intellectual property of interest. Lastly, small companies have embodied portions of their technology in limited R&D partnerships in which large companies invest with options to purchase the technology if it proves successful.

Large Companies

Large companies that invest in early-stage research often create new technologies that they elect—for one reason or another—not to commercialize. These technologies do not necessarily disappear. Instead, many are eventually licensed to other companies, become the basis for joint ventures, spinouts, and the like. Other technologies become the concepts around which small entrepreneurial companies are organized by former employees of the large company.

A large company will often sell or license a technology in which it lacks a strategic interest. It may also retain a technology for its exclusive use in its existing business, but license rights to it in other fields or territories in which it does not compete. This can be an effective way of leveraging the R&D effort. Finally, it may choose to license technology to competitors, reasoning that the transfer of funds strengthens its own financial position while weakening its competitors. Union Carbide practiced this innovative strategy with its Unipol polyethylene process, while its chief competitor, Dow, steadfastly refused to do the same.

Another common approach for large company technology providers is the joint venture, which provides technology in return for equity. This makes particular sense in nations such as Japan and China, where it is notoriously difficult for overseas companies to operate without local partners.

Universities

Universities are excellent sources of emerging technology. Their emphasis is on basic research that brings conceptual stage ideas to a point of intellectual maturity. At that stage, however, universities lose their advantages: federal subsidies and inexpensive graduate student labor, and their primary motive—building the academic reputation of the senior investigator—has been satisfied. At the same time, the incentive to work with an industrial partner increases, since further development requires progressively larger scale facilities and market linkages, which the university lacks entirely.

Industrial partnerships can be created at any stage in the university development process. A typical arrangement is for the industrial partner to fund early stage research in return for an exclusive license to technology created by the project. Such an agreement is highly speculative, but can be purchased essentially at cost plus negotiated overheads. Another arrangement is the option to license patents or technology already developed by the university. This is usually done through a competitive process whereby the technology is "shopped" by a university licensing officer to a number of interested parties. University-based technologies may also reach the technology marketplace through start-up companies created by entrepreneurial professors and venture investors, a path well worn in some technical fields.

National Laboratories

Major government laboratories, such as those run by the U.S. Departments of Energy, Commerce, and Agriculture, are important sources of licensable technology. Twenty years ago, these laboratories operated like universities, where good science was viewed as an entitlement and where scientific merit alone could ensure adequate government funding. Today, however, funding is scrutinized by Congress for its relevance to a spectrum of national needs, including national economic competitiveness ("corporate welfare" to its detractors). Under these pressures, there is greater incentive to develop industrial partnerships and gain access to industrial technology and funding. As a result, most national labs today have technology transfer offices that actively seek partnerships with private industry. Some are willing to negotiate on terms that are much more attractive than could be obtained 10 or 20 years ago, often using a CRADA (Cooperative Research and Development Agreement) format.

Consortia

A number of universities, federal laboratories, and individual companies have joined forces to fund consortia that aim to advance "precompetitive" technologies. Each consortium has its own rationale and structure.

Consortia have been reasonably effective in raising funds for early stage technical research, but it remains unclear whether they have on the whole created value for their industrial participants. Consortia offer prestige, technology intelligence, and an opportunity to rub shoulders with top academics and industrial players. But a divergence of motivations and commercial restraints may undermine some of the promise. In particular, industrial firms with established and leading technology positions view consortia with skepticism—seeing more to lose than gain. On the other hand, newcomers and foreign firms seeking a foothold in a new technology have little to lose and view consortia as an inexpensive introduction to a promising business. They are likely to be enthusiastic participants. In other words, *consortia tend to serve the weak better than the strong.*

Agreements Used by Universities, National Labs, and Consortia

Agreements with these technology providers typically take the form of licenses. With universities and national labs, the purchasing company often underwrites research, then obtains the right to license the technology exclusively in its field of interest in exchange for a royalty; the university or laboratory, in turn, retains actual ownership of the intellectual property. The royalty rate is usually negotiated to reflect the relative value of the background technology accessed at the university, offset by the fact that the industrial sponsor has already paid for the specific research being licensed.

There is an important caveat in engaging early stage providers. That is, if you educate someone in your business and you later hang him out to dry, he may likely either become a competitor or join a competitor. This is natural enough—whatever he has been doing the past few years is likely to be his most marketable skill. If you want to avoid the risk of creating your own competition, do the work in-house, or at least plan to continue to provide enough funding so that the outside technologist stays in your camp.

Later Stage Providers

Whereas the motivation for engaging early stage providers may be to reduce the costs of conceptual stage research or to jump-start innovation, the motivation shifts toward risk reduction and speed-to-market in the later stages. For example, a chemical firm that wants to make a new chemical product via a fermentation process can get that product to market much more quickly through a strategic alliance with a company that has fermentation expertise and an existing plant it is willing to share—for a fee or a piece of the profits. If it works for a fee, this partner is called a "toll" manufacturer. A business decision must be made as to whether the toller's fee is acceptable when traded off against the shorter period of time and reduced technical risk. On the marketing side, commercial vendors are available to assist in packaging, distribution, promotion, and other aspects of product rollout.

Specialized consulting firms also act as late stage providers. A market research firm may be engaged to confirm assumptions regarding price, volume, and product mix. In most cases, the plant design is contracted to a reputable engineering firm, often after an internal comparison of competing proposals and intensive dialogue with the house engineering staff. Likewise, the construction contract itself may be put out to bid, as will various specialized portions of the job, such as equipment, piping, structures, and electrical work.

A make-or-buy decision is implicit in each of these steps. Decision factors on the side of "buy" include the experience, reputation, and resources of the contractor. On the "make" side, one must be concerned with confidentiality, a clear understanding of internal capabilities, and the ability to keep the key learning from the project as trade secrets or even proprietary technology. These strategic factors must be weighed against cost and speed-to-market.

The acquisition of fully commercial technology is also an important case. This makes sense when a company with a strong commercial position wants to extend that position through vertical or horizontal integration. When speed-to-market is essential, strategic alliances can be far more nimble than the traditional go-it-alone approach. Consider a company whose core competence is refining. It may have options to expand upstream by pushing into exploration and production of crude oil, or downstream into the retail marketing of gasoline or petrochemicals, or laterally into other energy sources such as solar power. American oil companies have done all of these, sometimes successfully and sometimes not. Certainly, national oil companies in the Middle East and Far East, rich in raw materials and financial resources, but poor in experienced management and in technology, are doing exactly the same today.

The weak points of such a strategy are now apparent. For example, the oil company seeking to enter petrochemical markets must go through a long and vulnerable stage during which it is smaller than its main competitors and lacks economies of scale, a technological knowledge base, a sales force, and other prerequisites for business success. So, although the financial resources behind the venture are huge, expansion into the new field can be cut down at the knees by the existing competition, and the financial returns will not meet the cost of capital. Similar integration opportunities and temptations are found in computers, aviation, automobiles, and a host of other industries. In these cases, strategic alliances with global partners who can provide business skills and technology are often the key to success.

CHAPTER 14

R&D Metrics

Managing a company or its R&D portfolio for value creation is not easy. We have seen that identifying value-building processes requires sophisticated financial understanding and business experience. Real-life management requires the creation of achievable goals that create value, and incentives for individuals and organizations to achieve them. Metrics are quantitative tools that track progress toward those goals and are often the basis for rewards. Therefore metrics must be chosen carefully to reinforce, rather than undermine, a value mind-set. It is a common observation that the performance parameters we choose to measure, and the parameters used to base incentives, determine the type of performance we get. Thus, the choice of metrics is important. In a very real sense, metrics and incentives, and the culture from which they spring, define organizational performance, and as such will be deeply embedded in a company's products and technologies.

The information in this chapter can assist R&D managers in creating metrics that enhance value and in using those metrics to develop credible R&D budgets and project proposals. Other readers may be interested in the relationships between metrics and company financial performance.

An R&D organization should always be prepared to answer three questions:

1. What have you done for us in the past—has our historical investment paid off?
2. What are you accomplishing today?
3. What do you expect to produce in the future—and is it credible in light of past performance?

No single performance metric can answer all three questions—an organization must develop historical, current, and future-oriented sets of metrics and link them into a coherent whole. In this chapter, we consider examples of the following three metrics and then discuss how to link them into a coherent whole.

1. Historical:
   Patents.
   New product sales as a percentage of total sales.
   Costs and capital savings.
2. Current:
   New products commercialized as a percentage of the new product R&D budget.
   Percentage of projects advancing to next stage.
   Percentage of project milestones met.
   Charge-outs (internal marketplace).
   Licensing balance of payments.
3. Future:
   Projected sales of pipeline products.
   Direct valuation of current project portfolio.

Historical Metrics

The world of R&D management does not begin anew every morning. Today's technology pipeline is largely defined by yesterday's work. The contents of that pipeline are critical to both current and future performance and need to be carefully monitored. So, we will first consider several types of backward-looking metrics.

Patent Productivity

The number of patents that a research organization receives in a particular year is a traditional historic metric. It is easily obtained and has the advantage of facile benchmarking against competition, since patent statistics are often tabulated in trade magazines and are also available directly from the U.S. Patent Office. Because corporate R&D budgets are usually reported publicly, we can use these two pieces of information to determine the ratio of issued patents per million dollars of R&D spending, as shown in Exhibit 14.1. An alternative to this dollar-denominated ratio is the ratio of patents per scientific professional.

The first tier of the exhibit shows that our firm, PMI Corporation, has outperformed all but one of its competitors during the current year (labeled 2000). Competitor D appears to be highly focused on patents in its current strategy, and it may be wise for us to read those patents (if they have not already been analyzed) to see what is afoot. PMI's patent output is higher in absolute numbers than two competitors with larger R&D budgets, suggesting that creativity is alive and well. The lower tier indicates considerable fluctuation but no discernible trend.

Patents reflect the number of proprietary new ideas produced by the research organization when the patents were filed, on average about two years earlier. In this sense, it is backward glancing, telling us how well we were performing in the recent past. The time lag is caused

	Patents Issued	R&D Budget ($M)	Patents per $M
Year 2000			
PMI Corporation	102	150.2	0.68
Competitor D	40	38.2	1.05
Competitor C	92	141.4	0.65
Competitor A	80	173.8	0.46
Competitor B	213	525.8	0.41
Competitor E	64	208.4	0.31
History			
PMI 1996	67	104.0	0.64
PMI 1997	62	118.9	0.52
PMI 1998	91	130.5	0.70
PMI 1999	111	134.1	0.83
PMI 2000	102	150.2	0.68

Exhibit 14.1 Patent Productivity

by the interval between invention and patent filing, and the subsequent interval between the filing of an application and the issuance of the patent. The first interval is largely controllable and relates to efficient communication between inventors and patent attorneys. It can be shortened to a few weeks if necessary. The second source of delay, however, relates to the U.S. Patent Office and the backlog of its examiners, and also to the complexity of the patent process itself, which depends on the nature of the patent claims and the persistence of the parties in arguing their cases. Some classes of patents, such as those for biotechnology, may lag considerably more than two years due to a shortage of examiners. In interpreting the statistics in Exhibit 14.1, therefore, it is worth bearing in mind that the number of patents issued in 1998 may compare more closely with the 1996 R&D spending than the 1998 spending. Although patents may be generated at any stage in the R&D process, building patent positions is usually a key task in the conceptual stage. Therefore, the patent metric tends to be a lagged indicator of the vigor at the front end of the research pipeline.

Time lag is not the only weakness of the patent metric. Patents vary enormously in value, and most are of very low value. Some are purely cosmetic. A new ice cream carton design, for example, provides little real competitive advantage or added financial benefit for the firm. Other patents cover technology that will never see the light of day, although this outcome may not have been foreseen at the time of invention. Multiple patents with narrow claims are filed by some firms to defend a proprietary position, but they contain no real breakthroughs.

Whether patent productivity is a real measure of R&D productivity and a source of value creation is, therefore, debatable. Nevertheless, managers can use this metric to observe the internal trend, and to compare patent productivity and patent trends with competitors.

Should a company provide financial incentives for patents? This is a controversial question. Currently, patent incentives take these forms: recognition, moderate financial incentives, and significant financial incentives (including royalties). Some European countries have *legislated* significant compensation for inventors. Many (including the author) do not support the idea of using the patent metric for monetary incentives. The reasons for their opposition are straightforward:

- A monetary incentive linked to patent output encourages quantity over quality (i.e., value-creating power). Processing and maintaining lots of low-value patents can be enormously expensive. It is not unusual to incur initial costs of $10,000 or more per issued patent, and to see cumulative maintenance fees approach several hundred thousand dollars when patents are maintained in many countries for their full lifetimes.
- It is often possible to break a single patent application into several patent applications. Thus, judgments that should be made on business and legal grounds—better coverage, speed to issue, costs of prosecution and maintenance—may take a backseat to the parochial interests of inventors.
- In a well-managed technical organization, patents relating to the highest priority projects, based on perceived commercial value, normally move to the top of the list within the corporate patent department. Financial incentives would encourage inventors to lobby in favor of their patents, objective values notwithstanding.
- Financial incentives also create pressures to add supervisors and coworkers to the list of inventors, but since false claims of inventorship are grounds for invalidating the patent, these practices may jeopardize the patent itself. Teamwork among coworkers may also suffer.

Patent productivity is an important metric and its trends are worth analyzing, but financial incentives linked to patents may encourage behaviors that *destroy* value.

Percentage of New Products

Another common historic metric is the percentage of new products introduced in the past 5 years, the past 10 years, or some other meaningful time frame.

In process industries, there is ample evidence that it is desirable to have 15% to 30% of current sales coming from products introduced in the past 5 to 10 years. We have seen that new products can be an important source of corporate growth and hence of shareholder value.

Typically, they command higher profit margins and strengthen the image of the company in the eyes of customers.

3M Corporation has long been one of the worldwide new product leaders. To maintain high growth and its image for technology innovation, 3M requires each business unit to meet a criterion that 30% of product sales be derived from products introduced in the past *five* years. They also require that each business have at least one *pacing project*—a radical research project capable of transforming the basis of competition in its market.

The appropriate new product criterion for a company is often industry dependent. The demand for innovative products is extremely high in entertainment software but low in construction-related businesses, where innovation is impeded by building codes and product liability. This metric is inappropriate for commodities such as gasoline, rice, and timber.

The new product metric has several pitfalls. Unless a business is large and has many products, year-to-year fluctuations in new product statistics will be large. A single blockbuster innovation can distort the data and create a sense of complacency. Then there's the problem of definition—what is and what is not a new product is often unclear. Is a simple upgrade a new product? Does a derivative of an existing product ("new mint-flavored dental floss") fit the definition of "new"? A definition suitable for all businesses is unlikely; more likely, the standard must relate to the industrial sector to which it applies. In the end, new products should have sales characteristics that are *sharply differentiated* from existing products. This seldom occurs for small technical tweaks notwithstanding the ubiquitous "new and improved" label favored by marketers. The best criterion I have heard is: *If a product does not change customer behavior, it is not new.*

Exhibit 14.2 illustrates the new product sales metric for PMI. It tabulates a hypothetical history of the new products created by PMI's laboratory organization during the past 10 years (using a 10-year definition for a new product). It not only includes current sales of these products, but also offers a forward-looking five-year projection of their future impact. (For credibility, these projections should be obtained from the business unit accountable for delivering them.)

Year Introduced	Product	Market	Current Year Sales ($M) (Year 2000)	5-Year Sales Projection ($M) (Year 2005)
1990	Catalyst A	Refining	150	210
	Plastic B	Transportation	63	88
	Adhesive C	Construction	35	49
1991	Chemical D	Pharmaceuticals	41	67
	Chemical E	Food additive	50	82
	Film F	Packaging	120	196
1992	Sorbent G	Environmental	68	130
	Membrane H	Battery separator	112	214
1993	Film I	Packaging	50	112
	Ceramic J	Cutting tools	11	25
	Catalyst K	Environmental	95	212
	Chemical L	Pharmaceuticals	15	34
1994	Instrument M	*Licensed*	53	138
	Sealant N	Packaging	58	152
	Membrane O	Medical devices	40	104
1995	Adhesive P	Transportation	32	98
	Plastic Q	Transportation	42	128
	Ceramic R	Electronics	14	43
	Membrane S	Construction	50	153
1996	Chemical T	Coatings	25	76
	Catalyst U	Refining	38	116
	Catalyst V	Environmental	24	73
	Sorbent W	Medical devices	8	24
	Film X	Packaging	47	143
1997	Membrane Y	Battery separator	4	12
	Ceramic Z	Electronics	6	18
	Plastic A	Construction	67	204
	Adhesive B	Electronics	11	34
	Film C	Packaging	23	70
1998	Catalyst D	Environmental	39	119
	Adhesive E	Electronics	29	89
	Chemical F	Adhesives	25	76
	Software G	*Licensed*	6	18
1999	Plastic H	Transportation	8	24
	Ceramic I	Environmental	26	79
	Film J	Packaging	17	52
	Plastic K	Construction	28	85
	Catalyst L	Refining	14	43
	Chemical N	Pharmaceuticals	6	18
	Total		1,550	3,612
	Subtotal (1995–1999)			1,798
	All Sales		5,000	8,970

Exhibit 14.2 New Product Metric for PMI's R&D Organization

The exhibit has several interesting features. Quite typically, the older "new" products tend to have larger current sales than the recent entrants; this is because they have been on their growth trajectories longer. Today's midgets are often tomorrow's giants. Second, the laboratory has taken credit for two "licensed" products developed internally but not produced by the parent corporation. Both, nevertheless, are contributing to net income, just as if the sales were being booked by the parent company—and with no requirement for capital. Similarly, R&D may make contributions that enhance the value of businesses that are subsequently divested. These too should be included in the report card.

Adding the totals, we see that new products in the current year represent 31% of sales, slightly ahead of the company's target of 30%. The projection for the year 2005, however, indicates the company's expectation that the sales contribution of still new products, those introduced between 1995 and 1999, will fall far short of the 30% target. Many of the strong current products cannot contribute to this target since they, by definition, will no longer be new. PMI must look to its pipeline to make up the shortfall in the percentage of sales attributable to new products.

As noted, there are no surefire formulas for business success. The same applies to the new product metric. The virtue of a high percentage of new products has been touted to managers for the past half-dozen years or so, virtually without challenge. Here is the contrarian view: A high percentage of revenues coming from new products *may* indicate that the products being created are weak and cannot sustain themselves in the market for long, hence the low revenues coming from older products. They are either faddish, easily imitated, or simply fail to satisfy customers for long. Given the human effort and the cost of developing and introducing new products, revenues from old established products actually might be more attractive.

One example of this contrarian view is the trade book industry, where almost 75% of the average product's total lifetime revenues are captured in the first 4 to 6 months. Trade book publishers have high percentages of revenues from "new products" (as advocated) but must continually reinvent the wheel, since those revenues dissipate quickly. Not surprisingly, it is a very low-profit industry.

Costs and Capital Savings

In many industries, being a low-cost producer is absolutely imperative for survival and success. Low-cost production is essential for commodity products such as petrochemicals or metals—even for products that are becoming commodities, such as first-generation plastics. Process research that aims for incremental and radical production cost savings is often a major source of profits. Some cost savings and "debottlenecking" projects, which increase the capacity of existing plants, can be achieved with very low capital, resulting in very high returns on invested capital.

Despite their attractiveness in principle, incremental cost-saving projects get little respect. The promised cost savings are usually realized at the plant level, but don't fall to the bottom line, causing the business manager to ask, "If the estimated cost savings of the project were $1 million, why are my profits only up a few hundred thousand dollars?" In most cases, the answer is "competition." Competitors do not stand still, but conduct process improvements of their own. A competitive marketplace ensures that most of the cost savings will be passed on to customers. Even if weaker competitors do not advance their technology, they may drop price to maintain share and thus erode the benefits to the innovator. Debottlenecking projects have a similar drawback; the increased capacity does not produce profit if it is not used. So, if markets weaken, projected profits are deferred. Radical process innovation, on the other hand, when proprietary, can create great technical advantage—and profits—that competition cannot immediately match! Legendary examples of radical process innovation include the Oxirane process for making propylene oxide (a key precursor to urethane plastics), and the use of zeolites in fluid cracking catalysts in making gasoline.

One advantage of process-related metrics is that plant accounting systems are good at verifying them. Therefore, it is worthwhile collecting data on the costs saved or the value of debottlenecking in terms of imputed capital. Exhibit 14.3 shows such a table. Here we separate the value of operational improvements from the issue of whether the gains are fully realized on the bottom line. It must be recognized that business is intrinsically subject to diminishing returns,

Cost Savings

Year	Business	Project	Value[a] Realized
1990	Catalysts	Cogeneration project	$20.9
1990	Polymers	Yield increase	12.4
1992	Catalysts	Waste disposal	3.0
1993	Headquarters	Utilities (lighting)	2.2
1995	Ceramics	New raw material source	3.9
1997	Polymers	Royalties	2.3
	Subtotal		$44.7

Capital Savings

Year	Business	Project	Value[b] Realized
1991	Sorbents	DCS control system	$5.1
1995	Chemicals	Cheaper alloys in new plant	7.9
1995	Packaging	Debottleneck 3 film lines	6.2
1996	Polymers	Reactor throughput	3.8
1998	Polymers	Continuous prepolymer process	12.5
	Subtotal		$35.5
	Total		$80.2

[a] Cumulative annual cost savings less investment (current $M).

[b] Net Capital Savings on New Investments or Value of Incremental Plant Capacity (current $M).

Exhibit 14.3 PMI Cost and Capital Savings (1990–1999)

and if continuous improvement is not made, margins will inevitably deteriorate.

The value of cost savings projects, both projects that aim for one-time savings (of expense or capital) and continuing savings, is a useful metric. For the hypothetical example of PMI, they have averaged about $8 million per year over the past decade. These cost savings can be viewed as returns from the portion of the annual R&D budget spent on cost-savings projects. For continuing savings, the concept of *economic profit* can also be used to define the net present value of the improvements. This was covered in Chapter 10 and will be reviewed later in this chapter.

Current Metrics

So far, we have considered backward-looking metrics of R&D performance. They tell us how well R&D has performed in the past, often the recent past, but do little to indicate how well it is doing at the moment. Historical metrics are useful in helping managers identify and interpret long-term performance trends, but day-to-day management and control depend on real time indicators.

New Products Commercialized as a Percentage of the New Product R&D Budget

This metric is a variation of the new product metric described earlier. Its focus, however, is on the 5-year projected sales of products commercialized in the *current* year in relation to the R&D budget for new product research. We discussed the "macro" effect of this metric on growth and shareholder value in Chapter 6 and linked its use to the net present value of a model R&D portfolio in Chapter 11. Exhibit 14.4 illustrates the use of this metric for our hypothetical corporation, PMI.

In Chapter 11, our calculations showed that this corporation needed 1.43 lab units to achieve a growth rate of 12.4%, and that the cost per lab unit was $120 million. Accordingly, current R&D

Product	Market	5-Year Sales Projection ($M)
Membrane O	Medical devices	86
Chemical P	Construction	30
Film Q	Packaging	42
Catalyst R	Petrochemicals	11
Corrosion Inhibitor S	Water treatment	36
Sorbent T	Environmental	95
Plastic U	Packaging	48
Total		348
Current Year R&D Spending ($M) after-tax		103
Productivity Ratio		3.38

Exhibit 14.4 Current Year New Product Performance

spending for new products is $172 million pretax, or $103 million after-tax. In the current year, seven new products were commercialized, with sales projections 5 years out of $348 million, giving a productivity ratio of 3.38. This level of R&D productivity not only generates growth, but ensures that such growth will add value.

PERCENTAGE OF PROJECTS ADVANCING TO NEXT STAGE

A second current metric can be built off the project flow dynamics illustrated for a model steady-state portfolio in Exhibit 11.5. In this case (Exhibit 14.5), we track actual results and look for indicators of potential problems.

Low success rates in the development or early commercialization phase are especially serious and indicate that too many marginal projects may have advanced through earlier gates. This can be anticipated if success rates in the conceptual and feasibility stages are observed to be trending upward.

Is the current percentage advancing good or bad? This can only be answered relative to what competitors are doing and relative to the corporation's own past performance. Exhibit 14.6 indicates recent historical performance in rates of advancement, with the current year performance in the bottom row.

This metric is useful in three respects. Low numbers may indicate a drop-off in R&D productivity, and suggest the need for immediate managerial intervention. High numbers for conceptual or feasibility stage projects (such as for conceptual projects in 1999) may indicate that management was insufficiently selective in terminating projects,

Stage	New	Continue	Total Continuing	Advance	Kill	Prior Total	Percent Advancing	Cumulative Percent
Conceptual	57	59	116	19	37	115	33.9	10.0
Feasibility	20	21	41	9	10	40	47.4	29.5
Development	9	7	16	8	3	18	72.7	62.3
Early commercial	8	4	12	6	1	11	85.7	85.7
Commercial	6							
Total projects			185			184		

Exhibit 14.5 Percent Projects Advancing in Portfolio

	Percent Advancing			
Year	Concept to Feasibility	Feasibility to Development	Development to Early Comm.	Early Commercial
1995	30.0%	54.5%	80.0%	71.4%
1996	36.4	40.0	81.8	100.0
1997	21.8	42.1	72.7	83.3
1998	33.3	63.2	75.0	50.0
1999	40.9	52.2	58.3	100.0
5 year average	32.5	50.4	73.6	81.0
2000	33.9	47.4	72.7	85.7

Exhibit 14.6 Historic versus Current Project Advancement

or that long-term portfolio balance may be threatened by a larger number than usual of excellent projects advancing toward commercialization (a happy problem but a potential problem nonetheless). This may be an early warning indicator of a need to increase future R&D budgets, to prioritize, or to consider selling technology opportunities.

Finally, changes in these numbers may reflect changes in the probabilities of commercial success applied to the R&D portfolio, and should be translated into the probabilities factored into the future-oriented metrics to be discussed.

Percentage of Project Milestones Met

Milestones are measurable achievements in a project. Advancing from one stage to another represents a major milestone. So too does the resolution of a significant task or issue required for advancement. For organizations that use PERT or Gantt charts for project management, the completion of significant tasks on these charts represent project milestones. A "deliverable" is generally associated with a milestone. The deliverable may be the production of one kilogram of sample material, or it may be the issuance of a report or a clear decision on project direction: "We will produce this product by thermoforming rather than injection molding."

However defined, milestones are extremely useful for both planning and project evaluation. They are, for one thing, a metric for current R&D productivity. The percentage of milestones met during the current reporting period correlates with the percentage of projects on track. The percentage of missed milestones—those that failed to be met by their prescribed dates—indicate something else: poor execution, unexpected results, or poor planning. The last is the most common reason for missed milestones. Necessary equipment may not be readily available, key people may be on holiday, or unexpected results may raise new issues. Analyzing these problems with the aid of hindsight can be an important training exercise for both scientists and managers.

Who determines the milestones and their associated dates? If the project involves a cross-functional team, the team more often than not has responsibility for determining major milestones. In a conceptual or feasibility stage project, the milestones may be proposed by the principal investigator. But if a completion date for the project is mandated by business considerations, milestones may have to be determined in reverse to fit the overall schedule, with resources reallocated to ensure meeting them. This exercise often is stimulating to say the least.

The milestone metric is not without pitfalls, the principal one being the temptation to make them too easy to meet, or to propose milestones that have already been largely met. The latter practice, called "lying in the weeds," is encouraged when milestones are associated with either punishment or excessive rewards. Fortunately, we have already introduced tools that can alleviate the problem of unchallenging milestones. An overly long schedule or a project that consumes excessive resources will look less attractive than other internal opportunities when scrutinized through the lens of NPV. Quantitative analysis can be used to set milestones on a financially acceptable pace. When that pace is also technically sound, the project is on solid ground. Management can also do much to encourage a prudent degree of challenge and risk taking in the schedule; both stimulate superior performance.

The second problem with the milestone metric is the dynamic nature of R&D and the competitive environment in which it operates. Every day is a learning experience. The world changes quickly. New information and the actions of competitors can make a milestone inappropriate. For example, the discovery of a competitor's patent may successfully complete the "patent search" milestone, but render upstream milestones inappropriate. The discovery may also create a new milestone—"Invent around this patent"—or provide a reason to terminate the project.

What is a good number for the "percentage of milestones met" metric? The answer depends mostly on the R&D stage. Early stage misses are normal, but late stage failures are embarrassing and expensive. In the riskier conceptual and feasibility stages, below 50% of milestones met is problematic, but at the same time anything better than 90% should make us suspicious that the milestone hurdle is set too low. In the development stage, we would want 75% to 90%, and in early commercialization, better than 90%.

Charge-Outs: A Metric for Service Functions

Performance metrics for service organizations within R&D, such as analytical laboratories, patent services, technical information services, and specialized engineering support, are a challenging issue. Management needs to know that these functions are performing productively and meeting a competitive standard of excellence.

Corporate executives who are uncomfortable with making R&D decisions are often inclined to leave these decisions to the "internal" marketplace. In essence, decisions to fund the support centers are left to the profit centers likely to benefit from them. In these cases, the primary metric for the laboratory becomes the degree of funding, or *charge-out,* obtained. Budgets will over time be adjusted to move this metric into the desired range.

In a typical arrangement, top management agrees to fund a certain percentage (say 20%) of a service laboratory's budget from corporate funds to work on projects selected by the laboratory director, but stipulates that the rest must be obtained from the business units. The corporate funding would be used to create advanced capabilities that would eventually be used by the business units, or to address relevant problems that might not affect individual business units (such as a corporate response to a pressing environmental issue). The corporate allocation could also be zero, making the laboratory entirely dependent on charge-outs. This would put the central laboratory in a position of competing with each business unit's internal R&D groups *and* outside vendors for R&D contracts, and simultaneously put the business units in a position of treating R&D as a make-or-buy decision. It must become lean and mean.

There are some pitfalls to charge-outs. For example, business unit R&D managers may prefer independent laboratory facilities (which they control) to a central facility, even if this results in duplication of equipment and personnel or inferior capabilities. Never underestimate the motivation of technologists to build "empires." The charge-out paradigm also de facto eliminates the option of using advanced technology strategically in the areas governed by the internal marketplace.

A secondary metric for service functions is *backlog*. This metric is useful, but its interpretation may be ambiguous. A low backlog in service functions may indicate that supply of services exceeds demand from users; on the other hand, it may indicate that laboratory management is highly conscientious in providing timely services. A high backlog certainly indicates that demand for services exceeds supply, and that some users are not being well served. In patents in particular, backlogs associated with processing delays can cause damaging losses in intellectual property value.

Licensing Balance of Payments

Companies that practice a variety of technologies often pay, and receive, license fees. For example, PMI may pay license fees of $9.5 million per year to an inventor, a competitor, for the right to use a process for producing Polymer A. At the same time, it may receive $12.5 million from several licensees who produce Polymer B using technology developed by PMI. The licensing "balance of payments" in this case is +$3.0 million. (In many companies, R&D reports both a gross and a net budget; the net budget takes a credit for income received from licenses, contracts, and other income sources, such as products made in its pilot plants.) This approach is useful, because in licensing-in, there is an implicit make-or-buy decision. In principle, management may at one time have elected to have the technology developed internally, but at a penalty of cost and time. (We neglect the case where a radical innovation has been patented by a third party, and there is little choice but to license it.) On this basis, management may reasonably charge the license fees as R&D-related expenses, since the royalties might have been avoided if timely R&D had been performed. Certainly, a positive balance of payments indicates value creation by R&D; a negative balance and a recurring need to license technology is symptomatic of technical weakness.

Future Metrics

Backward-looking metrics help managers to see where they have been and to identify long-term trends that may benefit from their intervention. Current metrics provide something much closer to the real-time pulse of activities. Both are important tools for R&D management. But no tool kit would be complete without at least one forward-looking measure. The future is, after all, the object of all R&D work.

Projected Sales of Pipeline Products

The high mortality of research ideas makes it important to maintain a robust research pipeline. The pipeline represents the future of the company, and one filled with many diverse and promising projects is absolutely necessary for growth in shareholder value. But how can we estimate the timing and magnitude of contributions from the pipeline? Any estimate is fraught with pitfalls: researchers frequently underestimate the time required for commercialization because it is difficult to build time into the schedule for problems you have not yet encountered; also researchers are prone to overestimate the probabilities of success—again because the problems have yet to be identified.

To build credibility[1] future projections should be linked to past performance. We are all familiar with "hockey stick" forecasts: the business has been deteriorating for years, but the forecast shows it heading sharply upward from here on. Wise managers understand the pressures that create hockey sticks, and discount them heavily. Thus, R&D managers should resist the temptation to use rosy and unrealistic forecasts to sell their proposals.

For this reason, I prefer the historical approach to the timing and probability of future R&D business contributions. These should be based on internal data, but accept the idea that other systems, such as scoring projects by opportunity- and risk-related attributes,[2] can also be helpful. In the absence of historical data, create a model by benchmarking with similar companies or from industry statistics.

It is also worth repeating here why sales, rather than profits, are a simpler metric on which to base the value of projections. First, minimum acceptable profitability should be assured since few projects go forward that fail to pass the hurdle rate. Second, sales are more reliably measured and sales forecasts are standard operating procedure. On the other hand, profit measures create large uncertainties in allocations of overheads and other fixed costs.

Exhibit 14.7 demonstrates a process that links past, present, and future, and its key features. In many ways, it parallels the analysis of

	2000	2005					2010				
Revenues in $M	(1)	(2)	(3)	(4)	(5)	(6)	(7)	(8)	(9)	(10)	(11)
Stage V–Established products											
Commercialized 1975–1989	1,746	2,449				2,449	3,435	0			3,435
Commercialized 1990–1999	1,550		3,612	1,798	1,798	3,612	5,066	0		5,066	
Commercialized 2000			348		348	348	0	1,062		1,062	1,062
Subtotal established	3,296	6,061				6,409	8,501			1,062	9,563
Stages I–IV–R&D pipeline											
Stage I–Conceptual			0		0	0		4,345	3,892	453	453
Stage II–Feasibility			413	284	129	129		1,400	528	872	872
Stage III–Development			580	218	363	363		1,107		1,107	1,107
Stage IV–Early commercialization			679	113	565	565		1,726		1,726	1,726
Subtotal–Pipeline			1,672	615	1,057	**1,057**			4,420	**4,157**	4,157
Total–Products from R&D unit	3,296	6,061	3,818	615	3,203	7,466	8,501	9,639	4,420	5,219	13,720
Target new	1,500	2,691			2,691		4,828			4,828	
Total corporate revenues	5,000	8,970			8,970	8,970	16,093			16,093	16,093
% All products R&D unit	65.9%					83.2%					85.3%
Target–% new	30.0%				30.0%					30.0%	
% New	31.0%				35.7%					32.4%	
% Difference	1.0%				5.7%					**2.4%**	

Column Head Key:
(1) = Current Year Sales
(2) = Established Products in 2005
(3) = New Products in 2005
(4) = Contingency
(5) = New Products ex Cont.
(6) = All Products in 2005
(7) = Established Products in 2010
(8) = New Products in 2010
(9) = Contingency
(10) = New Products ex Cont.
(11) = All Products in 2010

Exhibit 14.7 Future Sales from New and Established Products (in $millions)

the R&D pipeline in Chapter 11, where the value of current projects was extrapolated from their pro forma business plans and historically based probabilities of success. What is added in creating a metric is the link to actual performance.

Consider the current year to be 2000. PMI has sales of $5,000 million. Of that, $1,550 million (or 31%) comes from new products introduced from R&D in the previous 10 years (1990–1999), meeting the company's 30% goal. The detail behind this number has already been presented in Exhibit 14.2. In addition, R&D notes for the record that an additional $1,746 million of sales came out of the company's labs in the previous prior years, but are no longer counted as new. The balance of corporate revenues come from products obtained by acquisition, from before the 25-year historical period, or from other non-R&D sources. If we turn to the year 2005, we see that the company has established a revenue goal of $8,970 million, of which $2,691 million should come from new products. Toward this goal, it anticipates a contribution of $1,798 million from products introduced in the period 1995–1999, but still qualifying as new, and from the year 2000 crop of projects, which is projected to grow to $348 million. The balance must come from the pipeline.

The pipeline is in good shape, however. We can see that products with projected year 2005 sales of $1,672 million (labeled "Subtotal—Pipeline") are in various stages of research. Because this hypothetical model assumed all projects are proceeding at equal speed over an 8-year interval with 2 years between each stage, early commercialization projects will begin to generate revenues in 2001 and 2002, development projects in 2003 and 2004, and half of the feasibility stage projects in 2005.

The next column discounts this expectation by the probability of failure in each stage, based on historical data. The discount is labeled as "contingency." This convenient term eliminates the problem of trying to separate winners from losers and allows acceptance of the financial targets of project champions, but still eliminates unrealistic projections.

The probability of failure is high (62.5%) for projects in the feasibility stage so the target 2005 sales of $413 million are reduced by $284 million to $129 million. After correcting for the contingencies for the other classes of projects, we expect $1,057 million in new sales from the pipeline. More than half of these sales come from early commercialization projects, indicating the critical importance of these to medium-term corporate performance. Conceptual projects make no contribution to the 5-year forecast. They cannot save the company if it is in difficulty.

Most importantly, we can see that the pipeline is realistically expected to produce an apparently comfortable margin of new product revenues versus the corporate target.

Year 2010 results are more speculative. Financial results in that year will be driven primarily by the growth of the new products expected to be in hand by 2005. Today, they are in Stages 3 and 4. But further growth will depend on the products currently deep in the R&D pipeline—Stage 1 and Stage 2 products due to be commercialized in 2006 through 2008. The year 2010 will also present opportunities not yet conceived—indeed, no projects currently on the books will generate initial revenues as late as 2009 or 2010—an artifact of our uniform 8-year research cycle. In the real world, some of our current conceptual projects would be delayed to this time (while creating valuable intellectual property positions) and the emerging technology portfolio would create abundant technical synergies with external developments emerging over this decade. The apparent drop in percentage of new products from 35.7% in 2005 to 32.4% in 2010 is simply because projects not yet in existence cannot be included in the plan and are not yet a concern.

Direct Valuation of Current Project Portfolio

The value of the R&D portfolio consists of five pieces:

1. The enterprise value of new product projects.
2. Cost savings projects.
3. Capital savings projects.

4. Licensable properties.
5. Embedded options.

Given the many intangibles associated with each of these, some may question the reliability of our attempt to determine their value. However, we have most of the tools required to do the job—the real issue is the quality of the input data.

New Product Projects

Chapters 9 through 11 developed methodology for calculating a net present value for each new product project in the R&D portfolio. It was based on the concept that each project represents a minienterprise with its own pro forma business plan. We know that the portfolio will consist of a relatively large number of conceptual and feasibility stage projects, each with a relatively low NPV, and a smaller number of developmental and early commercialization projects with higher NPVs. Much of the difference in value between early stage and later stage projects will be associated with time-discounting of the projected rewards and the low probabilities associated with early stage projects.

Cost Savings Projects

The methodology for using *economic profit* to approximate NPV for a continuing cost-saving project was outlined in Chapter 10. Basically, annual cost-savings can be treated as perpetuities; or, if they are expected to increase with time, the growth in perpetuity methodology of Chapter 5 can be applied. One time-cost savings in the current year can be taken at face value, and those that must be deferred should be discounted at the cost of money.

Capital Savings

Capital savings, like cost savings, are cash flow items and can be treated identically. The value of future one-time capital savings, such as the elimination of an operation in a proposed plant in 2003, should be discounted to the present. Continuing capital savings, such as a system to reduce working capital, can be valued using a standard DCF model.

Licensing Revenues and Costs

We have discussed how to use a licensing balance of payments as a metric of technical strength. Since licensing income and costs are generally quite predictable, at least for existing licensees and licensors, it is a fairly simple matter to determine the NPV of each licensing deal. Remember that many licenses expire, and unrealistic terminal values should not be assigned.

Technologies and patents that are not valued as minienterprises, presumably because there is no intent to commercialize, may also be valued as *licensable properties,* using DCF and discounted with appropriate contingencies to reflect the probability of successfully realizing such income.

Embedded Options

None of the values described include embedded options: options to spin off additional R&D projects, to create follow-on investment opportunities, or to link the firm's technology with strategic partners. In some cases, these options may be the highest value of all. In principle, financial methods exist to value such options (some are outlined in Chapter 12), although the "quality of assumptions" issue is extremely important.

Total Portfolio Value

The total value of the R&D portfolio is the sum of the various parts just described. Using the NPV approach, we determine value though the relationships shown in Exhibit 14.8. Over time, a major R&D organization creates a vast reservoir of intellectual property, some of which can be captured in the collective NPVs of individual projects. It is possible, in principle, to track net changes to that NPV as projects progress, are commercialized, or are canceled. At least for companies that compete in a dynamic market environment, the possibility

$$\begin{aligned}\text{NPV (Portfolio)} = &\ \Sigma\ \text{NPV (New product projects)}\\ &+ \Sigma\ \text{NPV (Cost-savings projects)}\\ &+ \Sigma\ \text{NPV (Capital savings projects)}\\ &+ \Sigma\ \text{NPV (Licensing revenues less licensing costs)}\\ &+ \Sigma\ \text{Value of embedded options}\end{aligned}$$

Exhibit 14.8 Net Present Value of R&D Portfolio

is real that the value hidden in technology options may exceed that which can be more or less rigorously identified by cash flow methods. The NPV approach can supplement other techniques, but is not a substitute for a strategic understanding of the potential of the R&D portfolio to sustain growth, to compete in the marketplace, and to create opportunity over the longer term.

Summary

R&D metrics are complex because the bottom line impact of R&D effort substantially lags the investment. In some cases—as in the pharmaceutical industry—that lag time is 10 years or more. Time is needed to discover a new drug in the laboratory, test it for safety and efficacy, scale up, and introduce the product to the marketplace.

Put another way, the impacts on shareholder value of changes in the R&D budget or in R&D productivity often seem gradual because the R&D boat steers slowly. Hidden from view, however, is much turmoil and urgency. Speed to market is critical. Patents must be filed ahead of competitors. Incorrect decisions can lead to irreversible losses of opportunity and talent. R&D metrics provide the first glimpse of the long-term future.

CHAPTER 15
Special Issues in Value-Based R&D

Much of this book has been devoted to examining how to maximize value by balancing risk and opportunity in the technology arena. This chapter addresses four R&D issues that impact value in unique ways and help to create favorable outcomes. These are:

- Patents
- Technology transfer
- Globalization
- Environment

Patents, globalization, and the environment are each in their own way important value drivers, since they multiply opportunity. Technology transfer, competitive patents, global rivals, and environmental regulation all contribute importantly to risk.

Patents as Value Drivers

Patents are written by lawyers for an audience of patent examiners and patent attorneys. They are highly technical. Claims can be voluminous and repetitious. There are millions of patents, and each can take many hours to read and understand. The vast majority are of no commercial value. And the real goodies are often deliberately obscured, since the patent-holder has no desire to reveal details that might benefit competitors. Easy reading they are not.

Still, people get excited about patents because they are licensed monopolies and fantastic value drivers. The best patents are valued at hundreds of millions or billions of dollars. They propel prices and gross margins upward. One of the world's most successful industries—pharmaceuticals—is based on patents.

The heart of a patent is its claims. If another company practices within your allowed claims, you will have a negotiating position: You can ask the competitor to desist or permit it to continue if it pays a satisfactory royalty for the privilege. Your claims do not, however, give you an unconditional right to practice your own patented technology—you must also be free of anyone else's patents.

The patent value chain has three key elements:

1. Writing the patent,
2. Obtaining and maintaining the patent, and
3. Enforcing patent rights.

All are important, and all involve quite different processes and people.

Writing the Patent

The objective of this process is to get valid claims allowed, particularly in those areas that would appear to be commercially valuable. The first step in the patent process is the basic invention itself. This must be documented, preferably in a signed and witnessed bound laboratory notebook. The next step for the inventor is the writing of a patent disclosure—a document that enables a patent attorney or agent to prepare a patent application, which is the third step.

The patent attorney must deal with both intellectual property and business issues. He knows that his patent application must meet three criteria: the invention must be (1) novel, (2) non-obvious, and (3) useful. Demonstrating novelty means finding and citing all relevant "prior art" (a term encompassing patents and publications related to the invention), and then distinguishing the invention from the prior art. (Missing some of the prior art can cause problems down the road in obtaining and enforcing the patent.) "Obviousness" can be a tricky issue—it can be argued that any combination of existing technologies are obvious to one skilled in the art—therefore it is important to show that something unexpected was discovered. "Useful" means that the invention has commercial value—the solution of a mathematical problem of no relevance to the real world is, by itself, not patentable. By reason of condition (3), patents fall in the domain of technology, rather than science.

The patent writing step can be a game of disclosure and concealment for the patent attorney. Obtaining the patent usually creates value; yet, the disclosures required to obtain the patent may aid competitors. Thus, the patent attorney must balance business issues against the technical issues of gaining the patent. He may not wish to disclose the most valuable known use of the invention to potential competitors when the patent is published. And he certainly won't describe the best and most detailed directions for making the patented material or employing the patented process for all the world to copy. Let competitors learn to do this at their own expense—he merely wants a claim allowed that *includes* the most favorable cases. Patents typically include a section of examples meant to illustrate the usefulness of the invention; however, the actual examples given may be decoys.

The inventor and the patent attorney have two other key tasks to perform to ensure that value is maximized. First, they must attempt to broaden the claims as much as possible to make it as difficult as possible for ingenious competitors to invent around the claims. Second, they will wish to "teach" downstream applications, to make it more difficult for others to show non-obviousness if they attempt to patent such applications. In other words, the inventor of polyarothene may claim the material and the process to make it, and then cite its possible application in a list comprising essentially all known uses of similar plastic materials.

After writing the basic patent, the attorney may write other patent applications to extend and defend the core invention. Patent writing involves on-going dialog between inventors and attorneys, since the attorney may suggest additional experiments to support extending the claims and to demonstrate that the invention has been reduced to practice. For this reason, considerable time may go by between time of invention and the filing of the patent application.

Care must be taken that publications by the company's scientists do not undermine the patenting process. Publications are the coin of the realm for scientists. They are the basis for advancement for academics and an absolute requirement for graduate students seeking Ph.D. degrees. For industrial scientists, publications are the only real measure by which one's status in the field is established and maintained. Scientists want to publish. (Giving a talk at a scientific meeting is equivalent to publishing for this purpose.) Unfortunately, publication preceding a patent filing may invalidate that patent by creating its own prior art and denying the condition of novelty. The rules and timing differ in different countries, but the principle is universal. As a result, businessmen usually argue against publishing any material for which a patent has yet to be filed.

While this issue is nontrivial, it can usually be successfully negotiated. Industrial scientists don't have much choice but to play by the company rules. Most research contracts between industry and universities have a section dealing with just this problem. However, very valuable technologies, such as the basic claims on hybrid monoclonal antibodies, have been "donated to the public" because the inventor disclosed the technology before filing for patents. In this case, the event had significant international trade significance, since the inventor was British but the technology, after it had been disclosed, was primarily exploited by American firms.

Obtaining and Maintaining Patents

The second step in the patent value chain is to actually obtain the patent. In its simplest form, this involves a negotiation between the patent examiner of a national patent office and the patent attorneys of the sponsoring company. Few patent filings are simply accepted. The examiner usually rejects or seeks to narrow some or all of the applicant's claims. The inventor's attorney attempts to counter these arguments until a final decision is reached.

Patent applications may result in multiparty proceedings. In the United States, a patent examiner may declare an *interference* if he believes two patent applications may have overlapping claims. The judgment is made in part by the seniority of the invention and the filing, but there are other criteria. Each of the parties may present evidence to support its priority. In several foreign countries, patent applications are published before the patent is issued, and anyone may argue against the issuance of that patent in a process known as an *opposition*. Some overseas competitors almost routinely oppose patent applications in their fields of interest.

The rules for obtaining patents are undergoing rapid change as a result of international agreements promoting "patent harmonization." You must keep current with these changes or consult with patent attorneys for more detailed and current information.

There are costs associated with obtaining and maintaining patents. National patent offices charge fees, usually several thousand dollars per application, to cover their costs. They may also charge additional fees for maintaining the patent—these often escalate as the patent gets older. These costs represent incentives for inventors to drop valueless patents. The second major cost is the inventor's patent attorney. Costs may range from a few thousand dollars to hundreds of thousands of dollars in the event of a major interference proceeding. These costs have increased significantly in the past decade.

Global Patent Protection

A number of different patent systems and patent philosophies are found around the world. All national patent offices, however, seek in some measure to favor local businesses and inventors.

For U.S. inventors, the decision to seek patent protection abroad must be made promptly. This is due to the requirement of absolute novelty. Novelty is particularly important in Europe. Commercialization or the publication of technical information in the United States will violate the novelty requirement and most likely destroy any change of obtaining overseas rights.

Unfortunately, a patent in some countries provides very little protection. Courts may favor local infringers. And law enforcement agencies may be reluctant to pursue local pirates, or be susceptible to influence or bribery. Some third-world politicians view patents as a form of neocolonialism, and proclaim that patented drugs should be available to their poor at cost. Thus, the prospects of patent enforcement in such countries are poor.

Businesspeople and inventors face a clear tradeoff between investing valuable dollars on patent protection in dubious places and risking future competition from an overseas location. A first priority is to cover one's bases in those European and English-speaking countries that respect patents and afford strong patent protection. The second priority is to seek patents elsewhere, usually on an opportunistic basis. Japan represents a special case, because it is very difficult to compete there, but if you intend to, it is extremely important to patent your products. One can anticipate that their patent system will be hostile to Western interests, and will move very slowly. Skill and dedication are needed to gain even minimal protection.

Advanced developing countries such as Brazil, Taiwan, China, and Korea are even more problematic. These nations have a reputation for very weak or unfair patent systems and a history of permitting locals to appropriate foreign intellectual property. However, many argue that these countries will be forced to play by the rules as they become more thoroughly integrated with the world economy, and in time patents will gain force. Until that time comes, local partners may give the outsider an edge in seeking value from patents.

Patent Enforcement

The final step in the patent value chain is enforcement. The need for this, and its costs, should not be underestimated; nor should the value of successful enforcement be discounted.

Some valuable patents have generated licensing cash flow values in the hundred million-dollar or even billion-dollar class. Searle's aspartame patent was valued at about a billion dollars when Monsanto acquired it. Patents on blockbuster drugs protect many hundreds of billions of dollars of shareholder value. However, even highly visible patents are but the tip of the value iceberg—most of the value to patent-holders comes in the form of additional gross margin.

With stakes this high, it is a fair bet that any important new patents will be carefully scrutinized by competitors and potential competitors. And many will be challenged. Challenges take several forms: challenges may occur in the patent offices of various nations; interference actions in the United States; or formal opposition procedures in Europe or

Japan. Later challenges may move to court, or competitors may knowingly infringe, saying, "sue me."

A patent challenge should not cause the inventor or the firm to panic—it represents business as usual. (If your income model includes extra gross margin based on patent protection, your cost model might well include some expenses for patent enforcement!) In fact, a challenge is a form of good news, a clear indication that the technology has real and substantial value and that the inventor is in a position to negotiate and capture value. Unfortunately, some business and financial executives draw the opposite conclusion—they perceive only an inconvenient, unbudgeted, and uncapped legal expense. The costs of a patent challenge in court can be nontrivial—$1 million or more. Yet this cost is all part of the process of validating a patent and extracting its full value for the benefit of shareholders.

One way to reduce the cost of enforcing the patent is to negotiate a licensing arrangement with the challengers. Often, if a respected industry player accepts a license, the rest of the industry will fall into line and the validity of the patent will be established for practical purposes. But to get that first licensee, the patent holder must be prepared to defend the patent, even if it means going to court.

Cross-Licensing

Some patent disputes can be settled through cross-licensing, instead of (or in addition to) paying cash royalties. One instance of this is the case in which one party's patents interfere with the other party's ability to do business. For example, if X invents a material, and Y invents a new use for the material, neither party can exploit the material in this use without a license from the other. They may resolve that dispute by cross-licensing the two patents—a value creating exercise for both parties. In other cases, a company that appears to be infringing a competitor's patent may grant rights to some of its own patents to the other party in lieu of a cash payment.

Patents on materials used in products are often enforceable. For example, it is a relatively simple matter to ascertain whether another drug company is copying your drug—which is why very few take that risk. Patents on *uses and applications* are murkier; and patents on *manufacturing processes* can be very difficult to enforce, since access to a competitor's plant may be required. If a patent is not enforceable, there is obviously little point in obtaining it—the information will be more valuable if maintained as a *trade secret.*

Patent disputes are basically manageable, and the decisions required by managers are similar in kind to those used in dealing with other R&D risks. When it appears that a patent lacks commercial value or is on shaky legal ground, it is wise to drop it. Attempting to enforce a dubious patent is a dangerous strategy and likely to backfire with expensive consequences. Effort and legal cost should be concentrated on that part of the patent portfolio that has financial and strategic value.

Technology Transfer

Technology transfer is the successful adoption of a technology package by a new organization. It occurs in every project in the inevitable transitions from the research lab to engineering and on to manufacturing—between different units in a single company. When technology is purchased or licensed between companies, technology transfer takes place. It also occurs between companies and consultants.

A familiar analogy to technology transfer is the purchase of desktop software. Presumably, the software worked just fine on the developer's machine when it was released. But it is a wholly different matter to get it to work on your machine, both because your system is configured differently and because the directions that seemed so clear to the person who wrote them may be incomprehensible to you. Most of the time you master the difficulties, perhaps after a call or two to the technical support desk, but occasionally you give up in frustration.

From a business viewpoint, technology transfer is primarily *an issue of risk*—a risk that must be managed and is often vastly underestimated. The "selling" organization, the organization that created the technology, is enthusiastic and anxious to see its technology implemented and has difficulty anticipating the receiving organization's problems. For a number of reasons, some of them shortsighted, technology sellers are prone to underestimate the issues that new technology creates for the receiving organization.

A case history will illustrate how difficult the technology transfer problem can be. This case, which had a happy ending, is one in which I was peripherally involved some 20 years ago. It involved the transfer of technology from a leading U.S. chemical company to the Dow Chemical Company—one of the three largest chemical firms in the United States. That technology had been practiced successfully for over 20 years in a full-scale manufacturing plant before it was licensed to Dow. Dow had first-class process engineering skills. The Dow plant was expanded slightly to improve economics of scale and redesigned in some sections to take advantage of what had been learned from the original plant. In principle, all of these changes were believed sound and the risks appeared minimal. Nonetheless, the plant didn't work. The start-up encountered very serious operational problems, among which were the plugging of pipes and process equipment. These problems were analyzed and eventually solved by crack technical teams, who had to make many changes in both procedures and equipment. Yet it took almost a year for this *demonstrated* commercial process to be transferred successfully between two very competent and experienced companies.

Companies take technology risks that are much greater than this one. For example, companies have taken licenses on processes that are only *partially* developed and, under pressure to show results, have engineered them directly into full-scale manufacturing. Because of the significant costs of operating pilot plants, there is a natural tendency among the inexperienced to avoid checking out each stage, or to do the process research in pieces with the assumption that if each piece works, the whole will work. This approach can be lethal! I'm familiar with at least two $100 million write-offs that occurred because of these shortcuts.

But technology transfer problems are not confined to transfers between companies; they also occur in various steps in the research process itself. Frequently the idea is conceived in a laboratory by an inventor, and its feasibility is first established in small-scale experiments. At the next step, engineering data must be gathered for scale-up of that work, and that involves communication between the scientists and the engineers. Items that are trivial in the laboratory, such as allowing a vessel to heat or cool, to dilute the reaction mixture, or to dispose of used solvent, can cause enormous economic and technical headaches in a commercial plant. A third step is to transfer that engineering process to a pilot facility. This may involve practical difficulties, particularly when the pilot operation may be constrained to using existing equipment that is somewhat different in design or concept than what was used in the bench-scale studies. The final stage is to move from that pilot operation to a full-scale plant.

Technology transfer issues are part of the risk involved in advancing projects through the four stages of R&D described in previous chapters, and how well they are handled counts significantly in the success rates and cycle time achieved by an organization.

Similar steps must occur in product introduction where the properties of the new product must first be communicated to marketing people, who must then translate that into terms that the salesforce can promote. The salesforce must further communicate to its customers and distributors. Serious impediments to successful implementation in the field can occur because of structural problems, such as the limitations of the customers' facilities or of commercial constraints, which were missed in the initial market research. Each of these transfer steps involves risk, and even if the risk of each step succeeding is as little as 10%, the cumulative risk in technology transfer may well approach 50%.

The competence of the receiving organization is extremely important in determining the risk in technology transfer. In fact, the technical competence of the receiving organization should be considered before any project is initiated. My own experience is that businesses that have a high degree of commercial ability, engineering skill, and which maintain high margins, enjoy much greater success than do entities that are undercapitalized, struggling, or financially unequipped to support scale-up and commercialization. I have seen situations in which a long series of good product ideas that were enthusiastically endorsed by the receiving organization all failed. In most cases, the cause of failure was an insufficient commitment of resources; the organization simply lacked the resources needed to overcome the barriers that stand between all commercial launches and eventual success.

Researchers must recognize that technology transfer risk is part of the process of evaluating a project, and may itself be sufficient reason for a "no go" decision. Similar reasoning applies to the licensing of technology. A highly competent licensee will have a far better chance of success than will an enthusiastic, hungry boutique that lacks the resources to execute a successful market entry.

Another barrier to successful technology transfer is found in the set of personalities and motivations of principle players on each side of the relationship. Technology transfer can fail if one of the initial project champions is transferred or otherwise leaves his post. A replacement player is prone to reexamine the merits of the project and to discount the hard-won learning already accrued by the transferring organizations. The newcomers skepticism is understandable, but it can lead to a breakdown of trust.

Differences in organizational culture represents another factor that can affect the success of technology transfer. These culture differences may be found within two divisions of the same company, between two different companies, or, as we increasingly find in a global business, between two companies representing different national outlooks. Cultural misunderstandings can arise between partners over issues such as who is the team leader, when to report problems, when vacations are taken, safety precautions, and a host of others. Personal and organizational rivalry may also come into play. Perhaps the wisest statement I've heard made about technical transfer is that it is a contact sport. It is an intensely personal exercise, and not something that can be done through documentation. All parties to the exercise must be prepared to spend significant time working together toward the desired result.

Finally, acquiring technology from the outside is no excuse for not having technology capabilities within. Successful transfer requires people of relatively equal competence on both sides of the transfer process. When things go wrong, the recipient organization cannot depend on the technology vendor or an engineering contractor to bail it out.

So far we have looked at technology transfer as a risk element, but there is a real potential for it to be used to add value. Organizations compete, and those that are consistently more skillful in technology transfer will rise to the top. There is increasing interest today in the concept of technology integration,[1] a process through which experienced individuals facilitate transfer by guiding a project through two or more stages of development to ensure that organizational strengths are integrated into the technology rather than added piecemeal at each stage. Improved product performance and reduced development cost, both key value drivers, have been documented in a number of instances in which technology integration has been used.

A persuasive case has been made that technology mirrors organization,[2] and that the best organizations imbed their organizational capabilities into their products and their technology. This is almost self-evident in products such as a Mercedes automobile, a McDonald's hamburger, or an IBM mainframe computer. In each case, the final product is the result of a development process that is a reflection of the culture of the firm that performs it. Because the process and culture are not easy to imitate, the product itself is unique. It follows that value creation will have a great deal to do with how organizational strengths are translated into commercial technology. Conversely, weaknesses in the finished product may stem from organizational weaknesses. We have learned that technology does not exist in a vacuum, but only in a context. Part of that context is organizational.

Globalization

Globalization has leveraged the value of technology and added new dimensions to the competitive and strategic equations of technology-based companies.[3] Today, new technologies flow virtually instantaneously around the world via fax machines and the Internet, in the form of text, data, and detailed engineering drawings. And the capital required to develop these technologies can be raised in global financial markets and deployed rapidly, even to remote locations.

The global networks of commerce have now overcome many of the cultural and infrastructure limitations of regions that a decade earlier were "off the map" of the modern economy. This change in circumstance, often referred to as the "rice paddy phenomenon," makes it possible to establish a world-class commercial operation, such as a state-of-the-art petrochemical complex, in a rice paddy in Southeast Asian or other third-world location. These complexes are being constructed with a variety of motivations: cheap labor, access to growing markets, access to raw materials, (until recently) low cost of capital, or national pride. This new reality has raised grave concerns in developed countries, particularly among unions, politicians, and others concerned with unemployment, but for technologists much of the news is good. Equalization among regions in access to labor, raw materials, and capital puts a higher premium on technology.

Let us consider how the globalization of technology has come about, first by reviewing historical trends in doing business globally, then by looking at the dynamics of today's global markets and the financial influences globalization has had on the creators of technology.

Changing Business Patterns

A clear pattern of evolution has emerged in how businesses have organized to operate on a global basis. Most companies had their origins as national companies. Many of these evolved into multinationals by cloning their core operations and planting them in different regions of the world. Today, corporations are moving beyond the multinational stage to that of the *transnational* company. In the former, business units are defined by national boundaries, while the latter have no borders. Companies are in different stages of this evolution, but it is clear where the game will end. Once a company's customers and competitors adopt transnational strategies, going transnational oneself becomes an imperative.

Faster and cheaper communications and a common language have facilitated the transnational trend. Today we communicate at the speed of light via fiber optic cables, microwave links, satellites, and so on. The bandwidth of this communication network is enormous and growing, and English has become its language. English has been adopted by the business and scientific communities of the world. Indeed, the national scientific journals of a number of European countries are now published in English—because scientists in Scandinavia, Hungary, and even France want their work to be readily available to the global technical community.

Implications for Technology

Globalization has tremendous implications for technology. Globalization affects each of the key attributes of good R&D: leverage, productivity, being close to the customer, and superior access to technology. A worldwide technology marketplace is being created, and its importance will advance in step with the global communications infrastructure.

Globalization has created opportunities to leverage the potential of new technologies by moving them quickly to those markets where they provide a competitive advantage—whether that be accomplished through direct sales, joint ventures, or licensing. In addition, globalization affects the cost and value of the R&D program. For example, if the U.S. market represents one-third of the global market and the global market can be served with mostly the same technology, a global presence can in principle triple the NPV of any R&D project. The effect on the R&D budget is also very powerful. Given the above scenario, a technology company can:

1. Perform the same R&D at one-third the cost as a percentage of sales;
2. Perform three times as much R&D at the same percentage of R&D to sales; or
3. Any combination of the above.

Clearly, such leverage can create enormous competitive advantage.

For a national company, leverage typically takes the form of earning royalties from licensing technology to other firms or to joint venture partners. These royalties can be used to offset domestic R&D spending in the home country laboratories. Multinational companies fall into two patterns, often within the same company. The first pattern is to support all technical activity from the home country. This is appropriate when there are few issues of technology transfer to customers, and few requirements for customization to local needs. An example might be commodity chemical products such as chlorine and caustic, where an overseas manufacturing plant can be readily supported from a laboratory anywhere in the world.

The second pattern is to establish a group of overseas technology centers. This is very useful when local customs differ substantially from those of the home country. Food packaging is an excellent example—local tastes, shopping habits, and the availability of refrigeration vary widely from country to country. A local laboratory customizes the product line to these varying requirements and provides a nucleus of technically skilled people to communicate to customers. A second advantage of this arrangement is *feedback*. The company can learn from its technical staff of new packaging concepts that originate abroad and that have potential value both in the home country and other regional markets.

Transnational companies, in contrast, establish global technology organizations and perform all phases of R&D in the countries where it is most logical to do so. Overseas laboratories may therefore assume prime responsibility for certain products, including exploratory R&D, product development, and initial commercialization. For example, the primary responsibility for alumina research in Alcoa is now centered in Australia.

The transnational company finds that it makes sense to have a technical presence in a country that has technology strengths comparable to or greater than the home country—ceramics in Japan, chemicals in Germany, and biotechnology and aerospace in the United States. Technologists with scientific roots in these countries can build contact networks through which promising new technologies can be identified and acquired. The local laboratory can take the lead in achieving technology transfer with the advantages of proximity and culture. No licensing officer can achieve such results. The creativity of overseas research staff and the value of their fresh approaches to the problem-solving skills of the global R&D team should not be underestimated.

Finally, a global technology organization can effect tremendous productivity gains in cost and speed to market. This is facilitated by the tremendous bandwidth capabilities of modern communication, the 24-hour day, and low salaries for technology-capable staff in some parts of the world. For example, there is a rapidly growing software industry in Bangalore, India, owing to the availability of skilled programmers and the ability to transmit new code to the home country instantaneously. Design engineering, another technical activity that can be carried out on workstations or terminals and communicated electronically, is also being rapidly globalized.

When speed is essential and much of the work is routine, transnational teams can carry out 24-hour-a-day programming. When the crew in California finishes its work, the task can be continued in East Asia, and transferred for the third shift to Europe, whose members hand the product back to California at the end of their day.

Environment

Environmental issues have had an impact on my career that was far greater than I might have ever anticipated. Their importance emerged when as a junior research manager, I was asked to help manage crises related to mercury and dioxins, and continued to the point where as a senior executive I became accountable for the environmental, health, and safety activities of a major corporation. Environmental issues as diverse as novel clean air technologies, bioremediation, radioactive waste, and superfund sites were important landmarks, as was my tenure on an advisory committee of the Environmental Protection Agency. These experiences taught me that environmental issues are important and complex value drivers in technology-oriented businesses and must be factored into both near-term investments and long-term strategic plans. Their dynamics are different from other economic issues. Managing them effectively and exploiting the opportunities they present creates competitive advantage.

We have generally focused in this book on the quantifiable manifestations of value, such as NPV and the market value of securities. This is too narrow a view of value when dealing with environmental issues, but it is a good place to begin. Our discussion of value drivers will be divided into four categories: *capital, costs, timing*, and *opportunities.* Much more deserves to be said about environmental issues and industrial technology, of course. Human health, the quality of human life, and the health of the planet earth and the species that inhabit it are all affected by the technologies we create and the ways in which we use them. But the broad acknowledgment of these issues and their importance is beyond the scope of this book.

Capital

For the industrial company, the most visible impact of environment and health on value is the greater fixed capital required to produce a product. It is not unusual today for 20% of the capital cost for a new plant to be tied to environmental factors, such as wastewater treatment, baghouses, stack gas scrubbing, scrap recycling, and on-line instrumentation. These also imply increased operating costs and depreciation.

Previous chapters have identified how sensitive NPV and IRR can be to a 20% change in fixed capital. Adding 20% to plant costs in a competitive work is enough to turn an attractive project proposal into a clear loser unless customers help defray the increased costs through higher prices. The higher cost plant *will* benefit from higher prices if the playing field is level and all suppliers must abide by the same environmental regulations, but if there are critical differences in regulations between localities, important elements of competitive advantage and disadvantage are created. Consider the patterns of manufacturing investment and plant closings in the United States in the recent past. Areas perceived to be expensive in terms of environment compliance, such as California and New England, have seen an exodus of industry, much of it moving to the Gulf Coast or the mid-South, where the perception is the reverse.

The dynamics of the process by which capital flows from one region to another is more obvious to an industrial manager than to a politician or regulator. The latter is likely to believe that a local plant is captive and has no choice but to comply. He probably does not appreciate the short useful life of industrial capital assets. He may be misled by the fact that most competing plants in the short run will comply with new regulations rather than shut down. However, the local manager knows that his plant must in principle compete economically with a potential new plant constructed elsewhere, or with a potential plant expansion at another site. If top management perceives it now has a high cost site, it will deny capital for further improvements and expansions and save that capital for higher value opportunities. The old plant will be "capped" and its fate sealed. Within five to ten years, when most of the equipment has been depreciated, it will be closed. In a sense, the real plant closing occurs when the plant manager's request for an important capital project is turned down and he is unable to modernize with the latest and best equipment. Shutting the gates and furloughing the employees is simply an epilogue to an earlier and less public decision to deny capital.

The fixed cost issue does not simply occur between regions of the country but is the stuff of international competition. If an overseas plant has less stringent environmental and health standards than a domestic operation, it gains a competitive advantage that can be just as important as lower unit labor costs. The issue of different CO_2 standards for developing and developed countries, for example, has the potential to cause massive location shifts in future capital investment from the latter to the former in industries relying heavily on energy and petroleum feedstocks.

Costs

We have already noted that increased capital costs for environmental control imply increased operating costs. There will be both variable costs such as operating labor, utilities, treatment chemicals, and outsourced environmental services. There will also be fixed costs at the plant level, such as depreciation and factory overhead. The latter may include the costs of obtaining and maintaining environmental permits, on-site environmental and safety managers, and environmental consultants. Other costs will occur at the corporate administrative level (the "A" of G&A on the expense portion of the income statement). These include legal staff, regulatory specialists, and the costs (often substantial) of cleaning up orphan and superfund sites, some of which may never have been owned by the company, but where the company's hazardous waste may have once been placed.

The third category of costs is in research and development itself. Many types of new products encounter broad regulatory requirements, not all related to environment. Some of these may be mandatory while others are at the discretion of the regulatory agency, for example, the Environmental Protection Agency (EPA), the Food and Drug Administration (FDA), or the Department of Agriculture (USDA). In some cases, several rounds of additional testing may be imposed by a regulatory agency and result in years of delay. Testing costs and approval cycles for drugs, pesticides, and potentially toxic chemicals have escalated dramatically in past decades—and have contributed significantly to the fact that the total development cost of a major new drug has risen to the $500 million level. It has also led executives to conclude that new drugs, chemicals, and pesticides aimed at small markets may under no circumstances be economically viable, since testing costs have little to do with whether a market is large or small. Perversely, new regulatory requirements have tended to entrench some existing materials with known deficiencies, such as pesticides, by reducing much of the incentive to research safer products.

The strategic consequences of these costs for U.S. researchers have been considerable. Many agricultural products, drugs, and medical devices are first introduced in foreign countries to give the developer some assurance he has a winner before making a costly and risky investment to gain U.S. approval. The result is that some of the technology base once concentrated in the United States is being exported. And inventors of technology, such as biotechnology start-ups, who cannot afford such costs, often have little choice but to auction their inventions to the small group of large firms that have the infrastructure and financial resources to underwrite the full costs of product development.

TIMING

Delay reduces the reward from past investment according to the laws of discounted cash flow. For example, a three-year delay caused by regulation could reduce NPV by 29% at a 12% cost of capital. In itself, this can easily be enough to turn a winning project into a net loser. In the case of a patented product the situation is worse, because three years of patent protection is also lost. In the case of a proprietary drug with an NPV of $1 billion or more, the lost value represents very serious money.

Opportunities

Environmental concerns have also created new markets, many of them substantial. The U.S. environmental industry alone has annual revenues just under $200 billion and employs 1.3 million people. The global environmental industry generates revenues of over $450 billion.[4] It is a major part of the world economy. However, not all is well. Growth rates have slowed from 10% to 15% in the late 1980s to 1% to 5% in the 1990s, and environmental companies are out of favor on Wall Street. In the equipment and technology sector especially, U.S. regulations have unintentionally favored overseas competition.[5] These issues may be expected to be highly visible in the coming decade.

The catalytic converter provides an interesting example of a product created by environmental regulation. It simultaneously created, expanded, reduced, and destroyed different markets. Some companies were big winners and some big losers in this process. The converter had the direct effect of creating new markets for precious metals, ceramic honeycombs, and metal housings. It also created an array of jobs at vehicle testing centers. Less obvious effects have also occurred. The market for tetraethyl lead, an octane booster (and a dangerous pollutant) that is incompatible with converter technology, was destroyed, while a new market for methyl tertiary butyl ether (MTBE), a more environmentally friendly octane booster, was created. The universal existence of price elasticity of demand argues that the automobile market itself, as defined by numbers of vehicles sold, was reduced, since converters increased the prices of cars quite substantially and probably reduced unit sales. The effect of fewer vehicles produced cascaded downward to automotive suppliers of tires, radios, mufflers, and hundreds of other components. The timing for these changes was driven largely by the scrappage of older vehicles that were generally "grandfathered" to avoid an additional economic burden on drivers, even though these clunkers caused a disproportionate share of the pollution. As a result, many years went by before leaded gasoline disappeared from the marketplace.

The new concept of *product stewardship* has become an important aspect of environmental marketing and product design. The phrase refers to cradle-to-grave concern for the environment during a product's life cycle.[6] The ability to assure the customer that a product will be made from environmentally friendly materials, that it can be manufactured scraplessly, and ultimately disposed of benignly, has become a source of competitive advantage, and is actively marketed to both industrial and retail customers. Product stewardship has created many research programs that aim to improve environmental benefits while sustaining performance and cost competitiveness. The use of plastics and aluminum to reduce vehicle weight, and thus increase fuel efficiency, is just one example.

On the opportunity side, manufacturers can gain competitive advantage by using materials that defer or eliminate the expenditure of new environmental capital. For example, high solids or water-based coatings can dramatically reduce emissions from a paint line, perhaps eliminating the need for an incinerator otherwise required to meet local air standards. A surprising amount of industrial R&D is targeted at capital-related environmental advantages.

Other aspects of the environmental business are not so clearly in the public interest. Given that environmentalism has spawned a massive industry, with large armies of experts, bureaucrats, scientists, engineers, and lawyers arrayed on each side of the issues, one cannot expect dispassionate analysis—in fact, the natural motivation for many of the individuals and institutions involved is to see the contests, and the costs, continuing. A rational perspective may have to come from outside the industry.

Value-Based Environmental Decision Making

Value-based decision making and market-based mechanisms may have the greatest long-term potential for resolving environmental issues. The direction of progress, in fact, points to this conclusion. These decisions require a basic understanding of key technical relationships, and the use of financial techniques to balance risk and reward. These methods are not unlike the tools discussed in this book. Risk assessment is increasingly accepted in health and resource allocation controversies and is beginning to emerge in the quality-of-the-planet arena.

At the same time, it would be simplistic to think that we can rely entirely on the marketplace to protect the environment, since there is no real financial proxy for many of its highest values: the habitat, the continued existence of rare species, the beauty of untrammeled natural vistas, and hundreds more. Recreational use of parklands and waterways, eco-tourism, and the reversal of deforestation can all be market-based—and have definite financial and measurable financial values—but these are insufficient in the face of planetary environmental threats. Any expectation that environmental problems can be solved in ways that make the most technoeconomic sense flies in the face of recent experience. Unlike most other business issues, environmental issues tend to be resolved by adversarial processes, and while science and economics carry weight, by themselves they seldom carry the day. In today's world, one cannot yet expect either the highest priority environmental threats to be at the top of the agenda, or for costs to be well-correlated with benefits.

Afterword

We have reviewed the complex and extremely important process through which science is transformed into technology, and technology is transformed into economic gain. The English language seems to have no single word for this process. The French, in contrast, have an excellent term—*valorization*. The Japanese likewise have a suitable term that translates as *techno-economics.*

Our understanding of how science and technology create value is relatively new. Only thirty years ago, investors and many leading business executives were more inclined to pursue value through the buying and selling of corporate ownership than through the slower, more deliberate approach that others have followed in laboratories and research centers. The "conglomerate kings" of the 1960s and 1970s sought economic value through financial sleights of hand—the trading of paper. That approach produced remarkable results—at least for a while. The market value of companies like Litton Industries, LTV, Gulf & Western, and ITT skyrocketed. Most of these gains, however, evaporated just as quickly. Corporations that created economic value through the slow but deliberate process described in the preceding chapters eclipsed the conglomerates in the pantheon of business leadership. Companies like Merck, Hewlett Packard, Dow, Intel, Microsoft, 3M, and dozens like them have generated enormous wealth for their shareholder through R&D, not through a financial shell game.

But what of the future? Will technology creation ossify or will it evolve into something better? From a technologist's viewpoint, the big picture is mixed: overall funding for R&D in the United States is growing, but that growth is unevenly distributed. Statistics indicate dramatic growth in the information, telecommunications, and health care industries, while R&D funding in traditional process and extraction sectors has remained flat. Direct government spending in the national laboratories is being curtailed as a result of the end of the cold war, and government support for university research, while rising slowly, is increasingly inaccessible to young researchers.

There is increasing evidence that the value mindset has propelled the American system of capitalism to an unprecedented level of prosperity, and that it is currently providing important competitive advantages. And R&D, on balance, is benefiting from the resulting prosperity, despite many examples of mindless cost cutting, and even of personal hardship.

It is particularly important to remember that the American form of public corporation is a relatively new institution, that corporate governance is continuing to evolve, and that corporations themselves exist only at the sufferance of society. Indeed, American corporations are hardly trusted repositories of our national wealth as evidenced by the popularity of many media exposés and politically motivated attacks targeting them. This occurs despite the fact that it is mostly the general public who owns these corporations through its pension plans, mutual funds, and insurance policies, not to mention the safe deposit boxes, desks, and mattresses that hold actual stock certificates. And it is our own demands for an ever higher return on our investments that drive the practices of money managers, CEOs, and profit-center managers, each in turn. Furthermore, the employees of corporations are mostly ourselves and our relatives.

There are alternatives to the modern American corporation: private corporations, partnerships, employee-owned corporations (ESOPs), "Asian capitalism," the bank-dominated financial systems of Japan and Germany, and myriad forms of socialism have the ability to survive and often, for a time, to prosper.

It seems likely that other national economies, which reflect different traditions and cultures, but where long-term economic performance is in some cases in doubt, will in time accept the value system. If they do, they will very likely become more competitive. Science in some of these societies is more sheltered from economic forces than in the American system, but that sheltering has not made scientists more effective contributors to industrial growth. Countries where the bulk of science had been carried out in governmental institutions—the former Soviet Union, China, and Eastern Europe—have been notably ineffective in harnessing technology for economic growth, although China and most of East Asia have done an excellent job in assimilating external technology. Certainly, the rapid demise of government-based science and engineering establishment of the former Soviet Union gives pause to the notion that the weight of support for science should come from the public rather than the private sector.

The American system, in which funding freely follows opportunity, appears to be successfully managing some difficult transitions. Like Adam Smith's "invisible hand," it is directing technical and scientific know-how—and capital—from maturing sectors of the economy to growing ones, and from defense to a balance of health, environmental, and security missions. The extent to which the decision makers who control that invisible hand understand the value creation process will determine the wisdom of those allocations. Hopefully, the concepts and tools presented in this book will improve that understanding.

Chapter Notes

Preface

1. Lawrence J. Lau, "Sources of Long-Term Economic Growth," in *The Mosaic of Economic Growth,* ed. R. Landau (Stanford, CA: Stanford University Press, 1996), 79–81.
2. Quote attributed to Donald E. Kash, "Taking the Measure of Basic Research," *Chemical and Engineering News,* 20 October, 1997, 30–33.
3. William G. Howard and Bruce R. Guile, eds., *Profiting from Innovation* (New York: Free Press), 45.
4. Tom Copeland, Tim Koller, and Jack Murrin, *Valuation: Measuring and Managing the Value of Companies* (New York: Wiley, 1995).
5. Ibid.
6. Simon Ramo, *The Management of Innovative Technological Corporations* (New York: Wiley, 1980).
7. Richard A. Brealey and Stewart C. Myers, *Principles of Corporate Finance* (New York: McGraw-Hill, 1996).

Chapter 1 Science, Technology, and Business

1. For an anthropological outlook on the culture of science, and its breeding ground, see Bruno Latour and Steven Woolgar, *Laboratory Life* (Beverly Hills, CA: Sage Publications, 1979).
2. Reference: C. P. Snow, *The Two Cultures* (Cambridge: Cambridge University Press, 1993).
3. See Jeffrey Sonnenfeld, *The Hero's Farewell* (New York: Oxford University Press, 1988).
4. Attributed to Louis XV.
5. Richard A. Brealey and Stewart C. Myers, *Principles of Corporate Finance* (New York: McGraw-Hill, 1996), 77.

Chapter 2 The Industrial R&D Process

1. Marc H. Meyer and Alvin P. Lehnerd, *The Power of Product Platforms* (New York: Free Press, 1997), 230.
2. Oscar Schisgall, *Eyes on Tomorrow* (Chicago, IL: J. G. Ferguson, 1981), 216–220.
3. For studies on lead users, see Eric von Hippel, *Sources of Innovation* (New York: Oxford University Press, 1988).
4. Greg A. Stevens and James Burley, *3,000 Raw Ideas = 1 Commercial Success* (Washington, DC: The Industrial Research Institute, *Research•Technology Management*, May–June, 1997), 16–27.
5. Robert G. Cooper, *Winning at New Products* (New York: Addison Wesley, 1993), 102–106.

Chapter 3 A Technologist's Guide to Financial Statements

1. Tom Copeland, Tim Koller, and Jack Murrin, *Valuation* (New York: Wiley, 1995), 160.
2. Nonoperating companies, such as R&D start-ups are in a different situation. They may pay no income tax, so they have no immediate income from which they may claim tax deductions for research expenses. Their R&D losses may have value, since they can be carried forward against the future as "NOLs" (net operating losses), which they or an acquiring company may be able to eventually recapture for tax purposes.
3. Data from *The Eaton Corporation, 1995 Annual Report* (dates have been altered).
4. Allen Michael and Israel Shaked, *Finance and Accounting for Lawyers* (Boston, MA: Legal Financial Press, 1996), 17.
5. Richard A. Brealey and Stewart C. Myers, *Principles of Corporate Finance* (New York: McGraw-Hill, 1996).
6. To calculate the term of the company's accounts receivable, multiply the percentage of total sales (from the income statement) represented by accounts receivable times 365 days.
7. Copeland et al.,*Valuation*, 172.
8. Ibid., 172.
9. Stanley Foster Reed, Lane Edson, and P. C. Edson, *The Art of M&A: A Merger Acquisition Buyout Guide* (Homewood, IL: Dow Jones-Irwin, 1989), 106 ff.
10. H. T. Johnson and R. S. Kaplan, *Relevance Lost, The Rise and Fall of Management Accounting* (Boston, MA: Harvard Business School Press, 1987), 200.

11. Ibid., 202.
12. S. Davidson, C. P. Stickney, and R. L. Weil, *Financial Accounting* (Chicago, IL: Dryden Press, 1988), 11; Financial Accounting Standards Board, Statement of Financial Accounting Standards No. 2, "Accounting for Research and Development Costs," 1974.
13. Form 10K, Iridium World Communications LLC, March, 1998. Development costs are included within system construction cost.
14. *Annual Report 1997*, Johnson & Johnson, New Brunswick, NJ.
15. In particular, see Copeland et al., *Valuation.*, 268.
16. Robert C. Higgins, *Analysis for Financial Management* (Chicago, IL: Richard D. Irwin, 1995), 40.
17. Brealey and Myers, *Corporate Finance*, 154.

Chapter 5 Calculating Values Using Discounted Cash Flow

1. See James M. Utterback, *Mastering the Dynamics of Innovation* (Boston, MA: Harvard Business School Press, 1994), 103–120.
2. Some people use the expression "DCF return" when describing the IRR.
3. Richard A. Brealey and Stewart C. Myers, *Principles of Corporate Finance* (New York: McGraw-Hill, 1996), 162.
4. Brealey and Myers, *Corporate Finance*, 207.
5. See Burton Malkiel, *A Random Walk Down Wall Street* (New York: W.W. Norton, 1996).
6. Brealey and Myers, *Corporate Finance*, 180.
7. Tom Copeland, Tim Koller, and Jack Murrin, *Valuation* (New York: Wiley), 268.
8. Brealey and Myers, *Corporate Finance*, 154.
9. Ibid., 206.
10. Alan C. Shapiro, *Modern Corporate Finance* (New York: Macmillan, 1989), G-4.
11. P. Morris, E. Teisberg, and A. L. Kolbe, "When Choosing R&D Projects, Go with Long Shots," *Research•Technology Management*, January–February 1991, 35–40.
12. Copeland et al., *Valuation*, 221.

Chapter 6 R&D, Growth, and Shareholder Value

1. F. Peter Boer, "Linking R&D to Growth and Shareholder Value," *Research•Technology Management*, May–June 1994, 16–22.
2. Simon Ramo, *The Management of Innovative Technological Corporations* (New York: Wiley, 1980).

3. The free cash flow will be either the difference between the gross cash flow and the gross new investment, or the operating profit less the net new investment. (Depreciation falls out of the calculation, as it is included in both terms.)
4. Tom Copeland, Tim Koller, and Jack Murrin, *Valuation: Measuring and Managing the Value of Companies* (New York: Wiley, 1995), 143.
5. Alden S. Bean, John B. Guerard, Jr., and Bernell K. Stone, "Goal Setting for Effective Corporate Planning," *Management Science*, 1990, 359–366.
6. Alden S. Bean and John B. Guerard, Jr., "A Comparison of Census/NSF R&D Data vs. Compustat R&D Data in a Financial Decision Making Model," *Research Policy*, Vol. 18, No. 4, 1989, 193–208.
7. Copeland et al., *Valuation*, 80.

Chapter 7 Strategy: Driving Value in the Competitive Arena

1. See David Matheson and James Matheson, *The Smart Organization* (Boston, MA: Harvard Business School Press, 1997).
2. Tom Copeland, Tim Koller, and Jack Murrin, *Valuation: Measuring and Managing the Value of Companies* (New York: Wiley, 1995), 44.
3. Philip A. Roussel, Kamal N. Saad, and Tamara J. Erickson, *Third Generation R&D* (Boston, MA: Harvard Business School Press, 1991), 97.
4. G. Hamel and C. K. Prahalad, *Competing for the Future* (Boston, MA: Harvard Business School Press, 1994), 23.
5. Steven Brandt, *Strategic Planning in Emerging Companies* (Reading, MA: Addison-Wesley, 1981).
6. J. S. Armstrong, "The Value of Formal Planning for Strategic Decisions: Review of Empirical Research," *Strategic Management Journal*, Vol. 3, 1982, 197–211.
7. Hamel and Prahalad, *Competing for the Future.*
8. For a discussion of core competencies, see Hamel and Prahalad, *Competing for the Future*, 23.
9. Robert Cooper, *Winning at New Products* (Reading, MA: Addison-Wesley, 1993), 310.
10. For a full description of the Xerox-L.L. Bean case, see Gregory H. Watson, *Strategic Benchmarking* (New York: Wiley, 1993), 153–167.
11. For a thorough treatment of competitor intelligence, see Leonard M. Fuld, *The New Competitor Intelligence: The Complete Resource for Finding, Analyzing, and Using Information about Your Competitors* (New York: Wiley, 1994).

Chapter 8 Marketing: The Top Line

1. Robert Cooper, *Winning at New Products* (Reading, MA: Addison-Wesley, 1993), 22.
2. MIT Professor Eric Von Hippel has shown how "lead users" are an important source of new product and technological innovation. See *Sources of Innovation* (New York: Oxford University Press, 1988).
3. For a complete discussion of the development of Sunbeam's "Global Iron," see Marc H. Meyer and Alvin Lehnerd, *The Power of Product Platforms* (New York: Free Press, 1997), 105–116.
4. Gary Hamel and C. K. Prahalad, *Competing for the Future* (Boston, MA: Harvard Business School Press, 1994), 26.
5. Cooper, *Winning*, 70–71.
6. P. Green, S. M. Goldberg, and M. Montemayor, "A Hybrid Utility Estimation Model for Conjoint Analysis," *Journal of Marketing*, 45, 1981, 33.
7. In many firms the authority to negotiate price is not granted to salespeople, but to the marketing function or general management. This is a good rule, since it prevents a narrowly interested individual's actions from changing industry relationships. A price change can cause a chain reaction of competitive pricing actions, risking a destructive full-scale "price war."
8. The first recognition of the experience curve was made in the late 1930s, in aircraft plants, where managers observed that the labor hours needed to assemble an airplane declined as production increased.
9. Peter H. Spitz, *Petrochemicals: The Rise of an Industry* (New York: Wiley, 1988), 390–417.
10. Richard N. Foster, *Innovation, The Attacker's Advantage* (New York: Summit Books, 1986), 102.
11. For the details of these particular discontinuities, see James M. Utterback, *Mastering the Dynamics of Innovation* (Boston, MA: Harvard Business School Press, 1994).
12. Ronald A. Mitsch, "R&D at 3M: Continuing to Play a Big Role," *Research•Technology Management*, September–October 1992, 22–26.
13. Foster, *Innovation*.
14. James R. Bright, *Practical Technology Forecasting* (Austin, TX: Sweet Publishing, 1978).
15. Ibid., 104.
16. Committee on Time Horizons and Technology investments, National Academy of Engineering, *Time Horizons and Technology Investment* (Washington, DC: National Academy Press, 1992), 13.

17. Note on more complex "top lines." In constructing our revenue model, I chose for simplicity to project dollar revenues. In most business plans, marketing first projects volume in units (which helps in projecting capacity), then projects average selling price and multiplies the two numbers to forecast revenues. In a complex product line, several different types of units, and their prices, are individually forecasted and then added together to obtain annual revenues. Revenue growth can then be a result of three factors: volume growth, price, and the *mix*.

Chapter 9 Building a Pro Forma DCF Model

1. Terrence W. Faulkner, "Applying Options Thinking to R&D Valuation" (*Research•Technology Management*, May–June 1996), 50–56.
2. S. Davidson, C. P. Stickney, and R. L. Weil, *Financial Accounting* (Chicago, IL: Dryden Press, 1988), 229–230.
3. One accounting refinement not included in our model is that most corporations (correctly) report returns on total capital and returns on equity using the average value of invested capital over the year. In most cases, invested capital increases from year to year, so this procedure generates a slightly higher number than the year-end figure. However, for transparency, we did not include this in our simplified spreadsheet.
4. Richard A. Brealey and Stewart C. Myers, *Principles of Corporate Finance* (New York: McGraw-Hill, 1996), 91.
5. Tom Copeland, Tim Koller, and Jack Murrin, *Valuation: Measuring and Managing the Value of Companies* (New York: Wiley, 1995), 172.

Chapter 10 Shortcuts and Market-Based Approaches to Value

1. Remember, however, that economies of scale can affect capital intensity. If you are looking at a larger market, you may want to use the 0.6 exponent rule or some other adjustment to reduce the key fixed assets/sales ratio. This factor can significantly enhance the value of upside cases (refer back to Chapter 4, Exhibit 4.2).
2. Joseph G. Morone, "Managing Discontinuous Innovation," talk presented at Industrial Research Institute Semi-Annual Meeting, Cleveland, Ohio, October 13, 1997.
3. Tom Copeland, Tim Koller, and Jack Murrin, *Valuation* (New York: Wiley, 1995), 149.
4. Gary Hamel and C. K. Prahalad, *Competing for the Future* (Cambridge, MA: Harvard Business School Press, 1994), 23.
5. William D. Bygrave and Jeffry A. Timmons, *Venture Capital at the Crossroads* (Boston, MA: Harvard Business School Press, 1992), 16–21.

6. Ibid., 67–93.
7. For a complete discussion of the stages of technology start-up financing, see John L. Nesheim, *High Tech Start Up* (Saratoga, CA: J.L. Nesheim, 1997).
8. Bygrave and Timmons, *Venture Capital*, 20.
9. Contact Recombinant Capital, San Francisco, CA, Mark Edwards, Managing Director.
10. Bygrave and Timmons, *Venture Capital*, 165.
11. Robert C. Megantz, *How to License Technology* (New York: Wiley, 1996), 55–69.
12. For simplicity all NPVs in this section are referenced to the date of commercialization. Time discounting to the present is straightforward and will not change the relative relationships.

Chapter 11 Managing Value and Risk in the R&D Portfolio

1. Using an arbitrary project unit with $48 million in year 5 sales would allow us to regard a smaller project, say with half these sales, as constituting 0.5 project units.
2. The term *net costs* is used in R&D. R&D is usually treated as a cost center in corporate accounting. However, in many companies, the R&D budget may be credited for revenues attributable to its operations. Hence the R&D department may run a gross and a net budget. Products produced by R&D pilot-scale operations and sold commercially are an example. Other offsetting revenue may be obtained from licensing technology, sale of technology, contract income, sale of services, or sale of assets.
3. Philip A. Roussel, Kamal N. Saad, and Tamara J. Erickson, *Third Generation R&D* (Boston, MA: Harvard Business School Press, 1991), 97.
4. Greg. A. Stevens and James Burley, "3,000 Raw Ideas = 1 Commercial Success!" *Research•Technology Management*, May–June, 1997, 16–27.
5. Robert G. Cooper, *Winning at New Products* (New York: Addison-Wesley, 1993), 102–104.
6. Financial analysis is based on after-tax costs, although this is an appropriate yardstick only for profitable companies. A tax deduction for an R&D boutique that has no profitable products is not applicable as long as it has no profits. Nevertheless, its accumulated nonoperating losses, or "NOLs" may subsequently be utilized by an acquiring company as credits against its own tax bill should it want to purchase this technology by buying the business.
7. The growth-in-perpetuity formula cannot be applied here because PMI's growth rate of 12.4% exceeds its cost of money 12%. Using the

perpetuity formula is also reasonable, since the calculation assumed there was zero growth in R&D resources.

8. Using the sensitivity analysis in Exhibit 10.1, we interpolated to find a gross margin corresponding to 20% IRR and interpolated again to obtain the corresponding NPV.

Chapter 12 Decision Trees and Options

1. Richard A. Brealey and Stewart C. Myers, *Principles of Corporate Finance* (New York: McGraw-Hill, 1996), 255–264.
2. Nancy A. Nichols, "Scientific Management at Merck: An Interview with CFO Judy Lewent," *Harvard Business Review*, January–February 1994, 91.
3. Crystal Ball™ v4.0, Decisioneering Inc., 1515 Arapahoe Street, Suite 1311, Denver, Colorado 80202.
4. Brealey and Myers, *Corporate Finance*, 590–592.
5. G. R. Mitchell and W. Hamilton, "Managing R&D as a Strategic Option," *Research•Technology Management*, May–June 1988, 15–22.
6. Brealey and Myers, *Corporate Finance*, 577–580. I have used the Black-Scholes model on several occasions and found it to be reasonably user-friendly. Financial textbooks offer tables of values to short-circuit the need for calculations.
7. I. M. Sokolnikoff and R. M. Redheffer, *Mathematics and Physics of Modern Engineering* (New York: McGraw-Hill, 1958), 650–654.
8. Brealey and Myers, *Corporate Finance*, AP12–13.
9. Nichols, *Scientific Management at Merck*, 91.
10. Gary L. Sender in Nichols, "Scientific Management at Merck," 92.
11. Ibid., 92.
12. In the case of operating systems, it is surprising that the operating system vendor (Microsoft) had the advantage of polarity when combining its technology with word processing and spreadsheet software. Existing software from long-established vendors such as Wordperfect and Lotus was excellent—but the superior compatibility features offered by Microsoft seemed sufficient to convince users to consider switching vendors as their earlier version software became dated. The point is made because polarization may be nonobvious (one usually assumes the existing suppliers have the advantage), and the value of capturing it through enabling technology can be extremely high.
13. Marc H. Meyer and Alvin Lehnerd, *The Power of Product Platforms* (New York: Free Press, 1997), xii.

Chapter 13 Creating Value through Diversification

1. Richard A. Brealey and Stewart C. Myers, *Principles of Corporate Finance* (New York: McGraw-Hill, 1996), 177.
2. Robert Cooper, *Winning at New Products* (Reading, MA: Addison-Wesley, 1993), 267.
3. Philip A. Roussel, Kamal N. Saad, and Tamara J. Erickson, *Third Generation R&D* (Boston, MA: Harvard Business School Press, 1991), 93.
4. Brealey and Myers, *Corporate Finance,* 67.
5. Richard N. Foster, *Innovation: The Attacker's Advantage* (New York: Summit Books, 1986), 102.
6. Deb Chatterji, "Accessing External Sources of Technology," *Research•Technology Management,* March–April 1996, 48–56.

Chapter 14 R&D Metrics

1. Tom Copeland, Tim Koller, and Jack Murrin, *Valuation: Measuring and Managing the Value of Companies* (New York: Wiley, 1995), 159. See also R&D Leadership and Credibility Steering Committee, Research-on-Research Committee, "Building R&D Leadership and Credibility," Industrial Research Institute, August 1991.
2. Philip A. Roussel, Kamal N. Saad, and Tamara J. Erickson, *Third Generation R&D* (Boston, MA: Harvard Business School Press, 1991), 97.

Chapter 15 Special Issues in Value-Based R&D

1. Marco Iansiti and Jonathan West, "Technology Integration: Turning Great Research into Great Products," *Harvard Business Review,* May 1, 1997.
2. Marco Iansiti, "Technology Integration," a speech given at Industrial Research Institute Semiannual Meeting, Tucson, Arizona, May 4, 1998.
3. For a broad discussion of the realities of global competition read Kenichi Ohmae, *Triad Power: The Coming Shape of Global Competition* (New York: Free Press, 1985).
4. David R. Berg and Grant Ferrier, "Meeting the Challenge: U.S. Industry Faces the 21st Century," *The U.S. Environmental Industry* (Washington: U.S. Department of Commerce, Technology Administration, Office of Technology Policy, 1997).
5. Ibid.
6. Product stewardship is one of the ten "Codes" of the Responsible Care™ program of the Chemical Manufacturers Association (Washington, DC), which is an excellent source for further information on the subject.

Index

Abbott, 233
Acceleration option, 290, 299–300, 301
Accounting. *See* Financial basics; Financial statements
Accounting *vs.* economic value, 54–55
Accounts payable, 58, 227
Accounts receivable, 55, 227
Accrued liabilities, 58–59
Accumulated depreciation, 56, 229
ADR, 255, 256–257
Ajinomoto, 233
Allocation of costs/resources, 71–72
Alternatives. *See* Strategy/strategic planning
American Can Company, xiv, 255
American corporations. *See* Corporation, American form of
American National Can, 24
Analyzers/synthesizers, 14–16
Angels, 256
Anheuser-Busch, 24
Apple Computer, 262, 317
Appraisal, technology, 272–273
APT. *See* Arbitrage Pricing theory (APT)
Arbitrage Pricing theory (APT), 109
Archimedes, 9
Arco Chemical, 290
Arthur D. Little, 160
Asian capitalism, 377
Aspartame, 57, 234, 235, 236
Assets, 55–58, 229
Assumptions, key, 243–247
  capital intensity, 246–247
  gross margin/pricing, 245–246
  market size, 245
  overheads, 245
  timing, 243–245
Assumptions, quality of, 214–216, 352
Astronomy, 5–6
AT&T, 18, 324

Background technology, 73
Backlog metric, 346
Balanced portfolio of businesses, 134
Balance sheet, 52–64
  accounting *vs.* economic value, 54–55
  assets, 55–58
  capital structure and investment performance, 63–64
  double entry bookkeeping, 63
  example, 62
  liabilities, 58–60
  shareholder equity, 61–63
Balance sheet, pro forma, 226–230
  accounts payable, 227
  accounts receivable, 227
  accumulated depreciation, 229
  gross fixed assets, 228–229
  inventories, 227
  net fixed assets, 229
  return on TCE (total capital employed), 229–230
  total capital employed (TCE), 229
  working capital, 227–228
Base/key/emerging technology, 160–161
Baxter, 233
Bean (L.L.), 157
Beef packaging patterns, 182
Bell Laboratories, 324
Benchmarking, 157–158
Best practices, 157, 158
Beta, 107–108
Black-Scholes model, 107, 110, 302–306, 307
Boeing, 144, 212
Bond-rating agencies, 64
Book value, 54, 61, 64, 72, 74
Boston Consulting Group (BCG), 193–194
Brand names, 72
Brazil, 359
Break-even analysis, 84–87
Bristol-Myers, 145, 201, 317
Burn rate, 34, 258

Business life cycle, 132–133
Businesspersons *vs.* scientists, 6–16. *See also* Science, technology, and business
  analyzers/synthesizers, 14–16
  approaches to data, 10–11
  close linkage of, 19–20
  forecasting, 11–14
  motivations, 7–9
  vision, 9–10
Business plan, 254
Buyback programs, stock, 50, 67

California, 26, 370
California, University of, 263, 272
Campbell Soup, 315
Capacity:
  excess, 87
  expansion projects, 228–229
  matching to demand, 89–91
Capital:
  cost of, 93–94, 96
  expenditures (pro forma cash flow statement), 232
  intensity (assumption), 246–247
  requirements (proactive use of R&D/engineering to reduce), 84–95
  savings (and project portfolio valuation), 351–352
  structure, and investment performance, 63–64
Capital, from operating viewpoint, 84–95
  break-even analysis, 84–87
  economies of scale, 87–89
  matching capacity to demand, 89–91
  new *vs.* existing plant, 91–92
  R&D effects, 92–95
Capital Asset Pricing Model (CAPM), 108, 109, 111
Capitalization, income (method of appraisal), 272
Capitalization, market, 51
Capitalizing R&D, pros and cons of, 75–77
CAPM. *See* Capital Asset Pricing Model (CAPM)
Case study: hypothetical (PMI—Polymers & Materials, Inc.), 217–218
  acceleration option, value of, 299
  balance sheet, 223
  cash flow statement, 231
  cost and capital savings, 340
  decision/event tree for, 295
  income statement, 220
  licensing balance of payments, 346
  licensing model for, 268
  modeling R&D pipeline, 276–281
  Monte Carlo analysis, 298
  new product metrics, 337, 341
  option value and risk, 309
  patent productivity, 333
  sensitivity analysis, 239
  technology option, and Black-Scholes theory, 307–309
  tornado chart, 241
Case study(ies): hypothetical, 23
  catalytic converter, 26, 30–31, 33–34, 37, 40
  genetic vaccines, 26, 30, 33, 36, 39
  ion chromatography, 26, 30, 33, 36, 39
Case study: real example, 233–236
  endgame, 235–236
  pro forma results, 234–236
Cash, 55
Cash cows, 134
Cash flow models:
  DCF [*see* Discounted cash flow (DCF) analysis; Discounted cash flow (DCF) model, pro forma]
  pitfalls, 124–126
Cash flow statements, 65–69
  example, 68
  financing activities, 67–69
  investments, 66–67
  operations, 65–66
Cash flow statements, pro forma, 230–233
  after-tax net income, 231
  capital expenditures, 232
  depreciation, 231
  free cash flow plus terminal value, 232–233
  operating cash flow, 232
  working capital, increase in, 232
Catalytic converter, 373
  fictional example, stages of R&D, 26, 30–31, 33–34, 37, 40
Change, 168
Charge-outs, 345–346
Chicago Board Options Exchange, 302
China, 327, 359, 378
Chrysler: K-Car platform, 313
Ciba, 180
Clinton health care proposal, 145
Coca-Cola Company, 73, 235
Commercial espionage, 159–160
Commercialization, early, 38–41, 281
Commodity products, industrial markets, 184
Competencies, core, 155
Competitor intelligence, 158–160
Conceptual research, 27–31, 280
Conjoint analysis, 189–190

Consortia, 328
Consulting firms/consultants, 156, 165, 188, 329–330
Consumer market, 171, 183–184
Continuing value, 115
Contribution margin, 85. *See also* Gross margin (GM)
Copernicus, 9
Copyrights, 72
Core competencies/technologies, 155
Corning, 98
Corporate culture, 216, 364
Corporate growth. *See* Growth, and shareholder value
Corporate reorganization, 141
Corporation, American form of, 17, 377
  alternatives to, 377
Corporation, value model for, 16–19
  diagram, 17
Cost(s):
  accounting for (*see* Income statement)
  and break-even analysis (*see* Break-even analysis)
  of capital, 59, 96
  and capital savings, 339–340
  of debt, 105–106
  of equity, 106–110
  fixed *vs.* variable, 84–85
  historic, 54
  of money, 83, 103–111
  and productivity/value in R&D portfolio, 286–289
  of sales, 47, 219–220
  savings projects, 351
Cost-plus pricing, 191
CRADA (Cooperative Research and Development Agreement) format, 328
Credit/debit, 63
Creditworthiness, 64
Cross-licensing, 360
Crowd behavior, 273
Culture, organizational, 216, 364
Culture of science, 3
Customers as source of ideas, 24–25

DCF. *See* Discounted cash flow (DCF) analysis; Discounted cash flow (DCF) model, pro forma
Debottlenecking, 93, 339
Debt, 49, 59–60, 78, 104–105
  on balance sheet, 59–60
  cost of, 105–106
  and equity, 78, 104–105
  interest payments on, 49
Decision(s)/decision-making:
  operating, and cash flow, 70
  support/tools, 84–95, 213, 247–248
  time for, in strategic planning process, 167
  value-based, 374–375
Decision trees, 114, 149, 213, 249, 290, 291–293, 295
Defense markets, 187
Delco, 325
Deliverables, 343
Dell Computer Corporation, 184
Delphi, 325
Demand, matching capacity to, 89–91
Demand, price elasticity of, 195–196
Department of Agriculture (USDA), 372
Depreciation:
  accelerated, 222
  accumulated, 56, 229
  of fixed assets, 56–57
  and pro forma balance sheet, 229
  and pro forma cash flow statement, 231
  and pro forma income statement, 222–224
  recapturing, 70
  of technology, 77–78
Development, 35–38, 280–281
Diamond, 173
Diapers, disposable, 24, 25
Digital Equipment Corporation, 255, 257
Dilution, 52, 80
Direct manufacturing costs (DMC), 85, 219, 246
Discontinuities, technological, 202–204
Discounted cash flow (DCF) analysis, 83, 97–126
  calculating reasonable gross margin, 191
  and comparison of strategic alternatives, 167
  cost of money, 103–111 [*see* Debt; Equity, shareholder; Weighted average cost-of-capital (WACC)]
  economic value added, 102–103
  examples, 100–102, 120–122
  net present value, 98–99 [*see also* Net present value (NPV)]
  pitfalls, 124–126
  rigor of, 149
  risk-weighted hurdle rates for R&D, 113–115
  and terminal value, 115–122
  time horizons, 122–123
  in valuations (market-based), 254
  valuing capital savings, 352
  valuing licensable properties, 352

Discounted cash flow (DCF) analysis *(Continued)*
valuing long-term growth of 1%, 130
and year consistency, 281
Discounted cash flow (DCF) model, pro forma, 213–236. *See also* Financial statements, pro forma
assumptions, 214–216
balance sheet, 226–230
case study (real case), 233–236
cash flow statement, 230–233
example (PMI), 216–218
formulas for pro forma financial statements, 221
income statement, 218–226
purposes, 213–214
Discount factors, 112
Discount rate, 96, 100, 103, 113
for royalty revenues, 269
Discretionary capital projects, 66
Distribution channels, 198–200, 244
Diversification, 43, 107, 284, 314–330
efficient portfolios, 314–320
external technology, 324–329
vs. productivity, 320
value proposition, early-stage R&D, 320–321
Dividend yield, 51
DMC. *See* Direct manufacturing costs (DMC)
DNA discoveries, 5, 6
Double entry bookkeeping, 63
Dow Chemical, xiv, 180, 192, 265, 327, 362, 376
Dow Corning, 18, 53
DRAM market, 194
Duke Energy, 315
Du Pont, 324

Earnings:
managing, 111
models (EPS/EBIT/EBITDA) *vs.* cash flow models, 124–126
quality of, 111
ratio [*see* Price earnings (PE) ratio]
retained, 51
Earnings before interest, taxes, depreciation, and amortization (EBITDA), 45, 70, 117, 225, 273
Earnings before interest and taxes (EBIT), 45, 47–48, 117, 225
Earnings per share (EPS), 50
EBIT. *See* Earnings before interest and taxes (EBIT)
EBITDA. *See* Earnings before interest, taxes, depreciation, and amortization (EBITDA)
Economic profit calculations, 252–254, 351
Economic value added (EVA), 102–103
Economies of scale, 38, 87–89, 140, 200
Efficient frontier, 316
Efficient portfolios, 314–320
achieving technology diversification, 317–320
liquidity, problem of, 316–317
80/20 rule, 25
Elasticity, price, 195–196
Electricity, as product, 174
Electric vehicles, 176, 311, 312
Electronics, 203
Emerging technology, 161
Empire building, 346
Employee-owned corporations, 377
End-use market segmentation, 178–180
Energy efficiency, 95
Environment, 26, 200, 369–375
capital, 370–371
costs, 371–372
opportunities, 373–374
timing, 372
value-based decision making, 374–375
Environmental Protection Agency (EPA), 369, 372
Equity, shareholder, 54, 61–64
on balance sheet, 54, 61–63
cost of, 106–110
and debt, 78, 104–105
return on [*see* Return on equity (ROE)]
ESOPs, 377
EVA. *See* Economic value added (EVA)
Evaluation fees, 269–270
Event tree, 291
Excel, 216
Excell, 182
Exclusivity, in licensing, 271–272
Exit strategy, venture capital, 256
Expenses, accounting for. *See* Income statement
Experience (or learning) curve, pricing ahead on the, 193–195
External technology, assessing, 324–330
External technology, early-stage providers of, 325–329
agreements used by universities, national labs, consortia, 328–329
caveat, 329
consortia, 328
large companies, 326–327

national laboratories, 327–328
small companies, 326
universities, 327
External technology, later stage providers of, 329–330
Exxon, 145, 263, 304

Factory. *See* Plant/factory
Failure, 169, 262
FDA. *See* Food and Drug Administration (FDA)
Feasibility, 32–35, 280
Federal Express, 9
Fiddler Magazine, 180
Financial basics:
accounting for intellectual property, 72–75
accounting issues, 71–72
capitalizing R&D, pros and cons, 75–77
depreciating technology, 77–78 (*see also* Depreciation)
financial statements (*see* Financial statements)
key concepts, 69–70 [*see also* Earnings before interest, taxes, depreciation, and amortization (EBITDA); Free cash flow (FCF)]
leverage, 78–81
measures of return, 81–83
value of technology is situational, 75
Financial burn rate, 34, 258
Financial information, and strategic planning, 146
Financial options, two characteristics of, 306
Financial portfolios *vs.* technology portfolios, 316
Financial statements, 45–83, 218–233
balance sheet, 52–64, 226–230 (*see also* Balance sheet)
cash flow statement, 65–69, 230–233 (*see also* Cash flow statement)
caveat, 46
and DCF [*see* Discounted cash flow (DCF) analysis; Discounted cash flow (DCF) model, pro forma]
income statement, 46–52, 218–226 (*see also* Income statement)
Financial statements, pro forma, 46, 218–233
balance sheet, 226–230 (*see also* Balance sheet, pro forma)
cash flow statement, 230–233 (*see also* Cash flow statement, pro forma)
formulas for, 221
income statement, 218–226 (*see also* Income statement, pro forma)
Financing, 256–262
activities, 67–69
failures, 262
initial public offering (IPO) (stage 4), 260–261
mezzanine round (stage 3), 259–260
private placement, round 1 (stage 2), 258–259
secondary offerings, 261–262
seed capital (stage 1), 256–258
stages in process, 256–262
Fisher-Pry equation, 206–207
Focus groups, 188
Food and Drug Administration (FDA), 33, 36, 39, 41, 97, 173, 174, 320, 372
Ford Motor Company, 313, 315
Forecasting, 10–14
hypothetical case, 12–13
methodology issues, 11
planning horizon, 11
projected sales of pipeline products, 347–350
scientists *vs.* businesspersons, 10–14
technology, 206–208
Foreground technology, 73
Forest slash example, 88
France, 366, 376
Free cash flow (FCF), 45, 66, 69–70, 115, 121, 128, 232–233
Funnel, R&D, 23

G&A. *See* General and administration expenses (G&A)
Gantt charts, 166, 343
Gates/gatekeeping, 22–23, 25, 159, 323
Genentech, 233, 255, 264, 290
General and administration expenses (G&A), 224–225, 371
General Electric, 9, 164, 317, 324
Generally accepted accounting principles, 54, 75
General Motors, 199, 313, 315, 325
Gene-splicing patents, 263, 271
Genex, 234
Geographic expansion, 164, 205
Geographic market segmentation, 182
Germany, 368, 377
Gillette, 274
Globalization, 176–177, 365–369
implications for technology, 367–369
patent protection, 271, 358–359
Goodwill (on balance sheet), 57–58
Government/institutional market, 171, 186–188

Government laboratories, 327–328
Gross domestic product (GDP), 139
Gross fixed assets, 228–229
Gross margin (GM), 85, 148, 191, 219–220, 239, 245–246
Gross profit, 224
Growth, and shareholder value, 127–145, 148
  balanced portfolio, 134–135
  business life cycle, 132–133
  caveats, 143–145
  corporate goals, and R&D, 274–276
  and product life cycle (mature/rapid), 204
  R&D: trading current profits for growth, 135–137
  sources of corporate growth, 138–141
  strategies for corporate growth, 141–143
  trade-off, growth and profitability, 129–132
  value of 1% in added growth, 128–130
Growth business, 125–126
Growth model, revenue, 208–212
Growth rates, 209
  and incremental value changes, 131
  intrinsic value of long-term sustainable, 130
Gulf & Western, 376

Harvard, 255
Headquarters expenses, 71
Heisenberg's uncertainty principle, 3
Hewlett Packard, 9, 376
Hockey stick forecasts, 347
Holding company, 317
Horizon value. *See* Time frame/horizon
Human resources, 153, 166
  personnel involved, stages of R&D, 25–26, 29, 34, 36–37, 41
Hungary, 366
Hurdle rates, 97, 113–115, 247, 275, 321

IBM, 18, 53, 189, 365
IBP, 182
Ideas, finding/screening, 24–27, 280
Imaging, 203
Income, after-tax net, 231
Income capitalization method of appraisal, 272
Income statement, 46–52
  analysis of, 51–52
  example, 50
  nonoperating expenses, 49–51
  operating income (EBIT), 47–48
  sales and operating expenses, 47
Income statement, pro forma, 218–226
  after-tax net income, 226
  cost of sales and gross margin, 219–220
  depreciation, 222–224
  EBITDA, 225
  EBIT/pretax profit, 225
  factory overhead, 221–222
  G&A, 224
  gross profit, 224
  R&D, 224
  return on sales (ROS), 226
  selling cost, 224
  taxes, 225–226
Incubation stage, 202, 204
Industrial market, 171, 184–186
Industrial R&D. *See* Research and development (R&D)
Industry, light/heavy, 228
Information overload, 25, 159
Initial public offering (IPO), 61, 256, 260–261
Innovation, sustaining, and diversification, 322–324
Intel, 200, 376
Intellectual property:
  accounting for, 72–75
  licensing of, 263 [*see also* Patent(s)]
Intelligence, competitor, 158–160
Interest on interest, 102
Internal rate of return (IRR), 83, 96, 100–101, 103, 113, 123, 216, 230, 232, 242, 243, 288, 297–298
  hurdle rate, 113
  and sensitivity graphs, 243
Internet, 163, 250, 306
Inventory:
  accounting, 72
  on balance sheet, 55, 227
  building up, 89
Investments:
  on cash flow statement, 66–67
  on balance sheet, 58
"Invisible hand," 378
IPO. *See* Initial public offering (IPO)
IRR. *See* Internal rate of return (IRR)
ITT, 317, 376

Japan/Japanese, 147, 177, 182–183, 194, 200, 327, 368, 376, 377
Joint ventures, 49, 58, 327
Justification, project, 248–249

K-Car platform, 313
Kepler, 9
Key technology, 161, 163
Korea, 194, 359
Kyowa Hakko, 233

Laboratory unit, model, 279, 284–286
Lead users, 24
Least-squares method, 11
Legacy building, 8–9
Leverage, financial, 78–81
Leveraged buyout (LBO), 70, 225
Liabilities, 58–60
Licensable properties, 352
Licensing, 235, 262–272, 326, 346–347, 352
  current metrics, 346–347
  discount rate for royalty revenues, 269
  exclusivity, 271–272
  fees, 269–270, 346
  negotiations, and pro forma analysis, 235
  revenues/costs, 266–272, 352
  royalties, minimum/running, 270
  up-front fees, 270
  and valuation, 262–264, 268–269
Licensing scenarios, 264–266
  patent rights only, 264, 266
  proven technologies, 264–265
  unproven or partially proven technologies, 264, 265
Life cycle of a business, 132–133
Lilly, 145
Linear extrapolation of current trend, 11
Liquidation, phantom, 117, 118
Liquidation value, 116–117
Liquidity, problem of, 316–317
Litton Industries, 317, 376
L.L. Bean, 157
Lotus 123, 103, 216
LTV, 376

Make-or-buy problem, 15, 325, 330
Manufacturing cost, 138, 140–141
Margin:
  contribution, 85
  cost of money at, 103
  gross [*see* Gross margin (GM)]
Market(s):
  capitalization, 51
  captive/merchant, 177
  definition of, 176–178
  growth, 138, 139
  penetration, 138, 139–140, 176
  research, 188–190, 284
  risk, 108
  segmentation, 178–183
  share, 138, 140, 200–201
  time to, 38, 330, 353
Market(s), categories of:
  consumer, 183–184
  government, 186–188
  industrial, 184–186
Market-based approaches to value, 254–255
Market-driven *vs.* technology-driven, 171
Marketing, 169–212
  assumptions, and value models, 215, 245
  defining "product," 172–176
  distribution channels and R&D, 198–200
  and pricing, 190–196
  processes, importance of, 174–176
  product life cycle, 204–206
  R&D perspective on, 170–171
  revenue model, 208–212
  technological discontinuities, 202–204
  technological performance and product life cycle, 201–204
  technology forecasting, 206–208
  value, delivering, 196–198
Marketplace for technology, 255–262
Market value *vs.* book value, 64
Mature year, 243
Maturity, 204
Maxwell's equations, 3
McDonald's, 365
McGaw, 233
Mercedes, 365
Merck, 145, 297, 306, 307, 308, 309, 376
Mevacor™, 173
Metallurgy, 6
Metrics. *See* Research and development (R&D): metrics
Mezzanine round, start-up financing, 259–260
Microsoft, 103, 193, 271, 290, 310, 315, 376
Milestones, 22, 31, 32, 343–345
Military markets, 187
MIT, 255
Mobil, 145, 163, 263
Monetary success as motivator, 7–8
Money, cost of, 103–111
  cost of debt, 105–106
  cost of equity, 106–110
  debt and equity, 104–105
  managing earnings, 111
  WACC, calculating, 110–111
Money, discounted value of, 112
Monsanto, 57–58, 359
Monte Carlo analysis, 213, 297–299, 321, 322
Montfort, 182
Moody's, 64
Moore's Law, 78
Morton Salt, 172, 173
Motivation, scientists/businesspersons, 7–9
"Myopia of the served market," 178

National laboratories, 327–328
Net income, after-tax, 226
Net present value (NPV), 83, 149, 216, 232
  approach to total portfolio value, 352–353
  calculated using spreadsheets, 103
  and comparison of strategic alternatives, 167
  and DCF, 96, 98–99, 230
  and economic break-even, 242
  negative, 248
  probability weighting of, 113–115
  project proposals and, 14–15
  R&D contribution, 287–289
  and time horizons, 123
Netscape, 291
New products. *See* Product(s)
Newton's laws, 3, 9
Niche marketing, 180, 201
Nielsen, 183
Normal distribution, 303
NOVA, xvii, 263
NPV. *See* Net present value (NPV)
Nutrasweet, 97–98, 235

Obsolete technologies, 203
Operating cash flow, 232
Operations, 65–66
Opportunity(ies), 323–324
  creation, *vs.* risk reduction, 35, 42
  defining strategically or economically, 18
  environmental concerns creating, 373–374
Option(s), 290, 300–309, 321
  acceleration, 290, 299–300, 301
  embedded, 352
  financial, 302
  follow-on investments, 290, 301
  and portfolio dynamics, 213
  technology, 306–312
  termination, 43, 290, 301
Outsourcing technology, 325
Overhead, 85, 221–222, 245
  assumptions, 245
  reduction, 138, 141

Pacing project, 203, 336
Packaging, 172
Patent(s), 29, 72–75, 158, 163, 263, 264, 266, 275–276, 332–335, 345, 353, 354–361
  accounting for, 72–75
  challenges, 359–360
  cross-licensing, 360
  enforcement, 359–361
  financial incentives for, 334–335
  global protection, 358–359
  harmonization, 358
  interference, 357
  life of, 275–276
  milestone, 345
  obtaining/maintaining, 357–358
  opposition, 358
  productivity, 332–335
  reduction to practice, 29
  rights-only licensing scenario, 264, 266
  time lags, 334
  as value drivers, 354–361
  writing, 355–357
Penicillin, 4
Pentium, 200
Pepsi-Cola, 9, 235
PE ratio. *See* Price earnings (PE) ratio
Performance limit, 202
Performance products, industrial markets, 185, 186
Perpetuity, 288
  growth-in-perpetuity method, 115, 116, 118–120
  terminal value as a, 117–118
PERT (Program Evaluation and Review Technique), 166, 343
Physical assets, 56
Pilkington Glass, 98
Plant/factory:
  on balance sheet, 56–57
  maintenance, 94–95
  new *vs.* existing plant, 91–92
  overhead (pro forma income statement), 221–222 (*see also* Overhead)
Platforms, technology, 312–313
PMI. *See* Case study: hypothetical (PMI—Polymers & Materials, Inc.)
Polarization of technology linkages, 312
Post-it notes, 24
Price, 138, 139
Price earnings (PE) ratio, 51, 82, 145
  tendency to revert to historical mean, 145
Price elasticity, 195–196
Pricing, 190–196, 245–246
  ahead on the experience curve, 193–195
  cost-plus, 191
  part of product, 173
  reinvestment, 191–192
  skimming, 192–193
  strategies, 191–195
Pricing umbrella, 192
Private placement, 258–260
  mezzanine round, 259–260
Probabilities of success, developing, 281–284, 291, 296

Probability tree, 296
Probability-weighted project outcomes, 295–297
Processes, importance of, 174–176
Process-related metrics, 339
Procter & Gamble, 24, 25
Product(s):
  current metrics, 341–343
  definition, 172–176
  future metrics, 347–350
  historical metrics, 335–338
  marketing (*see* Marketing)
  "new" (definition), 336
  pricing (*see* Pricing)
  and valuation of R&D portfolio, 351
Productivity factor, 135–137, 286
Productivity *vs.* diversification, 320
Product life cycle, 204–206
Product stewardship, 200, 373
Profitability:
  first derivative of shareholder wealth, 53
  and growth (trade-off), 130–132
  and hypothetical business life cycle, 133
  measure of (*see* Income statement)
Profit centers, 71
Profit & loss statement. *See* Income statement
Pro forma models, 237, 290
  discounted cash flow [*see* Discounted cash flow (DCF) model, pro forma]
  financial statements (*see* Financial statements, pro forma)
Projects. *See* Research and development (R&D)
Property/plant/equipment, 56–57

Quick and dirty valuation models, 166, 215, 242–247, 289
  assumptions, key, 243–247
  project justification, 248–249
  using for decision support, 247–248

Random walk theory, 108
Raw materials cost (RMC), 246
R&D. *See* Research and development (R&D)
Real estate appraisal, technology appraisal compared to, 272–273
Recombinant Capital, 257
Recruitment, 166
Refrigeration, 203
Reinvestment pricing, 191–192
Rennselaer Polytechnic Institute, 251
Research and development (R&D):
  accounting, 224
  capitalizing, 74, 75–77
  costs, and project duration, 279–281
  defining, 6
  effects on capital, 92–95
  and growth/shareholder value, 127–145
  and marketing, 170–171
  mode of converting cash into opportunity, 18
  productivity factor, 135–137
  science/technology context, 6
Research and development (R&D): managing value/risk in portfolio of, 250, 274–289, 350–353
  corporate growth goals and R&D, 274–276
  costs, productivity, and value in, 286–289
  direct valuation of current project portfolio, 350–353
  diversification, 314–330 (*see also* Diversification)
  modeling the R&D pipeline, 276–281
  project flow in a model laboratory unit, 284–286
  R&D contribution by net present value analysis, 287–289
  risk reduction (*see* Risk)
  success rates, 281–284
  valuation (*see* Valuation; Value)
Research and development (R&D): metrics, 331–353
  current, 341–347
  future, 347–350
  historical, 332–340
Research and development (R&D): process stages, 21–44, 280–281
  and costs, 280–281
  examples, 26, 30–31, 33–34, 36–37, 39–40
  financial/business implications, 26–27, 29–31, 34–35, 37–38, 41
  personnel involved, 25–26, 29, 34, 36–37, 41
  stages in, 22–41, 280–281
Research and development (R&D): special issues, 354–375
  environment, 369–375
  globalization, 365–369
  patents as value drivers, 354–361 (*see also* Patent(s))
  technology transfer, 361–365
Reserves, 60
Residual value, 115
Resource allocation, 147
Retained earnings, 51, 61, 63
  key link between income statement and balance sheet, 63
  warning, 63

Return, different measures of, 81–83
Return on capital, 84
Return on equity (ROE), 81
Return on invested capital (ROIC), 229, 252
Return on investment (ROI), 67, 82, 229
Return on net assets (RONA), 82
Return on sales (ROS), 226, 229
Revenue(s):
  accounting for. *See* Income statement
  licensing, 266–272
  model, 208–212
Revolver, 59
Risk:
  changing with time, 97
  free return, 106
  hurdle rates for R&D, weighted by, 113–115
  management *vs.* avoidance, 43
  market, 156
  parameter, *vs.* attractiveness parameter, 156
  premium, 106, 107
  principles/rules (five; Cooper), 41–43, 283–284
  reduction, and R&D, 114, 250
  reduction *vs.* opportunity creation, 35, 42
  vs. reward, and efficient portfolios, 315
  two types of, 107, 156, 317, 318
  variables related to (four), 115
ROE. *See* Return on equity (ROE)
ROI. *See* Return on investment (ROI)
ROIC. *See* Return on invested capital (ROIC)
Rolex, 180
RONA. *See* Return on net assets (RONA)
ROS. *See* Return on sales (ROS)
Rowenta, 176
Royalties:
  discounted cash flow of, 74
  licensing (*see* Licensing)
  minimum/running, 263, 270
  patent [*see* Patent(s)]
  rates, using for valuation, 268–269
Rubbermaid, 274

Sales:
  expenses (*see* Income statement)
  and operating expenses, 47
  projection, 277–279
  return on [*see* Return on sales (ROS)]
  sales-driven *vs.* marketing-driven, 170
Saturn, 199
Scale, economies of, 38, 87–89, 140, 200
Science, technology, and business, v–vii, 1–20, 376, 377–378. *See also* Technology
  defining technology, 4–6
  science *vs.* technology, 2–4
  valuation of technology, v–vii
  value model for the corporation, 16–19
Science-based start-up companies, 10
Science and government, 377–378
Scientific method/culture, v, 3–4
Scientists, 6–16, 237
  and analysis/synthesis, 14–16
  vs. businesspersons, 6–16, 19–20
  and data, 10–11
  "donating to public," 357
  forecasting, 11–14
  and marketing, 171
  motivations, 7–9
  vision, 9–10
S-curve, 132, 170, 201–202, 203, 205, 206, 209, 244, 322
Searle, 58, 234, 324, 359
Secondary offerings, 261–262
Second-generation products, option to create, 290, 301
Seed capital, 256–258
Selling cost, 224. *See also* Cost(s)
Senior management, role of, 43–44
Sensitivity analysis, 166, 213, 238–241
  key factors, 240
Sensitivity graphs, 242–247
Service functions, charge-outs, 345–346
Shareholders. *See* Equity, shareholder; Growth, and shareholder value; Wealth, shareholder
Shell, 180
Situational value of technology, 75–77
Skill gaps, 165–166
Skimming, price, 192–193, 195
Socialism, 377
Software, new financial, 4
Software, platform-based, 313
Software industry, Bangalore, India, 368–369
Soviet Union, 378
Specialty products, industrial markets, 184
Spreadsheets, 103, 112, 238, 281
Stages. *See* Research and development (R&D): process stages
Standard & Poor's, 64
Standard deviation, 106, 303, 315
Stanford University, 263, 271
Start-up companies, science-based, 10
Start-up financing. *See* Financing
Steady-state model, xviii, 127, 285–286
Stewardship, product, 200, 373
Stock buybacks, 50, 67
Stock options, 307
Stock price, and R&D expenditures, 144
Strategic alliances, 273

Strategic architecture, 150, 154
Strategy/strategic planning, 146–168
  case for, 151–152
  defined, 149–150
  for growth, 141–143 (*see also* Growth, and shareholder value)
  key points (three), 168
  and market segmentation, 182–183
  pitfalls in, 153–154
  resource allocation, 152–153
  role of value drivers, 148–149
Strategy/strategic planning, systemic eight-step approach, 154–168
  step 1: inventory core competencies/technologies, 155
  step 2: identify target markets, 155–156
  step 3: access competitive position in target markets, 156–163
  step 4: formulate strategy proposal and set of alternatives, 163–165
  step 5: identify gaps, resource requirements, time frames, 165–166
  step 6: use formal/informal valuation methods to select among alternatives, 166–167
  step 7: adopt, modify, or reject, 167
  step 8: set targets and implement, 167–168
Strings magazine, 180
Subsidiaries, 49, 71
Substitutions, 206–208
Success rates, establishing, 281–284, 291, 296
  attractiveness scoring, 282
  historical approach, 281–282
Sunbeam Corporation, 176–177
Sunk costs, treating, 250–252
Super Glue, 181
Swiss watches, 195
Synthesizers/analyzers, 14–16

Table salt, 172
Target markets, 155–156
  assessing competitive position in, 156–163
Taxes, on pro forma income statement, 225–226
Taxes, deferred, 60, 64, 76, 223
Teams, cross-functional/design, 92–93, 177, 344
Techno-economics, 376
Technological discontinuities, 202–204
Technological performance and product life cycle, 201–204
Technology:
  assessment, 160–163
  background/foreground, 73
  base/key/emerging, 160–161
  core, 155
  defining, 4–6
  diversification (*see* Diversification)
  forecasting, 206–208
  gaps, 165
  linkages, 15–16
  marketplace for, 255–262
  market segmentation by, 180–181
  option, 306–312
  platforms, 312–313
  vs. science, 2–4 (*see also* Science, technology, and business; Scientists)
  and time frames, 97
  transfer, 361–365
  translation of science into, 6 [*see also* Research and development (R&D)]
Technology, valuation of, v-xviii, 1–2. *See also* Valuation
  accounting methods inadequate, 76
  roadmap, xi
Telecommunications, 306
Terminal value, 96–97, 115–122
  growth-in-perpetuity method, 118–120
  liquidation value, 116–117
  terminal value as a perpetuity, 117–118
Terminating projects, 251–252, 290, 291, 301
Texaco, 145
Texas Instruments, 263
Theory of relativity, 3, 5
3M Corporation, 24, 203–204, 274, 336, 376
Time frame/horizon, 11, 38, 97, 115, 116, 122–123, 130, 165–166, 214, 243–245, 334, 372
Toll manufacturer, 329
Top-down *vs.* bottom-up leadership, 150
Tornado diagrams, 241–242
Total capital employed (TCE), 229
  return on, 229–230
Total capital *vs.* total assets, 64
Total quality management, 323
Total shareholder return, 82
Trade book industry, 338
Trademarks, 72
Transfer, technology, 361–365
Transfer pricing, 71
Transistors, 5
Turnover ratio, 228, 246

Umbrella, pricing, 192
Union Carbide, 263, 327
Unipol, 327
Universities, 327, 328–329

Up-front fees, and licensing, 270
Upside/downside cases, 292–293, 295, 297

Vacuum tube, 5
Valorization, 376
Valuation:
  in accounting, 72
  cash flow models *vs.* earnings models, 124–126
  and DCF [*see* Discounted cash flow (DCF) analysis; Discounted cash flow (DCF) model, pro forma]
  in eye of beholder, 1
  in general, 1–2
  market-based approaches, ix–x, 254–255
  of project proposals, 300
  purposes, 213–214
  quantitative models, counterarguments, 149
  revenue model for, 208–212
  of technology (*see* Technology, valuation of)
Valuation of R&D current project portfolio, 350–353
  capital savings, 351–352
  components of (five), 350–351
  cost savings projects, 351
  embedded options, 352
  licensing revenues and costs, 352
  new product projects, 351
  total portfolio value, 352–353
Valuation shortcuts/alternative techniques, 237–273
  economic profit calculations, 252–254
  licensing as, 262–264
  market-based approaches, 254–255
  quick and dirty models, 242–247
  sensitivity analysis, 238–241
  sunk costs, treating, 250–252
  technology appraisal, 272–273
  terminating projects, 251–252
  tornado diagrams, 241–242
Value:
  adding, 102–103, 196
  creating through diversification, 314–330
  delivering, 196–198
  managing [*see* Research and development (R&D): managing value/risk in portfolio of]
  in the marketplace, ix–x
  net present [*see* Net present value (NPV)]
  of 1% long-term growth, 130
  shareholder (*see* Growth, and shareholder value)
  situational, 75
  sources of, 311–312
  terminal/residual/horizon/continuing, 115–122
Value-based management, vii
Value drivers, 146, 166, 168, 273, 324, 369
  patents as, 354–361 [*see also* Patent(s)]
  prioritizing with tornado diagrams, 242
  role of, 148–149
ValueLine, 109
Value model for the corporation, 16–19, 323
  diagram, 17
Venture capitalists, 114, 255–256, 258, 320
Vision, 9–10
Vision statements, 154
Volatility, 307
  increasing value of option, 301, 305
Vulture capitalist, 257

WACC. *See* Weighted average cost-of-capital (WACC)
Warranties, implied, 172–173
Wealth, shareholder, xviii, 53
Weighted average cost-of-capital (WACC), 104, 105, 110–111, 242, 252
Western Electric, 324
Windows PC operating system, 310
Word processing, 202, 313
Working capital, 55–56, 67, 95, 227–228, 232
W.R. Grace, xv, 21, 160, 164, 197, 210, 211, 264
Write-offs, 49, 111

Xerox, 157
Xerox PARC, 324
X-ray diffraction technology, 6

Year, mature, 243, 244, 245
Yield, 93–94

About the Disk

Introduction

The worksheets on the enclosed disk are saved in Microsoft Excel version 97. In order to use the worksheets, you will need to have spreadsheet software capable of reading Microsoft Excel version 97 files. Please refer to the readme.txt file found on the floppy disk for information on the contents of each Excel file.

System Requirements

- IBM PC or compatible computer
- 3.5″ floppy disk drive
- Windows 95 or later
- Microsoft Excel version 97 or later or other spreadsheet software capable of reading Microsoft Excel 7.0 files.

  NOTE: Files are formatted in Microsoft Excel version 97. To use the worksheets with other spreadsheet programs, refer to the user manual that accompanies your software package for instructions on reading Microsoft Excel files.

How to Install the Files onto Your Computer

To install the files follow the instructions on page 402.

1. Insert the enclosed disk into the floppy disk drive of your computer.
2. From the Start Menu, choose **Run.**
3. Type **A:\SETUP** and press **OK.**
4. The opening screen of the installation program will appear. Press **OK** to continue.
5. The default destination directory is C:\BOER. If you wish to change the default destination, you may do so now.
6. Press **OK** to continue. The installation program will copy all files to your hard drive in the C:\BOER or user-designated directory.

Using the Files

Loading Files

To use the spreadsheet files, launch your spreadsheet program. Select **File, Open** from the pull-down menu. Select the appropriate drive and directory. If you installed the files to the default directory, the files will be located in the C:\BOER directory. A list of files should appear. If you do not see a list of files in the directory, you need to select **Microsoft Excel Files (*.XLS)** under **Files of Type.** Double click on the file you want to open. Use and edit the file according to your needs.

Printing Files

If you want to print the files, select **File, Print** from the pull-down menu.

Saving Files

When you have finished editing a file, you should save it under a new file name by selecting **File, Save As** from the pull-down menu.

User Assistance

If you need assistance with installation or if you have a damaged disk, please contact Wiley Technical Support at:

Phone: (212) 850-6753
Fax: (212) 850-6800 (Attention: Wiley Technical Support)
E-mail: techhelp@wiley.com

To place additional orders or to request information about other Wiley products, please call (800) 225-5945.

For information about the disk see the **About the Disk** section on pages 401–403.

CUSTOMER NOTE: IF THIS BOOK IS ACCOMPANIED BY SOFTWARE, PLEASE READ THE FOLLOWING BEFORE OPENING THE PACKAGE.

This software contains files to help you utilize the models described in the accompanying book. By opening the package, you are agreeing to be bound by the following agreement: